Research for Development

The series Research for Development serves as a vehicle for the presentation and dissemination of complex research and multidisciplinary projects. The published work is dedicated to fostering a high degree of innovation and to the sophisticated demonstration of new techniques or methods.

The aim of the Research for Development series is to promote well-balanced sustainable growth. This might take the form of measurable social and economic outcomes, in addition to environmental benefits, or improved efficiency in the use of resources; it might also involve an original mix of intervention schemes.

Research for Development focuses on the following topics and disciplines:

Urban regeneration and infrastructure, Info-mobility, transport, and logistics, Environment and the land, Cultural heritage and landscape, Energy, Innovation in processes and technologies, Applications of chemistry, materials, and nanotechnologies, Material science and biotechnology solutions, Physics results and related applications and aerospace, Ongoing training and continuing education.

Fondazione Politecnico di Milano collaborates as a special co-partner in this series by suggesting themes and evaluating proposals for new volumes. Research for Development addresses researchers, advanced graduate students, and policy and decision-makers around the world in government, industry, and civil society.

THE SERIES IS INDEXED IN SCOPUS

Giuliana Iannaccone • Eugenia Victoria Ellis
Editors

Towards a Nature-Positive Built Environment

Springer

Editors
Giuliana Iannaccone
Architecture Built Environment
Construction Engineering
Politecnico di Milano
Milano, Italy

Eugenia Victoria Ellis
Civil, Architectural and Environmental
Engineering
Drexel University
Philadelphia, PA, USA

ISSN 2198-7300 ISSN 2198-7319 (electronic)
Research for Development
ISBN 978-3-031-95291-3 ISBN 978-3-031-95292-0 (eBook)
https://doi.org/10.1007/978-3-031-95292-0

This Springer imprint is published by the registered company Springer Nature Switzerland AG
The registered company address is: Gewerbestrasse 11, 6330 Cham, Switzerland

Introduction

The built environment is at a turning point. Today, cities are the main contributors to the depletion of material and energy resources and the production of CO_2 emissions. Either we radically change the processes through which energy and materials are used to construct the world in which we live, or it will become increasingly difficult to remedy the direct and indirect damages our building practices have done to the natural environment and to humans, to the detriment of achieving the United Nations' Sustainable Development Goals.

The building stock needed to accommodate the functions required by society and people already exists in the countries of the north-western hemisphere and is mainly concentrated in urban areas. In this context, the long-term sustainability challenge is mainly based on the renewal and transformation potential of existing urban buildings, spaces, and infrastructures.

The goal of carbon-neutrality is now a global imperative needed to limit the future effects of climate change on the planet. However, within this imperative contains the acceptance of the unfortunate irreversibility of certain processes and the recognition that losses or degradation of the planet's available resources are inevitable to meet humanity's demand for food, energy, and materials.

However, the pandemic to which the entire world population has been exposed has highlighted how the impoverishment of natural resources and the loss of biodiversity represent a serious threat not only to the health of the planet, but also to all living species, including human beings. And, the most vulnerable areas are indeed the cities, where the damage caused by the progressive displacement of the natural environment is most evident.

The regeneration of ecosystems and therefore the adoption of nature-positive approaches represent complementary challenges to those of decarbonization because the conservation and preservation of natural systems makes it possible to ensure the necessary processes of removal and storage of carbon dioxide from the atmosphere.

Approximately 50% of natural ecosystems have been converted by humans for agricultural activities, urbanization, raw material extraction, and infrastructure development. Integrating approaches aimed at restoring and regenerating these ecosystems into the transformation processes of the built environment can generate positive impacts not only for nature but also, and above all, for safeguarding the well-being of humans.

This means looking beyond the areas of study traditionally included in the training of professionals in the field of architecture, engineering, and construction and introducing methods of systemic thinking capable of capturing the complexity of the relationships between the three pillars of sustainable development (society, environment, economy) that occur in the built environment at the area of their intersection.

This volume brings together the research and design thinking of a group of researchers, professionals, and decision-makers who, going beyond the limits of their own disciplinary fields, have explored areas in which it is possible to identify strategies to resolve the major challenges of the world of construction and at the same time to give back to ecosystems and human–nature relationships the space that has been lost to urbanization.

The premise is to encourage a dialogue between disciplines that intervene at different levels in the decision-making processes at the urban, infrastructure, and building scale and to elevate the point of view of professionals, including technicians, to a broader context.

The book is organized into essays that provide a scientific justification for the proposed approaches and case. The book is organized into essays that provide a scientific justification for the proposed approaches and case studies selected from some virtuous realities in Italy and the United States. To follow is a journey that begins by exploring the built environment as habitat and its impact on human health. Emphasizing the need to restore the bond between humans and the natural world, the first section takes the reader back to rediscover ancient design principles of the constructed world as a foundation for the ideas to follow. The second part delves into urban landscapes, highlighting the importance of preserving and restoring biodiversity in cities and demonstrating practical approaches to achieving this goal. The third section presents examples of nature-based solutions integrated into urban environments to enhance resilience, public health, and safety, fostering citizen engagement and collaboration with government agencies. The book concludes with strategies for cities decarbonization, aligning carbon-neutrality targets with the goal of conservation and preservation of natural systems to ensure a nature-positive built environment for the future.

This collection is also the story of a fruitful collaboration that for several years has united the Politecnico di Milano with Drexel University in Philadelphia in terms of both teaching and research, and which has been extended to professionals and colleagues from other realities and contexts who share the same objectives and whose contributions are reported here.

Department of Architecture Built Environment Construction Engineering
Politecnico di Milano,
Milan, Italy

Giuliana Iannaccone

Department of Civil Architectural Environmental Engineering
Drexel University,
Philadelphia, PA, USA

Eugenia Victoria Ellis

Contents

About the Authors

Tecla Caroli is an Architect and Sustainable Development Advisor in Arup. She joined the Arup team in 2022, achieving the same year the PhD with honors in Architecture, Construction Engineering, and Built Environment. She has been teaching Sustainable Technology Architecture in Politecnico di Milano since 2018. She is an expert on issues involving architectural technology, with a focus on Reversibility and Design for Disassembly, Circular Economy, Life Cycle Assessment, and Climate Adaptation.

Alessio Costantino Mirabella is a Senior Engineer in the Sustainable Development team in Arup, Milan Office, where he has been working since 2015. He also works with Politecnico di Milano as teaching assistant since 2013, and he is a freelance journalist for Arketipo, an engineering-architecture magazine. He is an expert in energy efficiency, detailed energy modeling, and renewable integration in the built environment, focusing on strategies to align with decarbonization pathways for 2030 and 2050.

Silvia Croce is a building engineer-architect, currently applying her research skills in the field of sustainable building consultancy. Her work aims to provide insights into solutions for the integrated design of the urban built and natural environment, with a focus on outdoor microclimate, thermal comfort, energy savings, and renewable energy production. At the same time, she aims to raise awareness of the interconnectedness of these issues and to activate different actors in the development of integrated and systemic solutions.

Andrea Giuseppe di Stefano, PhD is a Chartered Engineer and Licensed Architect with a specialization in Sustainable Integrated Design, with a robust background in sustainability-focused professional and research projects. His research interests are on developing data-driven tools for early-stage urban energy optimization. His work has been published in several international scientific journals. He holds a PhD and a Master in Architecture and Building Engineering from Politecnico di Milano. During his studies, he was an Erasmus mobility student at Universidad

Politécnica de Valencia and a visiting researcher at Drexel University. He gained over 4 years of experience at international design firms like Arup. He serves as a teaching assistant at Politecnico di Milano.

Kimberlee Douglas, ASLA, RLA, LEED AP is the Director of the Landscape Architecture Program at Thomas Jefferson University and the Director of the Lab for Urban and Social Innovation. As the Anton Germinshuizen Stantec Term Chair in landscape architecture, she researches the effects of contact with nature on communities. Her research includes the Park in a Truck project that provides a DIY open-source system for communities to design, build, and maintain beautiful, green, low-cost public spaces for healthy, sustainable living. Kim is also currently working on a Pollinator Network in southwest Philadelphia which will increase habitat and green space for the city as well as a Robert Wood Johnson fellowship to research the mental health effects of the Park in a Truck Ambassador program on 11–13-year-old teens.

Eugenia Victoria Ellis AIA is a Professor at Drexel University where she directs the dLUX light lab. A licensed, practicing architect at BAU Architecture, she holds degrees from the University of Illinois at Chicago, the University of Pennsylvania, and Virginia Polytechnic Institute and State University. Dr. Ellis' research includes daylighting, spatial visualization, visual perception, and altered states of perception, such as blindness and dementia. At intersections of nature, the built environment, and behavior, she investigates (eco)logical building systems, architectural theory and well-being with the goal of creating frameworks for the design of sustainable buildings and cities at the nexus of health, energy, and technology.

Sonja Gantioler linked to her enthusiasm in unravelling the big picture and combining views across scientific disciplines and stakeholders, in her research focuses on collaboratively identifying, clarifying, and solving transformational puzzles. This especially concerns the right to ecological space and environmental justice, in light of increasing socio-economic inequalities and spatial disparities. At Eurac Research, she currently leads work on nature-based solutions, environmental policy and governance of a just energy transition. In this regard, she is project coordinator of the Horizon 2020 project JUSTNature.

Aubrey Germ, MUP, AICP, CFM is a Senior Management Consultant with ICF where she helps state and local governments navigate the landscape of disaster resilience funding, planning, and strategy development. Previously, she was the Climate and Resilience Planner for the City of Baltimore where she spent 6 years managing the City's Disaster Preparedness Plan, Climate Action Plan, and Community Resiliency Hub Program. Ms. Germ has a Master's in Urban Planning from Harvard and a BS in Environmental Health Sciences from UNC Chapel Hill where she was a Morehead-Cain Scholar. She currently resides in Bend, Oregon.

Serena Girani is a licensed architect since 2010, joined Arup Milan's Masterplanning and Urban Design team in 2014. Working in Milan, Barcelona, and

London, she combines architecture, planning, and urban studies to support city projects. A former teaching assistant at the Polytechnic University of Milan, she's drawn to public spaces' societal roles. In 2013, she attended the LSE's Cities Programme, enriching her understanding of city-making. Serena's recent focus on urban resilience includes work with the 100 Resilient Cities for Rome and Milan, research on aging cities, Milan's rail yard regeneration, and the Reinventing Cities C40 competition.

Patrick Gurian is a Professor in the Department of Civil, Architectural, and Environmental Engineering at Drexel University. His research interests are in policy analysis of technological systems with particular attention to supporting regulatory decision-making under uncertainty. He has worked on planning efforts for greenhouse gas emissions reductions for Philadelphia, New York City, and Drexel University. His work has been recognized with awards from the Society for Risk Analysis, the American Water Works Association, and the journal Building and Environment.

Guillaume Habert graduated in 1999 from the Ecole Normale Supérieure in Paris, with a degree in earth, atmospheric and oceanic studies, after studying biology and geology. In 2004, he completed his PhD in Toulouse (France) on structural geology.

From 2007 to 2012 Guillaume Habert conducted research at the Laboratoire Central des Ponts et Chaussées in Paris (now University Gustave Eiffel). He worked on the development of sustainable concrete, which led him to address the environmental evaluation of building materials as well as the development of new binders such as geopolymers.

Since August 2012, Guillaume Habert holds the Chair of Sustainable Construction at the ETH Zürich and became full professor in 2023. He has lectured on sustainable construction and taught in various engineering and architectural schools over the world, published over 100 scientific papers and contributed to the development of a startup on earth construction and software to implement sustainability into design practices.

Drew Harris is a researcher, author, and lecturer in the fast-growing discipline of population health. As a member of the Thomas Jefferson University College of Population Health (JCPH) faculty, Dr. Harris presents regularly on population health and health policy to regional and national audiences. His current passion is finding ways to help lay and professional audiences understand the important role social determinants of health, green space, and the built environment play in population health outcomes.

Molly Hesson is an undergraduate student at Drexel University College of Engineering studying for bachelor's degrees in both Architectural and Civil Engineering with a concentration in Structural Engineering. She has done research on the effects of climate change on residential architecture and is interested in finding ways to combat those effects for current and future projects, specifically through the revitalization and reuse of existing structures. Molly is graduating in June 2024 and will be working as a construction consultant after graduation.

Simi Hoque, PE, LEED AP is a Professor and Head of the Architectural Engineering program at Drexel University. Her research scales from indoor environmental quality to urban livability. Simi was trained as an architect and an engineer, with degrees from the University of California-Berkeley, Carnegie Mellon, and Johns Hopkins. She teaches architectural engineering students how to design building systems that center on climate resilience, health, and well-being in the built environment. Simi is a 2024 Fulbright Scholar to Italy where she is exploring climate impacts on urban built environments. She is deeply invested in promoting diversity and equity in science and engineering.

Giuliana Iannaccone, PhD is Professor of Architectural Engineering and Head of the Building Engineering for Sustainability Graduate Program at Politecnico di Milano. Her primary research efforts have been to study and develop strategies and tools supporting the decarbonization targets of the building sector, exploiting the potential of cross-disciplinary, systems thinking approaches. Due to the long-term challenges of the built environment, she has more recently initiated collaborations to introduce futures thinking methods to support her research in the field. Since 2017, with Eugenia Victoria Ellis coordinates collaborative international learning activities on bio-inspired design and innovative sustainable urban environments.

Hugh Johnson has held several positions at Drexel in research development in the areas of energy, sustainability, and climate change. In his current role as the Senior Director of Research Development and Strategy within the Environmental Collaboratory, Hugh focuses on supporting community-based and co-lead research that seeks to align the climate transition and environmental justice with the needs of communities. Prior to Drexel, Hugh consulted and provided financial and technical due diligence on renewable energy and green building projects for private, municipal, and federal clients.

Heather Kostick received her BSc in Wildlife Conservation from Juniata College, her MES from the University of Pennsylvania, and is a PhD candidate in Environmental Science at Drexel University. She has research experience in biodiversity, resource management, conservation, and agroecology. Heather's current research focus is on biodiversity and urban ecology in urban green spaces. She works in climate change communication at the University of Pennsylvania with the Penn Center for Science, Sustainability, and the Media.

Joel Krakower holds a BS in mechanical engineering. He has worked on nuclear propulsion at NASA and currently works on turbomachinery and thermal fluids at Drexel University.

Franco Montalto is a licensed civil and environmental engineer with 30 years of research, planning, and design experience in the fields of ecological restoration, green infrastructure, and urban sustainability and resilience. He is a Professor at Drexel University, where he directs the Sustainable Water Resource Engineering Lab and is also the Founder and President of eDesign Dynamics LLC, an environmental consulting firm based in New York City, with an international portfolio of projects.

Eugenio Morello is an architect and Associate Professor in Urban Planning at Politecnico di Milano, Italy. He is also a coordinator of the "Laboratorio di Simulazione Urbana Fausto Curti" within the Department of Architecture and Urban Studies. His research focuses on the intersection of urban design and sustainable development, emphasizing environmental quality, climate design, and climate change adaptation. He explores urban transition solutions for designing sustainable communities, with a goal of closing energy and environmental cycles in cities.

Fatemeh Nasrollahi is an architect, engineer, and water resource modeler with over 10 years of experience in planning, design, and analysis in urban contexts. Her doctoral research in the Dept. of Civil, Architectural, and Environmental Engineering at Drexel University utilized hydrologic and hydraulic models to quantify the impacts of climate change on urban communities under different adaptation scenarios. She also has degrees from Iran University of Science and Technology, the University of Calgary, and George Washington University.

Francesco Pittau graduated in 2009 from Politecnico di Milano in Building Engineering-Architecture. He completed his European PhD in Building Systems at Politecnico di Milano with honors in 2013. In 2016, he was Visiting Professor at the University of Saint Joseph in Macau and the year after took the position of Research Associate at ETH Zürich, Chair of Sustainable Construction, and Adjunct Professor at Politecnico di Milano. From 2020 to 2025 he was Assistant Professor in Building Production at Politecnico di Milano, Department ABC, before becoming an Associate Professor. His research is focused on the identification and development of solutions and strategies for a rapid decarbonization of the construction sector. Particular interest is placed in the development and implementation of bio-based materials for the promotion of climate-regenerative solutions, both for new construction and retrofit.

Scott Quitel has practiced ecological restoration for 20 years—first as principal of Applied Ecological Services; next as founder of LandHealth Institute. Prominent Quitel-led AES restoration work includes Washington Avenue Pier, Philadelphia; Sycamore Island, Pittsburgh; and Chestnut Grove Natural Area, Lancaster. LHI work includes ProFESS (Program for Future Environmental Scientists and Stewards), LHI Native Plant Nursery, and nature-based coastal resilience planning in exposed urban neighborhoods. Scott also is a professor at Drexel University's Close School, where he co-teaches with urban ecosystems and leads efforts to grow a new BS Degree in Sustainability and Innovation.

Mahsa Safari Senior Associate at Cadmus, has more than 5 years of experience in energy efficiency and decarbonization projects. Mahsa has performed impact evaluations for utilities nationwide, including Code and Standard Evaluation, Workforce Education and Training Evaluation, and Prescriptive and Custom Program Evaluation. Mahsa also worked on the NYC Decarbonization Pathways Study, where she performed systemic reviews of deep decarbonization studies.

Leah Schinasi, MSPH, PhD is an Assistant Professor of Environmental and occupational health at the Drexel University Dornsife School of public health. She is also a Harvard TH Chan JPB Environmental Health fellow and a Strategic Council Member of the Drexel University Urban Health Collaborative. Leah's research investigates the structural determinants of climate-related health inequities. Her current projects include a NHLBI-funded study of links between housing characteristics, indoor temperatures, and mortality outcomes, and a JPB Foundation-funded study of associations of social and labor policies with occupational heat vulnerability. Leah received her BA in Environmental Studies from Vassar College, and her MSPH and PhD in Epidemiology from the University of North Carolina at Chapel Hill.

Ramyar Tajik is a PhD candidate in Architectural Engineering at Drexel University, focusing on indoor air quality (IAQ) and circadian lighting to enhance well-being. He earned his BSc in Architectural Engineering from the University of Guilan and an MSc in Building and Architectural Engineering from Politecnico di Milano. Ramyar is involved in key projects, including improving IAQ in Philadelphia's low-income households and aiding disaster response decision-making for Southeast Pennsylvania school districts, in collaboration with Temple University. As an awardee of the Kling-Lindquist Fellowship from Drexel University's Engineering School in 2023 for his dedication to research in Architectural Engineering, Ramyar is committed to advancing AI and machine learning technologies to mitigate health risks and tackle social injustice, he aims to develop equitable and healthy indoor environments.

Francisca Tapia a senior consultant and PhD candidate, applies her expertise in urban planning and architecture to research, sustainability, and innovation. Having experience in Ecuador, Argentina, Hungary, and Italy, her work includes implementing and evaluating NbS, managing integrated energy solutions during renovations, and influencing decision-making and policy in several Horizon 2020 projects. She is currently pursuing her PhD, with a research focus on combining NbS and user-centered principles to enhance engagement and collaboration for successful implementation. She is also involved with the German NGO, gaia-liNc, participating in sustainable development projects throughout Latin America.

Helena van Vliet, RA Dipl.-Ing. AIA is a Biophilic Architect with 30+ years of construction experience, a Consultant, Researcher, Educator, and Speaker on Health and Well-Being in the Built Environment. Helena is President of Helena van Vliet

LLC, President of BioPhilly, and a Steering Committee Member at Biophilic Cities International. She is a member of the AIA Academy of Architecture for Health. Helena teaches and lectures nationally and internationally at various universities, including Thomas Jefferson University; the University of Pennsylvania; Drexel; Pratt Institute; the Politecnico di Milano, Italy; and the Bauhaus University, Germany.

As an active Municipal Planning Commissioner since 2002, Helena has advised the governing body in planning, zoning, and conservation development towards sustainable low-impact growth for the well-being of her local community.

Dane C. Ward, PhD is an Assistant Professor in the Department of History and Anthropology, Director of GIS programs, and university cartographer at Monmouth University. Dr. Ward's research examines human–environmental interactions with a focus on ecosystems with extreme stress gradients. These systems are often marked by the absence of spatially explicit or spatially truncated transitional zones (i.e., ecotones) and include urban landscapes, urban-coastal ecosystems, and unique ecosystems such as the Pine Barrens. Dr. Ward is interested in understanding factors driving biodiversity, ecological stability, and urban resilience. Additionally, Dr. Ward partners with colleagues to examine the interaction between environment and human public health outcomes including mental health, nutrition, and access to green spaces.

Jin Wen is Professor of Civil, Architectural, and Environmental Engineering and Associate Dean for Research and Innovation at Drexel University. She has extensive experience in building energy simulation, including work on greenhouse gas emissions reduction plans for Philadelphia and New York City and serving as the Task Leader for International Energy Agency (IEA)'s Energy in Buildings and Communities (EBC) Annex 81 (Data-Driven Smart Buildings) Task C (Applications).

A Reference Framework Addressing Nature-Positive Goals in Cities

Eugenia Victoria Ellis and Giuliana Iannaccone

Abstract It is now accepted knowledge, not only in science, that the built environment plays a decisive role in the health of people and the planet. Data and scientific evidence increasingly support the development of research in the various fields of knowledge in which the built environment is articulated towards sustainable development goals, with a focus on the better use of resources, materials, and energy, the decarbonization of processes, and the health and well-being of occupants.

The complexity of the interdependencies between these objectives increasingly draws attention to the need to consolidate an interdisciplinary research model for the sector to achieve a convergence of objectives and responsibilities towards society and the environment.

The increasing pressures to which the urban built environment is subjected, e.g. because of demographic and climate change, push research efforts towards the need to find solutions to real, urgent problems capable of resolving or mitigating impacts at different levels.

This chapter discusses how interdisciplinary research approaches can offer the necessary breadth of purpose and vision to address the challenges at hand. In this context, the Nature-positive Built Environment initiative offers an example of how to address research in the field of the urban built environment through the lens of ecosystem management and investigates possible strategies to provide a complex adaptive system with the necessary flexibility and ability to respond to environmental change and increasing uncertainty.

E. V. Ellis
Department of Civil Architectural Environmental Engineering, Drexel University, Philadelphia, PA, USA
e-mail: eve22@drexel.edu

G. Iannaccone (✉)
Department of Architecture Built Environment Construction Engineering, Politecnico di Milano, Milan, Italy
e-mail: giuliana.iannaccone@polimi.it

G. Iannaccone, E. V. Ellis (eds.), *Towards a Nature-Positive Built Environment*, Research for Development, https://doi.org/10.1007/978-3-031-95292-0_1

Keywords Nature-positive · Built environment · Sustainability · Biodiversity · Carbon neutrality

Introduction. From a Human-Centered to a Nature-Positive Built Environment

Since the built environment was established as a recognized interdisciplinary field of study (Chynoweth 2009), its theoretical approaches have been initially characterized by a range of practice-oriented subjects concerned with the design, development and management of buildings, spaces and places (Griffiths 2004) with a focus on process- or product-related issues. Humans, as building users or occupants, were included in these approaches, but only later did they become central to them, while the natural environment was typically defined in opposition to the built environment, or something separated from it (Moffatt and Kohler 2008), that could be manipulated for human needs.

Generally, a distinction is made between user-oriented theories, where humans are a part or actor in the built environment system, and human-centered approaches, generally relying on the analysis of the relationship between the built environment and people's behavior, also with the aim to minimize the impact of the built environment on its users. In this framework, the user's experience of the built environment becomes central as it introduces a human perspective into both process- and product-related theories of the environment and the measure of its effectiveness or quality (Vischer 2008).

With the advancement of research in the field of sustainable buildings, the point of view of humans, as occupants or users of buildings, has been integrated into an operation-oriented approach to building design, where human factors—human behaviors, perceptions, and values—have been correlated with the building management and operation strategies to reduce energy consumption. Human factors are included when assessing comfort conditions (temperature, lighting, acoustic, humidity, etc.) and health necessities (air quality, water, etc.) in energy efficient building design, and the lack of consideration of the interactions between humans and the environment, as suggested by the literature, is regarded as the cause of performance gaps of sustainable buildings (Tao et al. 2020). Poor human–building interfaces lead to energy-intensive behaviors and violate the original intent of the building. To provide living environments with constant indoor comfort conditions regardless of season, latitude and climate, the imperative of energy efficiency has accompanied research toward the realization of buildings designed and constructed to accommodate these new comfort needs, while ensuring that building energy requirements are reduced. Demand-response strategies and smart building concepts have been introduced to predict and guide human behavior towards an energy-efficient management of buildings, in consideration of the desired comfort expectations. The human-building interactions have become central to the design of the latest generation of sustainable buildings and the human factor has increased relevance with the growing recognition of the role of the built environment on people's health since the last pandemic. In the COVID-19 years it has also become clear of how the destruction of biodiversity caused by human pressure on nature poses a threat to human life and undermines economies around the world. The more biodiverse an ecosystem is, the more difficult it is for one pathogen to spread rapidly from wildlife and captive animals to human population. At the same time, over half of the global GDP is moderately or highly dependent on nature and therefore exposed to risks when nature is degraded or lost entirely due to deforestation, encroachment on wildlife habitats, intensified agriculture, and acceleration of climate change (World Economic Forum 2020).

Most of the efforts introduced in Europe, as in the United States, to meet the decarbonization goals of the Paris Agreement (Kamel et al. 2024), primarily address the reduction of the operational energy demand of buildings and only partially the problem of reducing the overall impact of the construction sector concerning climate change and biodiversity (Röck et al. 2020).

This is the reason why in 2020 several global nature and development non-governmental organizations and business organizations set a Global Goal for Nature that could be integrated with other global goals to create an equitable, carbon-neutral, nature-positive world (Locke et al. 2021).

Through the research and projects presented in this book, emphasis is placed on a converging set of values and priorities toward the transition from a human-centered to a nature-positive built environment that would promote a shift in perspective (Du Plessis and Brandon 2015), prioritizing a harmonious co-existence between humans and nature within our built spaces, where buildings and infrastructure benefit from and contribute to healthy ecosystems.

The Need for a Nature-Positive Built Environment for Health

More than half of the world's population, more than 4 billion people, lives concentrated in urban areas (Ritchie et al. 2024). There are many recognized benefits of urban settings in so many areas, including high density of economic activity, transportation and shorter trade links, energy consumption, shared infrastructure, job opportunities and cultural offerings. However, disconnection from nature has emerged with increasing force as a major cause that is afflicting individuals and society at different levels (Beery et al. 2023).

Reversing the loss of nature due to the urban built area as we know it today is firstly crucial for people's health and well-being (Pinter-Wollman et al. 2018). Humans are subject to the many selective pressures associated with urbanization, which alters mortality, demography, disease transmission, air, water and soil pollution, hygiene, nutrition, and social relations. With nature comes biodiversity. Biodiversity is the variation among living organisms from different sources including terrestrial, marine and desert ecosystems, and the ecological complexes of which they are a part. Biodiversity describes the richness and variety of life on earth—and this range of variety is needed to sustain life on earth as we know it today. It is the most complex and important feature of our planet (Norton 2008). Without biodiversity, life would not sustain—life is dependent on the variabilities between plants, animals and microorganisms and their relative frequencies in an ecosystem. Biodiversity contributes meaningfully to human health and well-being globally and its positive effects on the city can be felt when biodiversity is considered as part of the city's urban plan for green public space. A biodiverse city is a city that works like a living system where economic, social and ecological functions work together in harmony. Biodiversity in the urban environment is important because biodiversity adds health to the equation of environmental, social and economic justice (Wolch et al. 2014).

Urban parks and their inherent biodiversity have been recognized as a new paradigm for community health and wellness (Africa et al. 2019). Studies have shown that access to green space promotes healthier eating habits in students and encourages exercise, both of which contribute to healthier youth and less obesity in children and adolescents. Studies have also shown that contact with nature and natural daylight improves cognitive functioning. Natural changes in daylight balance the body's circadian rhythm, which determines sleeping and eating patterns, brain wave activity, heart rate, cognitive functioning, and hormone production — basically everything having to do with human physical wellbeing (McEachron and Ellis 2017).

For example, just looking at nature, like walking past a park, can positively impact wellbeing by fostering a sense of nature relatedness, which has been associated with better cognitive, psychological, and physical health and encourages better eating habits (Milliron et al. 2022). Research has also demonstrated that contact with the natural environment increases mobility in older adults because green space encourages outdoor activity—even walking slowly through a park will provide health benefits. Green public space in the city encourages healthier eating habits, a healthier lifestyle and exercise, all of which decreases stress (Frumkin 2013) (Fig. 1).

Further, dynamic stress responses, using ambulatory heart rate, were monitored among neighborhood residents as they walked past vacant lots before and after a greening remediation project (Berman et al. 2008). These studies showed viewing a green space significantly decreased heart rate when compared to viewing a non-greened vacant lot, just by looking at the urban green space. Research has also discovered that patients who can enjoy the view of a garden outside their hospital window (Ulrich 1984), and those who can actually go outside and experience nature first-hand, heal more quickly and experience marked psychological benefits when compared to those not exposed to a natural environment. This symbiotic relationship has also been extended beyond the hospital room to suggest a multitude of cognitive benefits associated with interacting with nature (Stott et al. 2024). In addition to changes in individual-level behaviors, community gardens may also promote community cohesion, beautification, civic engagement, and social well-being.

Urban green space also helps to prevent crime. For example, farming in the city can engage young people not just with growing food, but also with the

Fig. 1 Biblioteca degli Alberi Milano (BAM). An example of a public-private partnership, between the City of Milan, COIMA and the Riccardo Catella Foundation, to secure a green lung in the heart of Milan and provide excellent maintenance, safety for citizens and an inclusive cultural program open to the city

entrepreneurship that comes along with harvesting the food and selling it—which in turn reduces teenage crime and improves eyes on the street, which results in increased feelings of safety and a reduction in gun violence.

A city with green public spaces can provide a fountainhead of ecosystem services, especially clean air and clean water, which keeps citizens healthy and prevents cities from flooding (Ahern et al. 2014). Ecosystem services are the natural benefits that diverse ecosystems provide to humans and the planet, such as food, water, climate regulation, and biodiversity. Biodiversity is critical for the sustainable development of cities (Opoku 2019). For example, as much as one third of the net reductions in greenhouse gas emissions required to meet the Paris Agreement's goals could simply come from Nature-based Solutions. Further, all food systems depend on biodiversity due to the broad range of ecosystem services that are required to support agricultural production, for example pollination, pest control and soil fertility. Healthy ecosystems also underpin the delivery of water supplies and water quality, and guard against water-related hazards and disasters.

Wetlands can be incorporated into the green public space of city riverbanks to clean hazardous materials from the water and soil in addition to supporting plant, animal and insect life (Zhang et al. 2022). Biodiversity itself supports the chain of life, starting with healthy soil with micro-organisms that can help support slightly larger creatures, diverse plant life, and so on. Overall, biodiversity leads to resilient cities because of its diversity and its ability to continue functioning due to the redundancy of nature that ensures some organisms will remain functioning even when others fail (Fig. 2). While wetlands can help control stormwater, stormwater management strategies can also be incorporated into urban design with green infrastructure. When green stormwater infrastructure is integrated with green public space it also improves air quality and reduces the heat island effect. The urban heat island effect is a typical phenomenon in urban areas. Buildings, roads and pavements absorb solar heat leading to higher temperatures especially in the summertime. This can be addressed by investing in urban forests, more trees in the streets and adaptive building practices such as green roofs and green walls (Figs. 3 and 4).

Biodiversity relies on the connectivity of green spaces, which is rare in cities (Dearborn and Kark 2010). Further, urban rivers and their corresponding riparian zones can also become important natural wildlife corridors. This is an important factor to recognize because pollinators are essential for the reproduction of many plant species, which in turn provide the vital ecosystem services needed to ensure

Fig. 2 The urban park along the Schuylkill River was opened to the public in 2000. The construction of the trail and greenway transformed formerly industrial land into a beloved resource for the people of Philadelphia. This stretch of trail is a vital commuter corridor, a valued recreational resource, and an oasis within the city. This project has also been essential in preventing erosion and protecting water quality

Figs. 3 and 4 The vertical forest in Milan (Stefano Boeri architect) and Palác Národní in Prague (Fiala Němec Architects). Two different examples of possible vertical green integration on buildings

sustainability. Because pollinators rely on natural vegetation as habitat, urban habitats, such as community parks, green roofs, and walls, offer considerable potential for supporting pollinators (Senapathi et al. 2017).

The Need for a Nature-Positive Built Environment for Climate Change

Parallel to the carbon neutrality goal defined by the Paris Agreement and the United Nations Framework Convention on Climate Change, nature conservation and sustainable use of natural resources should become as many nature-positive targets for governments and businesses (Locke et al. 2021).

Indeed, it is not enough to mitigate climate change without considering actions to restore natural systems to a healthy state, ensuring the conservation of existing carbon stocks and preserving the carbon sequestration function. The stability of the planet's temperature regime is ensured by nature's ability to absorb and store carbon dioxide on land and in the oceans, to regulate moisture fluxes from biomass to the atmosphere, and the solar radiation reflectivity of land surfaces and ocean ice. These functions are provided by biodiversity and the natural processes on which it depends. The decline of biodiversity is clearly visible and widely documented. And the resulting serious consequences for the health of the planet, human beings and all other living beings have been proven (WWF 2022). There is a considerable risk that we will very soon reach tipping points beyond which nature can never recover.

Ratified by 196 member states of the UN at COP15 in December 2022, the Kunming-Montreal Global Biodiversity Framework (GBF) is the equivalent of a Paris Agreement for nature, that established a global goal to halt and reverse nature loss by 2030, raising nature to the same level as climate on the global political agenda. Like net zero for carbon, this 'nature positive' goal articulates the level of ambition required from all parts of society if we are to prevent catastrophic nature loss (Nature Positive Initiative 2023).

This global goal for nature can be integrated with other global goals to create an equitable, carbon-neutral, nature-positive world. In the case of the built environment, the goal to become nature positive by 2030 is highly aligned with halving carbon emissions by 2030 and achieving net zero carbon by 2050. Indeed, many of the strategies that we need to implement to decarbonize the built environment also help to reduce nature loss and vice versa, for example: prioritizing building retrofit over new build, promoting circular design, and practicing sustainable procurement (Allen-Sader et al. 2023).

The Need for a Nature-Positive Built Environment for Business

Nature loss is already impacting business especially for those sectors—over half of the global GDP—that are moderately or highly dependent on nature: indeed the World Economic Forum puts biodiversity loss among the top five risks to global economy (World Economic Forum 2020). Three socio-economic systems represent over one third of the global economy and provide up to two-thirds of all jobs with significant opportunity and responsibility to reverse nature loss: "Food, land and ocean use", "Energy and extractives" and "Infrastructure and the built environment". For this reason, aligning the global economy with planetary boundaries is fundamental to generate a more effective economic growth and build more resilient societies.

The built environment is highly dependent on nature, being the largest global consumer of raw materials. Indeed, it is not only responsible for 40% of greenhouse gas emissions, but also for the consumption of 75% of natural resources and the production of 50% of global waste.

According to the IPBES (the Intergovernmental Science-Policy Platform on Biodiversity and Ecosystem Services) only in 2019 urban areas were responsible for the extraction of sixty billion tons of renewable and non-renewable raw materials, of which 40% was stone, sand and gravel, and 25% was virgin timber (IPBES 2019). As with other building materials, the scarcity of resources can lead to a huge problem, also in relation to the demand that will increase due to population growth, with almost 10 billion people expected to inhabit our planet by 2050 (United Nations Department of Economic and Social Affairs, Population Division 2022). For instance, sand is an essential aggregate for concrete and glass production but is being mined at a far quicker rate than is being naturally replenished.

With populations growing, cities are expanding, areas are being redeveloped, and new cities are being built, this growth will intensify pressures on nature, leading to increased water usage, pollution and waste production. Population growth, increased frequency of extreme weather events, and the resulting need for repair, and rising sea levels are accelerating the demand for construction and resource use, especially in urban areas, where more than two-thirds of the world's population is expected to live in the next 30 years. Recent studies also show that built-up areas are expanding even faster than population growth: by 2050, cities are expected to cover two to six times the area they did in 2000; this could increase material use to 90 billion tons by 2050, nine times more than in 2010, putting increasing pressure on urban systems such as water, energy and waste (Kissinger et al. 2013). Biodiversity loss is widely recognized as a systemic risk that threatens not only our prosperity but our very future as a species. Our 'take-make-waste' economy is increasingly being recognized as the main underlying cause of the biodiversity crisis. Surprisingly, more than 90% of this biodiversity loss is due to the extractive, polluting, and wasteful way we use resources. To halt and reverse biodiversity loss, we need to fundamentally transform our production and consumption systems (Di Sacco et al. 2021). The increased conservation and restoration efforts currently underway might slow this down for us. However, we need to take much larger strides in this effort and the circular economy offers an actionable framework for this transformative change. The circular economy strategy would fundamentally transform our production and consumption systems first of all by eliminating waste and pollution which would then reduce threats to biodiversity (Ellen MacArthur Foundation 2021). This would leave room for biodiversity to exist by recirculating recycled products and materials instead of continuing to deplete the earth's resources. This transformation would lead to the regeneration of nature which would enable biodiversity to thrive. It is also necessary to consider the in-situ (on-site) vs ex-situ (off-site) impacts along the construction value chain. Many processes of the built environment system depend on functioning water flows and water reserves, such as for extraction of raw materials, construction of building materials and water use during operations. Not to mention the increased demand for water, needed to produce concrete, which will increase in those very regions where water scarcity is expected to increase by 2050 (Miller et al. 2018).Taking mining as an example, key for the built environment, on-site activity causes significant air, water and soil pollution, which can have severe impacts on human health. Additionally, mining has a detrimental direct impact on nature through land-use change, with the sector driving 7% of all global deforestation (Bradley 2020).

Taking Action: A Global Commitment

By acknowledging the responsibilities of the built environment for the degradation of the natural environment, and the impact of nature loss on the built environment system, the World Business Council for Sustainable Development (WBCSD) is developing roadmaps to nature positive businesses. They offer specific guidance for the high-impact socio-economic systems which delve into the potential to move beyond reduced harm and achieve net positive benefits for nature. In particular, the roadmap for the built environment system, by highlighting the construction industry's enormous footprint, identifies the opportunities for a transformative and positive change across the entire value chain.

Six key priority actions have been identified to ensure the built environment system plays its role in halting, healing, and reversing nature loss by 2030 (WBCSD 2023):

1. Avoid further terrestrial, freshwater, and marine habitat conversion.
2. Prioritize re-use and retrofitting of buildings above demolition
3. Introduction of nature-related criteria in the selection of raw materials
4. Reduce raw material demand with circular design.
5. Address climate change and nature simultaneously.
6. Integrate Nature-based Solutions (NbS).

The following chapters provide concrete examples of how these priorities can be centered at different levels in the transformation processes of the built environment and what the main benefits are.

First, a new imperative is needed for construction: avoiding building new, whenever and wherever possible. The loss of natural habitats should be avoided by locating new buildings in brownfields, i.e. previously-impacted areas, or extending the service lives of existing buildings, often underutilized, which would reduce the need for new resources for construction (including land use) and hence mitigate both climate change and biodiversity loss (Kuittinen 2023). The built environment's footprint on nature also includes the impact of the material choices in the sector. True transformative changes can be achieved by careful selection of construction materials and by their maximum reuse. Circular economy principles should be adopted across the construction industry to reduce the use of virgin raw materials and greatly lower the levels of building waste, which can be a significant proportion of the weight of the building materials in the construction and demolition phases. In contrast to the current linear model, the circular economy is based on sharing, leasing, reuse, repair, refurbishment and recycling, in an (almost) closed loop, where products and the materials they contain are highly valued. In practice, it implies reducing waste to a minimum. This model takes its lesson from nature: nature uses only the resources it needs and recycles everything with no waste, for example harvesting nearby resources as birds do for their nests. These cities aim to eliminate the concept of waste and keep their assets at their highest value. Adopting circular economy approaches in a high-growth, high-waste sector like the built environment presents a tremendous opportunity for business, governments and cities to minimize structural waste and realize greater value from existing assets.

Adopting a whole life cycle approach to the assessment of nature impacts on buildings and infrastructure is essential to halt and reverse nature loss by 2030. Whole life carbon assessment (WLCA) could be a starting point to measure embodied nature impacts along the construction value chain: from raw material extraction and production to design and construction, and operations and maintenance, then finally to demolition and waste. Indeed, over the past 10 years there have been substantial advancements in the measurement and understanding of embodied and whole life carbon of buildings. We can leverage our knowledge of whole life carbon to accelerate our understanding of the whole life nature impacts of buildings. By using the advances that have been made to assess the whole life cycle carbon of buildings and leveraging areas of commonality wherever possible, a common language for the nature impacts of the construction value chain can be introduced, even if the whole life assessment of nature impacts associated with buildings diverges significantly at the point of measurement and metrics which are far more complex than for carbon.

It is particularly critical for new buildings and infrastructure to be nature positive as decisions on existing infrastructures are locked for many years to come. As a matter of fact, 90% of buildings currently in use in the European Union will be still in use in 2050. This transition would not be easy and would require collaboration between architects, engineers, ecologists, medical professionals, and policymakers. But the potential benefits are vast, fostering biodiversity, improving air and water quality, and promoting human health and well-being.

A growing interest has developed in recent years regarding the range of skills that will be needed by future professionals to meet the global and complex challenges of tomorrow's world. While technical knowledge is essential to a successful career in architecture, engineering and construction (AEC) professions, “soft skills” are playing an increasingly important role as emphasized by the current demands of the labor market and society: they include problem solving and critical thinking, communication skills, teamwork, social responsibility, emotional intelligence, and creativity (de Campos et al. 2020).

In addition, within the AEC industry, the need for new approaches and tools has become crucial to ensure sustainability at different levels, especially regarding the global objectives of protection of the planet, sustainable use of resources and people's health and well-being.

With this respect, the AEC professions have a critical role to play in the making of a sustainable human development which will require skills, tools, and education to address the global problems that our planet is facing or will be facing within the next 20 years (Mintz et al. 2014). A such, it is becoming more and more apparent that universities have a critical role in providing knowledge, testing, solutions and innovations to sustain and support the success of the Sustainable Development Goals (SDGs) that have been identified by the United Nations Department of Economic and Social Affairs (Gómez-Martín et al. 2021).

The essays collected in this volume are the result of many years of experience in global learning that places students, teachers and professionals into contact with students, teachers and professionals in other countries, demonstrating that problems can be solved in more than one way and also carrying the learning one step further: learning through the collaboration with others to effectively engage in different understandings and ways of thinking, which also makes more visible the non-technical human dimensions of the profession that stand alongside the technical dimensions as essential core features (Downey et al. 2006).

References

Africa J, Heerwagen J, Loftness V, Ryan Balagtas C (2019) Biophilic design and climate change: performance parameters for health. Front Built Environ 5:28. https://doi.org/10.3389/fbuil.2019.00028

Ahern J, Cilliers S, Niemelä J (2014) The concept of ecosystem services in adaptive urban planning and design: A framework for supporting innovation. Landsc Urban Plan 125:254–259. https://doi.org/10.1016/j.landurbplan.2014.01.020

Allen-Sader C, Gillhespy S, Schweizer Liégeard C (2023) Exploring nature positive buildings. In: WBCSD. https://www.wbcsd.org/Overview/News-Insights/WBCSD-insights/Exploring-nature-positive-buildings

Beery T, Stahl Olafsson A, Gentin S et al (2023) Disconnection from nature: expanding our understanding of human–nature relations. People Nat 5:470–488. https://doi.org/10.1002/pan3.10451

Berman MG, Jonides J, Kaplan S (2008) The cognitive benefits of interacting with nature. Psychol Sci 19:1207–1212. https://doi.org/10.1111/j.1467-9280.2008.02225.x

Bradley S (2020) Mining's impacts on forests. In: Aligning policy and finance for climate and biodiversity goals. The Royal Institute of International Affairs

Chynoweth P (2009) The built environment interdiscipline: a theoretical model for decision makers in research and teaching. Struct Surv 27:301–310. https://doi.org/10.1108/02630800910985090

Dearborn DC, Kark S (2010) Motivations for conserving urban biodiversity. Conserv Biol 24:432–440. https://doi.org/10.1111/j.1523-1739.2009.01328.x

de Campos DB, de Resende LMM, Fagundes AB (2020) The Importance of Soft Skills for the Engineering. Creat Educ 11:1504–1520. https://doi.org/10.4236/ce.2020.118109

Di Sacco A, Hardwick KA, Blakesley D et al (2021) Ten golden rules for reforestation to optimize carbon sequestration, biodiversity recovery and livelihood benefits. Glob Change Biol 27:1328–1348. https://doi.org/10.1111/gcb.15498

Downey GL, Lucena JC, Moskal BM, et al (2006) The globally competent engineer: Working effectively with people who define problems differently. J Eng Educ 95:107–122. https://doi.org/10.1002/j.2168-9830.2006.tb00883.x

Du Plessis C, Brandon P (2015) An ecological worldview as basis for a regenerative sustainability paradigm for the built environment. J Clean Prod 109:53–61. https://doi.org/10.1016/j.jclepro.2014.09.098

Ellen MacArthur Foundation. (2021) The nature imperative: how the circular economy tackles biodiversity loss

Frumkin H (2013) The evidence of nature and the nature of evidence. Am J Prev Med 44:196–197. https://doi.org/10.1016/j.amepre.2012.10.016

Gómez-Martín ME, Gimenez-Carbo E, Andrés-Doménech I, Pellicer E (2021) Boosting the sustainable development goals in a civil engineering bachelor degree program. Int J Sustain High Educ 22:125–145. https://doi.org/10.1108/IJSHE-02-2021-0065

Griffiths R (2004) Knowledge production and the research–teaching nexus: the case of the built environment disciplines. Stud High Educ 29:709–726. https://doi.org/10.1080/0307507042000287212

IPBES. (2019) Global assessment report on biodiversity and ecosystem services of the intergovernmental science-policy platform on biodiversity and ecosystem services. https://zenodo.org/record/3831673

Kamel E, Pittau F, Dal Verme LM et al (2024) A parametric integrated design approach for life cycle zero-carbon buildings. Sustainability 16:2001. https://doi.org/10.3390/su16052001

Kissinger M, Sussmann C, Moore J, Rees WE (2013) Accounting for greenhouse gas emissions of materials at the urban scale-relating existing process life cycle assessment studies to urban material and waste composition. Low Carbon Econ 04:36–44. https://doi.org/10.4236/lce.2013.41004

Kuittinen M (2023) Building within planetary boundaries: moving construction to stewardship. Build Cities 4:565–574. https://doi.org/10.5334/bc.351

Locke H, Rockström J, Bakker P, Bapna M (2021) A nature-positive world: the global goal for nature

McEachron DL, Ellis EV (2017) The human eye: a window to health and wellbeing. In: 105th ACSA annual meeting proceedings, Brooklyn Says, "Move to Detroit". ACSA Press, Washington, DC

Miller SA, Horvath A, Monteiro PJM (2018) Impacts of booming concrete production on water resources worldwide. Nat Sustain 1:69–76. https://doi.org/10.1038/s41893-017-0009-5

Milliron B-J, Ward D, Granche J et al (2022) Nature relatedness is positively associated with dietary diversity and fruit and vegetable intake in an urban population. Am J Health Promot 36:1019–1024. https://doi.org/10.1177/08901171221086941

Mintz K, Talesnick M, Amadei B, Tal T (2014) Integrating sustainable development into a service-learning engineering course. J Prof Issues Eng Educ Pract 140:1–11. https://doi.org/10.1061/(ASCE)EI.1943-5541.0000169

Moffatt S, Kohler N (2008) Conceptualizing the built environment as a social–ecological system. Build Res Inf 36:248–268. https://doi.org/10.1080/09613210801928131

Nature Positive Initiative (2023) Nature positive. A global goal for nature. https://www.naturepositive.org. Accessed 8 Mar 2024

Norton BG (2008) Toward a policy-relevant definition of biodiversity. In: Askins RA, Dreyer GD, Visgilio GR, Whitelaw DM (eds) Saving biological diversity. Springer US, Boston, MA, pp 11–20

Opoku A (2019) Biodiversity and the built environment: implications for the Sustainable Development Goals (SDGs). Resour Conserv Recycl 141:1–7. https://doi.org/10.1016/j.resconrec.2018.10.011

Pinter-Wollman N, Jelić A, Wells NM (2018) The impact of the built environment on health behaviours and disease transmission in social systems. Philos Trans R Soc B Biol Sci 373:20170245. https://doi.org/10.1098/rstb.2017.0245

Ritchie H, Sambrorska V, Roser M (2024) Urbanization. In: OurWorldinData.org. https://ourworldindata.org/urbanization

Röck M, Saade MRM, Balouktsi M et al (2020) Embodied GHG emissions of buildings—the hidden challenge for effective climate change mitigation. Appl Energy 258:114107. https://doi.org/10.1016/j.apenergy.2019.114107

Senapathi D, Goddard MA, Kunin WE, Baldock KCR (2017) Landscape impacts on pollinator communities in temperate systems: evidence and knowledge gaps. Funct Ecol 31:26–37. https://doi.org/10.1111/1365-2435.12809

Stott D, Forde D, Sharma C et al (2024) Interactions with nature, good for the mind and body: a narrative review. Int J Environ Res Public Health 21:329. https://doi.org/10.3390/ijerph21030329

Tao YX, Zhu Y, Passe U (2020) Modeling and data infrastructure for human-centric design and operation of sustainable, healthy buildings through a case study. Build Environ 170:106518. https://doi.org/10.1016/j.buildenv.2019.106518

Ulrich RS (1984) View through a window may influence recovery from surgery. Science 224:420–421. https://doi.org/10.1126/science.6143402

United Nations Department of Economic and Social Affairs, Population Division (2022) World population prospects 2022: Summary of results

Vischer JC (2008) Towards a user-centred theory of the built environment. Build Res Inf 36:231–240. https://doi.org/10.1080/09613210801936472

WBCSD (2023) Roadmap to nature positive. Foundations for the Built Environment System, Geneva

Wolch JR, Byrne J, Newell JP (2014) Urban green space, public health, and environmental justice: The challenge of making cities 'just green enough'. Landsc Urban Plan 125:234–244. https://doi.org/10.1016/j.landurbplan.2014.01.017

World Economic Forum (2020) New nature economy report II—the future of nature and business. Cologny/Geneva

WWF (2022) Living planet report 2022—building a nature-positive society. WWF, Gland, Switzerland

Zhang X, Zhang L, Wang Y et al (2022) Pollinators and urban riparian vegetation: important contributors to urban diversity conservation. Environ Sci Eur 34:78. https://doi.org/10.1186/s12302-022-00661-9

Reconnecting Humans with Nature

Buildings as Habitat: Adaptive Investments in Public Health

Helena van Vliet

Abstract Buildings kill an estimated 1 billion songbirds each year in the United States alone. Apart from the heartbreaking destruction of beautiful, innocent life, this catastrophic loss of biodiversity represents a public health emergency that tragically illuminates an all too literal clash between nature and culture. If evolution is any guide, culture must urgently realign with the biology of life. As the pervasiveness of human inflammatory conditions suggests, our hard-edged glass, steel, and concrete-dominated environments do not support our own biology either. In addition to biodiversity loss, chronic stress from extreme heat, noise, light, and air pollution degrades public health and well-being in many of our neighborhoods. From anxiety and loneliness to dementia, from hypertension to cancer, daily stress conditions corrode our immune response, reduce productivity, and fray our social fabric. The problem illuminates a science-informed path from opportunity to obligation: all buildings, new and existing, must adapt to become restorative, transformative, nature and body-positive infrastructure investments in public health. Local ecology must become intrinsic to each ecotonal building-envelope's design, supporting physiological well-being by mitigating air, noise, and light pollution, naturally cooling interiors and surrounding neighborhoods, sequestering carbon, and fostering urban micro-habitat biodiversity in symbiotic realignment with the biology of all urban lives.

H. van Vliet (✉)
Thomas Jefferson University, College of Architecture & The Built Environment, Philadelphia, PA, USA

University of Pennsylvania, Environmental Innovations Initiative, Philadelphia, PA, USA

Biophilic Cities, Charlottesville, VA, USA
e-mail: helena@helenavanvliet.com

G. Iannaccone, E. V. Ellis (eds.), *Towards a Nature-Positive Built Environment*, Research for Development, https://doi.org/10.1007/978-3-031-95292-0_2

Keywords Resilience · Health · Adaptation · Biophilic design · Biophilic urbanism · Neurobiology · Ecotonal building envelopes · Urban biodiversity · Micro-habitats

> *This is a fundamental view of the world: It says that when you build a thing, you cannot merely build that thing in isolation but must also repair the world around it so that the larger world at that one place becomes more coherent and more whole;*
> *and that the thing which you make can take its place in the web of nature as you make it.*—Christopher Alexander, *A Pattern Language*

In December of 2021, I had the distinct honor of participating on a panel assembled by the *National Museum of Natural History* in Washington, DC as part of their exhibit "*Unsettled Nature: Artists Reflect on the Age of Humans.*" Our webinar, titled "*Seeds of Potential: Art Re-imagines How Birds and Buildings Collide,*" sought to highlight artist *Andrew Yang's* inspirational project *The Flying Gardens of Maybe* as the starting point for a discussion about the deadly impacts of our buildings on the more-than-human world. What "Seeds of Potential" could we imagine turning what is currently a catastrophe into an opportunity to re-commit our buildings to supporting life? The National Audubon Society estimates that buildings, in the US alone, kill 1 billion songbirds each year: 56% are killed by buildings between 4 and 11 stories high, which constitutes the majority of urban structures, and an additional 1% by taller buildings (National Audubon Society 2023), with migration mass killing events that break into the news cycle. There is something uniquely horrific in the soft bodies of our feathered fellow beings, so destined to be free in weightless flight, slamming into walls of glass. The sheer number of innocent lives lost feels like a gut punch, so painful that we tend to look away. Yang sought to pierce our apathy by honoring the dignity and personhood of each individual bird, the abrupt silencing of each beautiful voice. He honored both grief and possibility, illuminating the disastrous loss of future potential and biodiversity this daily slaughter represents. Our beautiful songbirds' deaths amount to an all too literal clash between nature and culture, and since we cannot redefine nature, it is our culture we must redefine. We know that for us as well, there will be no survival, let alone well-being, without a vibrant biosphere. And yet, many of our buildings accelerate the already rampant biodiversity destruction, where, among many other species, insects, and birds are suffering a decline of unfathomable magnitude. If evolution is any guide, culture must once again be placed into the service of survival by realigning our built environments with the biology of life.

Birds experience our buildings as urban cliff formations, hunting and searching for food in our gutters and under our eaves. They try to raise their families on and around our buildings, only too often with tragic results. Many of our cities are built along rivers, essentially highways for birds who arrive hungry and depleted after long migration routes only to encounter hot, dry, and windy deserts dominated by concrete, steel, and glass and the many perils associated with those. Contemporary buildings have dramatically increased glass surfaces to please human users inside, neglecting to consider the danger this represents to other urban residents. At the risk of stating the obvious, buildings that regularly kill hundreds of birds cannot be considered sustainable buildings. They may well carry the important net-zero label but seem conceived as energy-efficient machines limited by a quantitative reductionist approach. Certainly, bird-friendly glazing solutions (see also Audubon and the American Bird Conservancy guidelines) must be obligatory for any project considering itself to be sustainable. Forward-looking cities, such as San Francisco and New York, have already passed bird-friendly glazing ordinances, which must become industry standard worldwide. Beyond that, we can do more. As the designated representative of the building industry on the Smithsonian panel, a decidedly humbling position, I sought to focus on the 'Seeds of Potential.' Every project, new and existing, represents an opportunity to get something essential right: namely, the responsibility of each building to be a benevolent community member that actively supports the well-being of *all* urban lives. Buildings that fall short of meeting this obligation must be updated via *nature-positive interventions*. Changing course to commit to an urgent re-alignment of our built environment with the biology of life, including our own biology, is the obvious path forward. The massive scale of building-bird collisions, which we as an industry apparently accept as part of 'the cost of doing business,' appears to be an unfortunately literal metaphor for the collision course we ourselves are on with nature in general and, considering public health data, with our own nature. Our reductionist approach turns out to be broadly nature-negative. It serves neither the bodies of birds nor the bodies of humans nor the bottom line. Still entrenched in a mechanistic definition of 'efficiency,' by what neuroscientist Antonio Damasio refers to as "the Cartesian idea of the disembodied mind" (Damasio 2005), we have produced built environments our ancient body-brain struggles to associate with life-sustaining shelter, and they most certainly are not designed to shelter anyone else. A culture accepting itself to be unmoored from its own body-nature sees fit to build machine-like environments modeled on the efficiency of disembodied software programs, adding strain to populations already immunocompromised by the corrosive effects of various other stressors, including biodiversity destruction. Why, then, do we continue to build buildings that do not support the well-being of people and, for that matter, of birds as the beautiful representatives of biodiversity? Like birds, we are biological organisms, open systems that internalize our environment with each and every breath. The COVID pandemic painfully awakened us to the fact that the same buildings intended to provide shelter can be uniquely perilous if conceived as hermetically sealed, mechanically controlled boxes. Because the weather patterns we created are killing us, we decided to live without them (Kennicott 2023), locking ourselves away indoors and unleashing public health inefficiencies that are impossible to ignore. As we discovered, our self-inflicted dependence on technology will offer no reliable safe haven from a virus, especially with an immune response already weakened by pre-existing inflammatory conditions. Have we forgotten that our biology requires fresh air, connection to nature, and dynamic sunlight? Hardly breaking news since Florence Nightingale already promoted this with great success as a tool both for the prevention and cure of illness in the 19th century. As Dr. Joseph Allen, a healthy building expert at Harvard's T.H. Chan School of Public Health, flatly stated, "Our building infrastructure (…) we know is not designed for health" (qtd. in Anthes 2023), a refreshingly frank admission, which begs the question: if not for health, what then *are* they designed for and what *is* our obligation as designers if not the well-being of building occupants? Daily contact with nature-rich environments is urgently needed to support a healthy immune response, especially in early childhood, so that our gut microbiome can remain the resilient protector of our health that nature designed it to be. Indoors, we may be sheltered momentarily from perils such as extreme heat by mechanical air conditioning, but at the same time, we contribute to worsening the deadly problem of extreme heat outdoors, where birds and other non-human fellow residents must contend with rising temperatures. Their livelihoods unsheltered, their suffering unconsidered, they exist as increasingly imperiled mere decorative props viewed from our sound-proof reflective towers. Ensconced behind mechanically cooled floor-to-ceiling plate glass, we are profoundly disconnected from birdsong, from nature, and we suffer the consequences.

In 2022, the US spent an astonishing $4.4 Trillion on healthcare costs (American Medical Association & American Medical Association 2025). Even more astonishingly, only 3.5% of that was spent on preventative services (Hargraves 2022), a perplexingly inefficient healthcare model indeed. We spend vast sums fixing things after they break instead of preventing illness in the first place. The building industry, by contrast, has had a tradition of focusing on prevention, at least when it comes to preventing buildings from collapsing, leaking, catching on fire, etc. More recently, we have concerned ourselves with protecting the public from indoor air toxins due to molds, glues, formaldehyde, and other unpronounceable chemical compounds lurking in construction materials. We have focused on reducing a building's operational and embodied carbon footprint to tackle climate change. And yes, all building enclosures must absolutely aspire to passive house standards. But beyond that, must we not also extend our care and efforts to protect the public from buildings that are physiologically and neurologically degrading because they represent "a mismatch between our ancient genes and our modern building habits" (Sussman and Hollander 2021). Yes, as a species, we are adaptive. However, judging from the high levels of anxiety, depression, loneliness, violence, and alienation, not to mention suicidal ideation in the population at large, things don't appear to be going that well. Evolutionary adaptation is slow, too slow, for the many profound ways in which we have departed from living conditions we are innately primed to require for health and well-being. Undeniably, the reasons for widespread inflammatory conditions are complex and multi-faceted, but the role of the built environment in exacerbating the problem has not been sufficiently considered, certainly not by the majority of architects and developers. Among mental health professionals, however, persistent findings that urban environments appear to be the source of an epidemic of schizophrenia and other severe psychotic conditions have drawn increasing attention (Kelly et al. 2010; van Os et al. 2010, qtd. in Architecture, the Urban Environment and Severe Psychosis: Aetiology n.d.). It stands to reason that the further we remove ourselves from environments that integrate the physiologically nourishing place ingredients we evolved in and with, the more separation anxiety we seem to experience. A profound misalignment with our innate biology has occurred in our day-to-day environments. Simply put, many built environments do not offer place conditions that our ancient body-brain associates with shelter and flourishing. While a fortunate few live in leafy, tree-covered neighborhoods or aside roof terrace gardens, the less fortunate majority struggles daily against overwhelming and literally unnatural because de-natured odds. So do thousands of our beautiful songbirds, who perish each day and each night in the built environments we have deemed fit for life. They are the hapless 'canaries in the coal mine,' illuminating the folly of our deadly constructs but also the path to redemption. As a matter of public health policy, we must commit to designing resilient places that make all of us – humans and other fellow travelers—more resilient by meeting our physiological and neurological needs on an everyday basis. Nature-positive places that support urban biodiversity are equally nature-positive in supporting our own biological nature. Beyond net zero, only *nature-positive/body-positive* built environments that function akin to preventative healthcare can truly be deemed sustainable, effective, efficient, or even just plain functional in the context of public health and well-being.

Nature-Negative/Body-Negative Aspects of Current Built-Environments Contributing to a Corrosion of Public Health

> *The facts of nature cannot, in the long run, be violated. Penetrating and seeping through everything like water, they will undermine any system that fails to take them into account.*—
> C. G. Jung, *Psychoanalyst*

Courtesy of evolution, we are designed to cope well with short stress events. They can even be energizing or, in the case of sports, reduce inflammation and eliminate toxins. Prolonged and repetitive stress, however, is another matter entirely. Few mechanisms impact our immune response more negatively. Even if not consciously registered as such, low-grade chronic stress corrodes our immune defense, leading to inflammatory conditions, which underlay most, if not all, major physical and/or psychological ailments. We may survive for a time, but we will not flourish. While certainly not the only source, evidence suggests that nature-negative built environments contribute significantly to the problem of chronic stress, the health consequences of which are not currently covered by construction warranties. Approximately since the end of WWI, we have enshrined the paradigm of cognitive dualism in built environments that are profoundly out of step, and quite literally out of touch, with our embodied animal nature, creating neurobiologically mismatched places that generate in us something akin to cognitive dissonance. Accelerating technological dependencies have only deepened our disconnect from nature, including from our own nature. In her provocative book *How to be Animal: A New History of What it Means to Be Human,* Melanie Challenger alludes to the dissonance that permeates our entire nature-of-body-negative culture:

> We are animals when we embrace and when our bloodied newborns slide from the bodies of women, but not when we make vows. We are animals when we bite into the flesh of our meal, but not in the workplace. (Challenger 2021).

Should we continue: "We are animals when we breathe, see, hear, smell, and touch, but not in built environments."? Cognitive dissonance is readily apparent in this question, and yet, from this same dissonance are born design processes and 'solutions' that "direct precious little attention to considering the emotional or livable qualities of our human habitats" (Mallgrave 2013). Cognitive dissonance seems to have become normative, enshrined in concrete, steel, and glass in the excessively noisy, hot, dry, over-lit, and de-natured environments so many of us attempt to call our home, exerting relentless underlying pressures on public health and therefore on societal equity, peace, and prosperity. In the context of public health, these are profoundly in-efficient places contributing to what Richard Louv refers to as "Nature Deficit Disorder" (Louv 2013). Our body-minded brains, ever alert to anything "that could threaten the integrity of the organism and compromise life" (Damasio 2010), have noticed the misalignment, the lack of physiological nourishment. Unsurprisingly, our bodies react with an entirely appropriate biological response pre-installed by evolution: fear. Melanie Challenger writes: "How do the aspects of the immune system come to know something unknown, let alone classify it as danger? What immunologists discovered was that there are capacities to probe and identify almost on a molecular scale" (Challenger 2021). I would hold that daily existence in an environment that forces misalignment with our neurobiological needs represents a level of chronic stress akin to trauma and may only be manageable by a degree of dissociation from one's body self. As Dr. Jan Golembiewski, a leading voice in architectural design psychology and a fellow in the *Centre for Urban Design and Mental Health,* writes: "This is where the built environment may become psychotoxic. Designers deliberately seek salience, and not only for the greater good. Design purposefully elicits responses; to impress, to spur on commerce, to alter behaviour" (Architecture, the Urban Environment and Severe Psychosis: Aetiology n.d.), resulting in an overload of aggressive sensory triggers from which there seems no physical escape. Dr. Golembiewski continues:

> The fact that designers are generally unaware of the effects that their inventions have on the brain and therefore take no responsibility for this role is deeply worrying. The world of design is like a highway, where each and every driver is asleep at the wheel (Golembiewski 2015).

In his 2023 national campaign to raise awareness about what he calls "Our epidemic of loneliness and isolation" (Office of the Assistant Secretary for Health (OASH) 2023), US Surgeon General Dr. Vivek Murthy elevates the re-design of physical elements in the built environment to foster community and connection to the first position in his framework of "foundational pillars." Many of our built environments exacerbate, if not outright foster, our archetypal fear of having been abandoned by our tribe, a fate foreshadowing death as far as our ancient body-brain is concerned. Considering the cost of chronic stress-related illness in populations, fear-inducing built environments should be considered akin to professional malpractice. Such places are characterized by a line-up of egocentric attention-seeking, nature-negative structures, unconcerned with their responsibility to support physiological well-being, community, or the play of children. These are not places where urban wildlife can make a living or informal and spontaneous encounters between all of the above can flourish. Instead, we are presented with hermetically sealed and mechanically conditioned constructs that leave us fundamentally out of touch with soils, seasons, climate, weather, and each other. Such places will slow our evolutionary adaption to the altering climactic conditions by creating technological dependencies that current global energy systems cannot sustainably satisfy. They force increased usage of mechanical ventilation and cooling, further exacerbating the problem of *extreme urban heat.* They also do nothing to absorb *fine particle pollution,* nothing to mitigate glare and circadian disruption from *light pollution,* and nothing to mitigate *excessive noise*; quite the contrary. Heat and noise are reflected and amplified relentlessly. Additionally, those buildings *do nothing to support urban biodiversity.* The catastrophic quantity of bird collisions serves as devastating testimony to our failure in this regard. There is a deep physiological connection between a bio-habitat's resilience and our own body's micro-habitat resilience (Mills et al. 2019; Tasnim et al. 2017), a connection established eons ago. Living outside this connection compromises our health and manifests as a sense of loss, loneliness, and unease, emotions not necessarily available to consciousness but powerfully corrosive to our physiology, nevertheless. To enjoy a robust immune response, the environmental conditions surrounding us must remain within the range of dynamic homeostasis, the ancient system in charge of optimizing efficient energy management of our body-brain processes.

> *On an urban level, one of the insights of newer emotional models is the recognition that our emotional responses are strongly integrated with our peripheral autonomic nervous system - that is, the working of our sympathetic and parasympathetic nervous systems.*—Harry Francis Malgrave, *Mind in Architecture*

We know that each of the major urban stressors has serious consequences for public health. Combined, they amplify each other, amounting to a devastating juggernaut that results in significant societal costs due to loss of well-being, loss of productivity, and even loss of life. While all of us are affected in one way or another, the distribution of stressors is profoundly inequitable, with some urban neighborhoods suffering from far greater heat, light, noise, and air pollution than others, leading to alarming discrepancies in life expectancy more determined by zip code than genetic code. Even low-grade stress, if unrelenting and repetitive, can activate ancient self-protective evolutionary responses. Triggered by daily sensory overstimulation, the somatic systems inform a physiological response, preparing us for fight, flight, or freeze by continuously pumping high levels of adrenaline and cortisol into our bloodstream. Repetitive and prolonged neurobiological stress moves from the neural into the hormonal realm, reaching deeper and deeper as "cortisol shuts down immune cells' responses" (Sternberg 2001). This is especially damaging to the developing bodies of children (Gunnar and Quevedo 2007). Numerous symptoms, including chronic basal hypersecretion, mitochondrial disruption, and adrenal exhaustion, can result from chronic stress-induced activation of the hypothalamic-pituitary-adrenal (HPA) axis (Herman et al. 2016). Prolonged stress activation supercharges ancient alarm responses that leave us chronically overwhelmed, anxious, perhaps even aggressive, and/or depressed. Heart rate and blood pressure can become chronically elevated. Cognitive functioning and memory become impaired (Westerink et al. 2020). "Autonomic and circulatory nervous systems carry distress signals to the gut" (Madison 2019), leading to dysbiosis, further weakening immune response. In short, the sympathetic nervous system is operating in overdrive in an attempt to protect. We are 'on edge.' Our body feels the need to be hypervigilant, which interferes with our ability to sleep, leading to circadian disruption. We know that a stressed and fearful mind narrows our ability to problem-solve creatively, to resolve conflicts constructively, trapping us in a vicious cycle of feeling unable to cope. We may become so alienated from our physical environment that we disengage completely into a kind of numb or frozen state. Neural inhibitory mechanisms become overwhelmed, and 'directed-attention-fatigue' sets in (Kaplan 1995). It is, perhaps, but a small step from here to seeking refuge in disembodied virtual worlds. Data suggest that parasocial fora tend only to heighten anxiety, potentially even foster anomic tendencies in some, by dissociating us further from the birthright of our animal-body selves (Office of the Assistant Secretary for Health (OASH) 2023). We have made progress in developing methods that can quantify exactly how corrosive chronic stressors are to human physiology. Stress biomarkers such as skin conductance response (SDR), heart rate variability (HRV), brain MRIs, and blood cortisol levels can provide definitive information. Evolutionary biology, cognitive neuroscience, biopsychology, neurobiology, and chronobiology provide important tools designers can use to curate all built environments as intentional well-being places informed by evidence-based biology and public health data. For the most part, the diagnostic systems we developed only serve to prove to our minds what our bodies already know: our nervous systems were designed for life in nature-rich environments, a fully immersive life that engages our many senses effortlessly. Data suggest designing urban environments that align with our physiological needs, that restore and sustain us on a neurobiological level, are the only efficient places. De-natured built environments offer no neurological sustenance, no respite for our biology. They carry significant negative externalities, societal costs that render them profoundly uneconomical and, in the context of public health and productivity, unaffordable. Let us examine the main nature and body-negative stressors more closely:

Loneliness & Biodiversity Loss: In his May 2023 public health advisory, US Surgeon General Dr. Vivek Murthy establishes as fact that loneliness and isolation represent a form of chronic stress with public health consequences akin to smoking, ranging from anxiety and depression to dementia, hypertension, heart attacks, and even death (Office of the Assistant Secretary for Health (OASH) 2023). Here again, a misalignment with our very nature has occurred. Our body-brain registers fear, ushering in susceptibility to inflammatory conditions. Isolation from the tribe represents a fundamental departure from evolutionary patterns that have evolved over hundreds of thousands of years to enhance our survival: isolation spells doom. The

COVID-19 pandemic certainly worsened the trend of isolation. However, many of our built environments have long been notorious for being "depressing, brutal, ugly, unhealthy, and spiritually degrading," as Howard Kunstler already pointed out in "Geographies of Nowhere" (Kunstler 1994), and indeed, Lewis Munford already lamented thirty years prior. As Dr. Murthy's urgent advisory indicates, we have not made sufficient progress in designing interventions that reassure and comfort our neurobiology and soothe the sense of isolation many of our urban and suburban environments seem to curate.

A related kind of loneliness afflicts us in places that are devoid of living nature. Biologist E.O Wilson argues that on a genetic basis, we are hard-wired for emotional connection with other forms of life, that our innate tendency towards *biophilia, a brotherly/sisterly kinship* with other animals, with birds, bees, and butterflies, must be considered an essential aspect of evolutionary optimization (Wilson 2009). In the presence of biodiversity, we feel quantifiably restored. The absence of nature has the opposite effect. Our careless destruction of the biosphere notwithstanding, there lives in us an underlying longing for connection with other living systems. And yet, we have had to subsist separated from nature for so long that we have difficulty putting our consciousness finger on the exact source of this particular loneliness. Richard Louv and others refer to it as "*species' loneliness.*" Until cognitive neuroscience only recently updated the paradigm, we spent some 400 years under Descartes' edicts, which declared (other) animals to be *res extensa*, material "things" devoid of consciousness, indeed, devoid of dignity or soul. Insisting, erroneously as it turns out, that the mind is of a higher order than the body, we have elevated ourselves above all other creatures only to find that it is lonely on the hard, dry, and windy tip of the pyramid we have invented for ourselves, a near perfect metaphor for a great many buildings we have constructed to take us higher and further away from earth. Perhaps we are longing to partake in the gravity-defying freedom of birds. Tragically, however, too many buildings intercept *their* freedom in midflight instead, sending millions in death spirals crashing to earth. Pretend as we might, we are not unaffected by this suffering. In her beautiful book "Braiding Sweetgrass," Robin Wall Kimmerer writes:

> Philosophers call this state of isolation and disconnection "species loneliness"- a deep, unnamed sadness stemming from estrangement from the rest of Creation, from the loss of relationship. As our human dominance of the world has grown, we have become more isolated, more lonely when we can no longer call out to our neighbors….We have built this isolation with our fear, with our arrogance, and with our homes, brightly lit against the night (Kimmerer 2020).

It is certainly hard to 'call out to our neighbors' from within sealed glass towers. The more we discover about the various forms of consciousness in other sentient life, the more absurd our invented hierarchy becomes. We know that "there are interdependencies between biodiversity, (…) and public health, which lead us to argue that human health outcomes could be improved by increasing contact with biodiversity in an urban context" (Mills et al. 2019). Residents of areas burdened by low biodiversity are more likely to suffer from decreased microbial gut diversity, which, especially in children, is "a key factor in determining disease susceptibility patterns later in life" (Tasnim et al. 2017). Since 70% of our immune cells are located in the gut, microbial deprivation can usher in any number of inflammatory diseases including allergies, setting the stage for zip-code-related inequities even before day one of a child's life.

Extreme Urban Heat: Among the deadliest side effects of planetary warming, extreme heat hits the top of the list. Most of our cities and buildings are not designed to cope. On the contrary, they generate tremendous amounts of heat due to mechanical systems, lights, and the density of heat-absorbing materials such as brick, stone, concrete, steel, and heat-reflecting glass. Instead of design techniques geared toward passive survivability, our near total commitment to sealed surfaces and technology has left parched cities vulnerable to expected brown-outs and, as we discovered, to viruses, which render many a building uninhabitable under those conditions. As extended extreme heat periods multiply, soaring mechanical air conditioning usage—already responsible for the largest increase in building energy usage—creates a perilous cycle by leaking increasing quantities of refrigerants with global warming potential 2000 times greater than carbon dioxide (Kennicott 2023). Many cities lack sufficient natural ventilation corridors, adequate quantities of greenery, and bodies of water to facilitate the ventilation, shading, evaporative cooling, and air purification they require to offset existing hardscapes and air conditioning heat exhaust. Without adaptative measures, deaths from extreme heat, already one of the leading causes of death, could increase by more than six times in the United States. As a result of extended periods of daily average temperatures above 90°F, more than 8500 deaths would be anticipated in a typical year under baseline climate and current demographic conditions. By 2050, this is anticipated to more than multiply by six, reaching 59,000 (Atlantic Council 2021). Birds fare equally badly or worse since we do not afford them the luxury of air-conditioned cooling centers. Like many other animals, birds are unable to cool themselves via sweat and, without access to cooling vegetation or water, struggle to survive. In 2023 extreme heat events in the US South-West, desperate volunteers tried to save dehydrated city birds, many baby birds among them, who were near death on sizzling pavements and without access to shade or water in temperatures in excess of 100°F for weeks on end. For humans and birds alike, sleep is disrupted by prolonged heat because natural summer night cooling has become impossible in many cities. In their excellent and comprehensive paper 'Biophilic Design and Climate Change: Performance Parameters for Health' (Africa et al. 2019), the authors offer a detailed analysis of our bodies' inability to adapt to rising temperatures and the devastating public health consequences and societal costs, from cognitive and immune system breakdown to aggression, and social unrest, that arise therefrom. Due to our overreliance on mechanical rather than passive cooling strategies, rising temperatures are predicted to increase the potential for multi-day electrical grid failures, which would further "amplify mortality and morbidity risk" (Stone et al. 2023). Authors describe this as an "infrastructure failure," which will place large percentages of urban populations at serious health risk, with hospitals unable to cope (Stone et al. 2023). Inner cities are warming twice as fast as surrounding suburbs, often with temperature differences in excess of 10°F. It is in our cities that the problem is most keenly felt and must urgently be solved.

Fine Particle Pollution: In addition to interior air pollutants from molds or potentially toxic building materials, the concentration of fine particle pollutants at urban street levels consistently exceeds public health recommendations. In fact, we increasingly understand that there is no "safe threshold for exposure" (Pugh et al. 2012). From gasoline car and diesel truck emissions to the microscopic particles generated by the abrasion of tires on asphalt, pollution lingers in poorly ventilated street canyons where the height of buildings often exceeds the width of streets by multiples. Low wind speeds in those canyons enable increased deposition of pollutant particles on horizontal and vertical surfaces on both the exterior and interior of buildings. Human exposure to $PM_{2.5}$ has been linked to inflammatory respiratory conditions, including lung cancer, chronic obstructive pulmonary disease, asthma, and interstitial lung damage (Pryor et al. 2022). Since fine particle pollutants pass directly into the bloodstream, they can trigger a systemic inflammatory response, affecting tissues and organs as well as causing blood clots and cardiac and placental damage (Pryor et al. 2022). Convergent epidemiological data present strong evidence that exposure to $PM_{2.5}$ affects brain network functions and is linked to decreased cognition, dementia, and a heightened risk of developing depressive disorders (Li et al. 2021; Air Pollution May Increase Risk for Dementia 2023). Other 2023 research found that "significant correlations between $PM_{2.5}$ and antibiotic resistance were consistent globally in most antibiotic-resistant bacteria" (Association Between Particulate Matter (PM)2·5 Air Pollution and Clinical Antibiotic Resistance: A Global Analysis 2023). It stands to reason that other biological organisms living in polluted urban centers, such as birds, small mammals, and insects, suffer silently in similar, if not identical, ways.

Light Pollution: As we transition to energy-saving LED lighting systems, it is an unfortunate fact that our cities are becoming brighter at night. The magic of starry skies has gone missing from our lives. Blue-white LED bulbs of 4000–5000 K customarily used in streetlights turn night into day. The detrimental health consequences they cause are now well understood. Our ancient body-brain associates these color temperature ranges with morning and floods the body with adrenaline and cortisol, increasing heart rate and generally marshaling energy resources for the demands of the day, except it is midnight. Melatonin, the sleep hormone essential for healthy immune system activity, is suppressed. Chronic circadian disruption is a serious form of stress that ushers in systemic inflammatory conditions, making us vulnerable to any number of problematic conditions such as hormonal dysregulation, diabetes, heart disease, obesity, depression, premature cell aging, dementia, and even cancer, specifically melatonin-sensitive breast and prostate cancer (Lin and Farkas 2018; Shafi and Knudsen 2019). Also negatively affected are impulse control mechanisms and conflict resolution skills. Even relatively small amounts of blue-white light exposure while sleeping affect us below the level of conscious cognitive perception, keeping the body in a semi-alerted state. In addition to the obvious public health costs, the cost of productivity loss induced by circadian disruption represents a serious problem in schools, healthcare systems, and businesses alike.

Our non-human neighbors suffer as we do, if not more. The majority of birds migrate at night, including sparrows, warblers, thrushes, and many others. The glare of our cities and electromagnetic radiation at close range interfere with their internal magnetic orientation mechanism, and the beacon effect of our night-illuminated glass towers lures tens of thousands of night-migrating birds to their deaths. There is much evidence that circadian disruption affects urban birds as well. They have been observed to feed at night in these daylight-simulated conditions, often making poor life-jeopardizing decisions or falling prey to night-marauding cats due to the stress of sleep deprivation. Similarly, night pollination is seriously affected, rendering night moth populations, of crucial nutritional importance for 60% of our songbirds, stressed and defenseless against bats in our streets, parks, and gardens flooded with blue-white light all night. Many night predators, such as owls, require darkness and quietude to hunt and simply cannot survive.

Noise Pollution: As is the case with disruptive night illumination, the corrosive impact of chronic noise on cardiovascular homeostasis does not appear to require cognitive perception. In fact, it is believed that subcortical brain areas like the hypothalamus, which has connections to the limbic, endocrine, and autonomic nervous systems, are involved in the activation of the fight-flight and defeat reactions. *Directed attention fatigue* is to blame (Kaplan 1995). Even at low decibel levels, persistent noise is a public health hazard, disrupting sleep and contributing to heart disease, stroke, metabolic disturbances, the escalation of psychological disorders, and premature morbidity (Münzel et al. 2014). Noise hinders learning and cognition, exacerbating behavioral issues and lowering productivity and achievement. Acoustic stress is known to negatively impact 'ideational originality, a key indicator of creativity' (An Ear For Nature 2018).

Children are among the most vulnerable, as more than 100 million Americans' health is at risk. Continued exposure to noise does not lead to habituation. Rather, the effects are cumulative. Chronic noise is associated with a higher risk of hearing loss, which in turn increases the likelihood of social isolation and depression—see also loneliness as a stressor above—and possibly of neurogenerative diseases such as Alzheimer's and Parkinson's Disease (Noise as a Public Health Hazard 2022). Here again, the distribution of stressors is profoundly inequitable. As of date, the United States has no federal non-occupational noise exposure standards. Like us, birds struggle to hear each other in the cacophony of our cities. Forced to sing louder, they, too, have to expend more energy, and mating cycles can become disrupted. The same is true for all other animals trying to make a living in our noise-burdened urban centers.

The Paradigm Shift

Design of the Built Environments Is a Healthcare Profession

We intuitively evaluate the homeostatic value of a constructed environment in the same way that we might be attracted to the pleasant smell of a rose or put off by a foul odor. We cannot step outside or override the valence of our emotions without some potential consequences because our genes and a few million years of evolution have made us who we are.—Harry Francis Mallgrave, *Architecture and Embodiment*

Restoration: *The process of renewing, recovering, or reestablishing physical, psychological, and social resources or capabilities diminished in ongoing efforts to meet adaptive demands (Hartig* 2017*)*

Restorative Environment*: An environment that promotes (and not merely permits) restoration (Hartig* 2017*).*

In his seminal book "The Body Keeps the Score," psychiatrist Bessel van der Kolk, M.D. writes:

> If an organism is stuck in survival mode, its energies are focused on fighting off unseen enemies, which leaves no room for nurture, care, and love. For us humans, it means that as long as our mind is defending itself against unseen assaults, our closest bonds are threatened, along with our ability to imagine, plan, play, learn, and pay attention to other people's needs. (Van der Kolk 2014).

A great many of our built environments seem to keep our ancient body-brain in a state of chronic overstimulation, requiring crisis-level physiological energy expenditures on a daily basis. Harry Francis Mallgrave writes: "A building can arouse our metabolic systems and demand high energy expenditure, or a building can provide a place for relaxation and comforting sociability." He asks: "…what happens in an urban environment where most buildings are designed to be active, aggressive, or even abrasive to our senses?" (Robinson and Pallasmaa 2015). Sadly, many of us daily live the answer to that question. Both van der Kolk and Mallgrave focus on the *expenditure of physiological energy,* energy spent by the sympathetic branch of the autonomic nervous system attempting to manage repetitive high-arousal stress scenarios, energy then not available for productive, creative, or social pursuits. Energy management in the context of maintaining homeostasis is also a major theme in the work of neuroscientist and Professor of Psychology Antonio Damasio. Homeostasis is understood to be dynamic, a constant moment-to-moment internal body negotiation, a balancing and rebalancing mechanism between the sympathetic and the parasympathetic branches of the autonomic nervous system. The goal of this internal negotiation, Damasio explains, "is the formidable enterprise of managing energy" (Damasio 2018). Information arrives via sensory and psycho-neural pathways. Cognitive neuroscience has established that much remains below the radar of consciousness. Some information rises to the level of cognitive perception as feelings, affording us diagnostic tools for conscious intervention. Feelings of well-being signify "that homeostasis is in the effective range" (Damasio 2018). Conversely, feelings of distress, fear, and exhaustion signify that homeostasis is *not* in the effective range. The body is unable to optimize energy management toward flourishing. Damasio states:

> The brain mapping of states in which the parameters of tissues depart significantly from the homeostatic range in a direction not conducive to survival is experienced with a quality we eventually called pain and punishment. Likewise, when tissues operate in the best part of the homeostatic range, the brain mapping of the related states is experienced with a quality we eventually named pleasure and reward. (Damasio 2010).

Judging from our stress reactions, nature-negative environments engage the “pain and punishment” side of things, resulting in health-compromising inefficiencies: too much energy is being drained away in basic life maintenance, in fortification against the background infrastructure of our lives. Physiological depletion is followed by increasing inflammation and, finally, breakdown. If our goal is to shift the unaffordable paradigm of the current status quo, we must follow the science and embrace the evidence-based ‘seeds of potential,’ a kind of allostatic approach utilizing adaptive, nature-positive designs that, akin to preventative medicine, strengthen our natural resilience by mediating and mitigating this unsavory cocktail of environmental stressors. We must ‘treat’ the ailments of our built environments as the public health epidemic they represent. This is sometimes referred to as *salutogenic design*, a design that specifically supports the causes of health and well-being. “Our buildings,” Dr. Allen of Harvard is quoted to say, “should be seen as a public health tool” (qtd. in Anthes 2023), or to paraphrase a quote by Will Rogers, president of the Trust for Public Land (TPL): Let’s take sustainability out of the emergency room (qtd. in Louv 2013). This establishes as an obvious priority that on an everyday, everywhere basis, built environments must facilitate physiological restoration. Nothing else makes sense. To our current quantitative goals of optimizing a building’s operational and embodied energy, we must add the *qualitative goal* of supporting building occupants’ neurobiological energy, specifically *supporting the parasympathetic branch*, which stimulates energy restoration, relaxation, and emotional bonding. To think along with Damasio’s work, we must redesign built environments to *foster neurobiologically and cognitively restorative pleasure/reward experiences* via place ingredients that signal evolutionarily relevant resource abundance, thereby keeping building occupants “in the best part of the homeostatic range,” a range that optimizes body energy management by reducing heart rate and raising vagal tone. Only together will quantitative *and* qualitative goals represent a comprehensive approach toward curating truly *sustainable shelter*. While architecture as a profession is a fairly recent phenomenon, our fundamental biology has not changed since the days of the mammoth hunters. Then, as now, our place experience is governed by ancient survival imperatives via embodied cognition. The experience of place is not intellectual but primarily visceral and completely immersive. This, unfortunately, is still news to many institutions of higher learning, especially those teaching design of built environments! Since at least the 1970s, science has worked with the updated paradigm of non-dualistic embodied cognition (Lakoff 2012), returning to the somewhat obvious understanding that the mind exists courtesy of the body and that the biological condition of our body strongly influences cognition, reasoning, and psychological well-being. This changes everything about how we approach the design of the built environment! So how, then, do we renew and transform built environments to be homeostatically effective and viscerally compelling, to be restorative rather than aggressive, and to be un-demanding of negative physiological energy expenditure? What do we know about transforming built environments to optimize physiological energy management toward cognitive restoration and well-being? What place ingredients, per Damasio, curate this effective homeostatic range? Indeed, we have all voted with our feet on this. Feeling anxious, stressed, or overwhelmed, all of us long to escape to the same kind of places. Let's call them 'vacation places.' Informed by a multi-dimensional richness of sensory perceptions, our body-brain seeks to guide us toward environments most conducive to flourishing and most likely to offer conditions with the potential to correct homeostatic imbalances (Damasio 2010). Judging from the vacation places we collectively choose, our shared physiology agrees that leafy groves, meadows, mountainous overlooks, and the edges of sparkling bodies of water are restorative places to be. We call these places 'beautiful' because embedded here are sensory patterns and place elements that offer enhanced survival opportunities via resource abundance. Protected and elevated perches promising sweeping views and refreshing breezes, flowing water, and vibrant plant life are key ingredients we seek. We agree that places offering dappled sunlight, ample birdsong, the sound of a waterfall or the ocean, and the presence of colorful flowers and butterflies are preferable to north-facing or indeed windowless rooms with stagnant air or rooms with glare-inducing quantities of fixed glass for example. No debate is needed. As Damasio writes: "The Diagnosis requires no special expertise but merely the fundamental process of consciousness" (Damasio 2010). We follow each other on the well-worn path to humanity's favorite vacation environments. Our preferences are validated by the quantifiable physiological restoration response those places elicit. These are *salutogenic places that support the causes of health***.** Our bodies know them when we see, hear, smell, and feel them. They are effortlessly neurologically legible and perceptually fluid. Like musical harmonies, they require no explanation, translation, or education. It is as though we tune into innately familiar frequencies that our bodies instantly recognize and 'understand.' The parasympathetic branch is strengthened. We experience joy, pleasure, and playful delight. The data speak for themselves: Place patterns and place ingredients that engage that 'vacation place feeling' reduce our heart rate, raise vagal tone, activate NK cells, restore our cognition, and lift our mood. Negative air ionization must be relatively high for us to love a place, suggesting the intuitively and evolutionarily obvious: the presence of water and vibrant plant life is essential for our well-being. Research indicates that places richly biodiverse in colorful plant and animal life support a healthy microbiome, strengthening our immune response (Mills et al. 2019), leaving us physiologically restored and cognitively refreshed. We feel gently stimulated but also relaxed. We feel well, indicating that *physiological energy management is optimized* because place ingredients curate settings *within the homeostatic effectiveness range* (Damasio 2010). Optimal ranges are experienced as pleasurable and valuable in health but also in monetary terms. Consider the real estate prices of properties with water, mountain, or woodland views. As Damasio explains: "Goods and actions that induce optimal life regulation will be regarded as most valuable" (Damasio 2010). The task then is to curate built 'goods' of value for all, environments that "improve the efficiency of the adaptive response to stressors while minimizing overactivity of these systems, since such overactivity results in many of the common diseases of modern life" (McEwen 2005). To my students, I say only half in jest: Let's design everything from schools to hospitals, offices to airports, and certainly homes to be within the quite obviously effective range of embedded vacation places ingredients. If this sounds extravagant or unaffordable, consider the inefficiencies and cost of continuing with business as usual. We already know these costs and can add increasing pressures from climate change-related extreme weather events, particularly extreme heat. Given the geographically inequitable distribution of stressors in many urban neighborhoods, it is also the only equitable path forward. Leading companies such as Google, Interface, and many more have for years, and especially in their post-Covid effort to entice workers back into offices, sought to blur the lines between work and 'leisure time,' in effect embedding vacation-style place ingredients overtly and covertly to increase productivity, an efficient strategy which aligns worker satisfaction with the company's bottom line (The Economics of Biophilia 2023). In so doing, they leaned deeply into the principles of *Attention Restoration Theory* (ART), which focuses on neutralizing 'directed attention fatigue' (Kaplan 1995) by introducing nature-rich place ingredients that engage the restorative sweet spot between relaxation and stimulation. Human habitations of any kind must be infused with living plants, flowing water, dynamic daylight, high-grain wood, and other natural materials to reduce stress responses such as anxiety, adrenal exhaustion, and impaired cognitive performance (Kaplan 1995). Considering the economics informing Fortune 500 companies' well-being-by-design strategy, it would make sense to implement this ART-based approach at the public health scale in cognitively stressful settings of any kind, especially urban settings. In the words of Dr. Jan Golembiewski: "We must design for genuine respite if we are to achieve an environmental antipsychotic effect through design" (Golembiewski 2015) (Fig. 1).

Fig. 1 The Nature-Diet Pyramid, reprinted with permission from Timothy Beatley, Biophilic Cities | www.biophiliccities.org

Soft Edges for Soft Bodies

From Thigmotaxis to Ecotonal Building Envelopes

> *Architects are still designing the 'it,' and seldom the edge, even though it is at the edges, or ecotones, where the richest exchanges and interactions take place.*—Sim van der Ryn, Stuart Cowan, *EcoLogical Design*

Like birds, we are soft, relatively small, and vulnerable bodies, equipped with ancient body-brains that have evolved over millennia for an immersive life in nature. Inexplicably, however, architects are frequently trained to focus on other matters, such as intellectualized spatial concepts or abstract minimalist forms, too often conceived from a disembodied viewpoint courtesy of the latest design software. As biologist E.O. Wilson observed, ignorance of human biology, of the body's very architecture, and lack of awareness of the full range of human sensory experience are oft-observed traits in even the most celebrated creative artists (Wilson 2017). As architects, whose work arguably leaves an indelible day-to-day imprint on the public, we are obligated to educate ourselves about the biologically ancient and visceral nature of human place experience and the physiological impact of our work on the public we serve. We understand how to design and retrofit thermal building envelopes that meet net-zero operational & embodied carbon goals. However, as the disastrous biodiversity loss caused by the perimeter of our buildings proves, we must give equal and urgent consideration to the exterior surfacing of those building envelopes: are they restorative or, in fact, destructive? It is, after all, the hard and unforgiving edges of our buildings that kill all those beautiful songbirds. It is the mirrored reflections, the beacons of light our buildings become at night, the see-through illusions of corner glazing, glass railings, and glass atria, especially if filled with interior greenery, which birds seek to reach in vain.

It turns out we, too, are deeply affected by edge conditions. Like other animals, we experience the world through somatic sensations, yearning for soft comfort, nourishment, and shelter. Much focus is on the *physiological quality of enclosures*. As a mom, I have fond memories of tucking my children into their beds while other babies, those of the feathered kind, were tucked into *their* nests, ever so skillfully woven by their parents into the leafy ivy curtain on the other side of the very same wall. As a biophilic architect, I have given a great deal of attention to the design of edge conditions that shelter human, plant, and bird life. Like other animals, we experience the stress of disorientation, a sense of exposure, or of feeling trapped when edge conditions are ambiguous, reflective, forbiddingly closed, or defy supervision such that danger could lurk in dark crevices. Such edge conditions create inefficiencies in terms of physiological energy expenditure. They elevate cortisol and are consequently guaranteed to be unpopular. As such, they are reliable predictors of economic failure in building projects. In *Cognitive Architecture,* Ann Sussman and Justin Hollander discuss the example of Boston's City Hall, whose brutalist, scary-desolate edge conditions have single-handedly wiped out a vibrant historic downtown: "The plaza's urban renewal history illustrates the high cost we and future generations pay when planners and architects do not appreciate their clients as evolved mammals with embedded reactions to place" (Sussman and Hollander 2021). The example of Boston's City Hall and similar places makes abundantly clear that the economic success of any built environment is directly proportional to its successful alignment with innate biological preferences. Damasio writes: "My hypothesis is that objects and processes we confront in our daily lives acquire their assigned value by reference to this primitive of naturally selected organism value" (Damasio 2010). To put it simply, places that make us feel trapped, exposed, or unprotected do not permit physiological energy optimization. Instead, they represent an unsustainable waste of human energy and material resources. Edge conditions are where much of this plays out.

In biology, the term *Thigmotaxis* refers to "The motion or orientation of an organism in response to a touch stimulus" (Oxford Dictionary). In design, Thigmotaxis refers to our relationship to horizontal and vertical edge conditions. Sussman and Hollander write: "When it comes to edges, biologists classify humans, exactly like other mammals of prey, as thigmotactic, or a 'wall-hugging' species" (Sussman and Hollander 2021). Sussman and Hollander describe this in evolutionary terms as another "energy-conserving strategy and a survival one," a common biological trait that can be observed in organisms as ancient as bacteria and as recent as mammals. All of us orient on edge conditions, and it turns out we favor a particular kind: I call them *engaged edges* because they permit us to *engage neurologically* in a manner aligned with biological flourishing. They engage our body-brain, which is constantly scanning for survival-related place information with the goal of physiological energy conservation. Primarily, we seek to establish a neurological relationship with place that promises life-sustaining shelter. It appears that nature has designed us to instinctively seek out 'soft' and permeable edge conditions that offer the potential for retreat (*refuge*) as well as opportunity for observation (*prospect*). Research indicates that *refuge* generates a higher relaxation response than *prospect.* Just like our fellow animals, we feel exposed in wide-open, unprotected spaces and prefer to move along edges, hedges, or walls, assuming those walls have intermittent and readily available openings that permit us to take cover in case of inclement weather or perceived danger. Think thickened wall conditions with variations on the theme of porous spatial layering, such as shady colonnades, plant-covered arbors, trellises, porches, balconies, verandas, pergolas, or the like. In other words, we like our edges body-positive: enriched, even embellished, offering a measure of protected depth and the flexibility, variability, and refuge of *transitional* conditions between inside and outside depending on usage, weather, and circumstances. We like options! Places with 'soft' edges are *neurologically legible*. More likely than not, they support a reduction in our heart rate because they simultaneously address our need for refuge & prospect (Appleton 1975), personal safety, and community (also related to safety). We are innately drawn to these places because we experience a relaxation (pleasure/reward) response. *Our physiological response indicates that place conditions are within the homeostatically effective range and likely to enhance flourishing.* Understanding our biological preferences is key to creating restorative, meaning physiologically efficient, design solutions. Taking a biomimetic approach, let us examine how nature designs edge conditions that support, rather than destroy, life.

On both a macro and micro level, nature favors permeable edge conditions. We refer to such places as *Ecotones (Kark 2017). Ecotones* are transitional spaces, open boundaries that facilitate life and energy transfer in both directions, fostering especially biodiversity-rich conditions. Van der Ryn and Cowan write: "An ecotone is a soft overlapping of very different regions. Like patches of watercolor on wet paper, different regions intermingle in an ecotone to create a new spectrum of colors" (Van Der Ryn and Cowan 1995). Fast forward to our cities: those that are successful in translating the qualities of 'soft edges' into built environments tend to be beloved and memorable. Those unsuccessful are experienced as bleak, uncomfortable, and inhospitable. Our biological needs are not supported here, and our body informs our consciousness to that effect. We do not like these places. They feel stressful because they quantifiably are (Ellard 2024). Neither health nor commerce will reliably thrive. If we live here, it is often not by choice. To be well, we require built environments that are vibrant and feel alive, buildings that surround us with the equivalent of a 'second skin': a transitional layer that protects as it connects, affording experience of the ever-changing experiences of seasons, weather, community, and more-than-human life.

How does nature develop ecotonal edge solutions akin to building envelopes? Let us consider the bark of a tree: it shields from wind, weather, and ultraviolet rays while also protecting the interior 'climate' of the tree in terms of moisture, nutrient levels, and fungal growth. Additionally, bark functions as a micro-ecotone, providing food and shelter for insects and birds, as well as serving as habitat for a rich diversity of epiphytes such as mosses, liverworts, lichen, and some ferns. Epiphytes do not trap moisture or harm the tree. As a symbiotic system, this living layer enriches the tree by providing additional soil nutrients, establishing a microhabitat on each trunk and branch, which sequesters carbon. Epiphytes achieve microclimatic improvements by reducing daytime vapor pressure deficits (VPD) and moderating temperature variations (Stanton et al. 2014) (Fig.2). What if we took a similar approach to make our buildings more resilient? Sussman and Hollander write: "People use buildings as orientation devices and protection screens" (Sussman and Hollander 2021). So, let us future-proof our buildings by adding 'protection screens' with qualities similar to the bark of a tree, albeit more transparent to suit human visual and daylight needs. Let us take our favorite *thigmotaxis* pattern, *refuge & prospect*, and add the dynamic sensory experiences of dappled sunlight through a leafy curtain of vegetation swaying in a cooling breeze. Add the daily experience of seasons and weather, the chirping of birds, the humming of bees and butterflies, and the scents and negative air ionization of plants, and you have the 'vacation ingredients' of a vertical garden that can be implemented just about anywhere, even conceived as a vertical rain garden if conditions require. Clothed in verdant *Ecotonal Building Envelopes* (EBEs), formerly hard-edged buildings are transformed into soft and vibrant places, alive with place ingredients offering enhanced well-being conditions that mitigate air, noise, and light pollution,

Fig. 2 Images courtesy of Sophia BenJeddi

naturally cool interiors as well as surrounding neighborhoods, and foster micro-habitat biodiversity in symbiotic realignment with the biology of all urban lives.

From the hanging gardens of Babylon to the ivy and grapevine-covered walls of historic houses including my own, vegetated buildings are hardly a new building typology but are still altogether too rare. Considering how many urban stressors they mitigate, this is perplexingly out of step with the economics of resilience. Much misinformation persists regarding potential façade damage, leakage, or bug infestations. As successfully completed projects prove, all of these can be managed by having experienced experts from various disciplines on board from day one of the design process, focusing on technically correct detailing, positive drainage, climate and elevation-appropriate plant selections, and state-of-the-art irrigation, and filtration systems. What if we learned from local hills and mountains about the kinds of vegetation that thrive in high, dry, and windy places? Remembering that birds nest and hunt from tall buildings as though they are cliffs, we can follow their lead. In this context, it is interesting to consider biologist Jeremy Lundholm's work on "The Urban Cliff Hypothesis" as a "Habitat Template" (Urban Habitats—Green Roofs and Facades: A Habitat Template Approach n.d.). To serve the twin goals of fostering public health and biodiversity, the human designer assumes a posture of humility, first learning from nature, then stepping back to permit nature to provide the design language of a façade that is seasonally changing and beautifully alive. As van der Ryn and Cowen write: "By designing ecotones, rather than hard edges, we intensify interactions. We bring together a greater variety of life…. we facilitate the flow of materials, energy, and information that can catalyze design processes." (Van Der Ryn and Cowan 1995). Well beyond mere imagining, many beautiful vegetated tall buildings have been built worldwide. Renowned botanist Patrick Blanc spear-headed a version of EBEs he calls "vertical gardens" (Mur Végétal), providing cliff-like landing pads and nesting opportunities for urban residents, such as birds, squirrels, and pollinators, and well-documented health and community-building benefits for humans. In describing *One Central Park* in Sydney, Australia, currently the tallest vertical garden in the world designed by Patrick Blanc in concert with the design teams of Foster & Partners, Ateliers Jean Nouvel, and PTW Architects, University of Virginia's Professor of Sustainable Urban Communities, and Founder and Executive Director of *Biophilic Cities* Timothy Beatley writes:

> The green walls increase energy efficiency, serving as natural insulation for the building, as well as an air filtration system that transforms pollutants into useful plant fertilizers. Providing jobs and utilizing local plant species, economic stimulus and educational opportunities coincide, providing a connection between city dwellers and the natural flora of their habitat (Beatley 2016).

Blanc refers to his work as 'Living Architecture' and asks us to "imagine a building that grows, sprouts and flowers, changes colour with the seasons, or sheds leaves in winter as though magically alive with petals, vines, and buds" (|Vertical Garden Patrick Blanc, n.d.). Italian architect Stefano Boeri designed and built the popular "Vertical Forest" (Bosco Verticale) buildings, such as the Bosco Verticale towers completed in 2014 in Milan and several more. Boeri explains that the Bosco Verticale houses the nests of over 20 different bird species, in addition to a plethora of plant species, whom he views as the building's main occupants (Lubell 2020). Architecture firms such as *WOHA, BIG, MVRDV, Ateliers Jean Nouvel, Vincent Callebaut, Smyrlis Architects, Ingenhoven Architects, Enrique Browne, Sheppard Robson*, and others are designing and building vertically vegetated designs. Considering the urban stressors we increasingly face, we must urgently scale up this effort, replacing any lingering luxury label with a public health mandate (Wood et al. 2014) (Fig. 3).

Fig. 3 Images by author

EBEs include both *extensive* and *intensive Vertical Greening Systems* (VGS) as well as *extensive and intensive Living (Green) Roofs.* Living roofs support urban biodiversity and are important stormwater management and urban cooling tools. Likewise, VGS can be designed as vertical rain gardens. Well-documented examples of large "stormwater-powered" living walls designed in direct collaboration with water engineers are the 21-meter-high wall in the Victoria neighborhood of London (Andrews and Andrews 2022) and the façade of the Enni Headquarters in Moers, Germany (Turner 2023). Elaborate intensive wall-bound VGS (Living/Green Walls) are fairly complex to install and maintain. A focus on the greater simplicity and affordability *of extensive ground-based VGS (Living/Green Façades),* as far as installation and maintenance, might be the most adaptive path to meet the goal of wrapping the existing stock of naked buildings in cooler, softer, more resilient clothing. These *gound-based systems* consist of climbing vegetation either (a) self-climbing on building facades (*direct/two-dimensional Green Façade*) or (b) supported by a secondary trellis *(indirect/double-skin/three-dimensional Green Façade*) (Medl et al. 2017; Bustami et al. 2018; Susorova et al. 2013). In both cases, vegetation grows from ground-based (including sidewalk-based) planters or planting strips that are watered via rainwater directly and/or rooftop or parking lot collection cisterns, making them drought-resistant. Root systems are protected by native surface plantings. Several cities, such as San Francisco, have already implemented 'Grey2Green' initiatives, issuing sidewalk greening permits after any potential conflicts with utility infrastructure have been resolved. *Non-deciduous, direct/two-dimensional Green Façades* create a thermal layer that both warms building interiors in winter and cools them in summer. In the Northern Hemisphere, *deciduous indirect/three-dimensional Green Façade Systems* add a soft double-skin-screen of vegetation that is especially effective in lowering interior glare and summer temperatures on South and West-facing glass walls, otherwise difficult to shade due to the low sun angle in mid to late afternoon.

All VGS provide noticeable evaporative cooling benefits through plant leaves for both building interiors and surrounding areas that other shading systems do not offer, which "improves the urban microclimate…and acts as *near-natural air conditioning systems*" (Kg 2022). VGS also provide protection against UV radiation and pollutants, may thereby inhibit material deterioration, and serve as effective deterrents to graffiti. The lightweight trellis structure of ground-based VGS consists of either stainless steel cables, mesh, rope, or wood trellis, depending on height, climactic conditions, and wall surface materials. Even the most mundane buildings can thus be transformed into a *vertical micro-habitat.* Modular cable and mesh solutions can easily be adapted to follow biomorphic building forms (Perez and Perini 2018; Greenscreen 2023). To protect the birds we invite, Audubon Pennsylvania's Connie Sanchez recommends the distance between building wall and *Green Façade System* not to exceed 3 feet and to add bird-safe glass as a matter of course (Fig. 4).

An interesting example of a *ground-based, double-skin/three-dimensional VGS in combination with intermittent plant troughs* is the award-winning South Elevation of the Institute for Physics at the Humboldt (Fig. 5)

University in Berlin, Germany, designed by Augustin & Frank and completed in 2002. As an ecological model, this VGS system's performance was studied and

Fig. 4 Vertical Greening Systems (VGS). Image: Green Wall Basics | EFB (n.d.) https://efb-greenroof.eu/green-wall-basic

Fig. 5 Institute for Physics at the Humboldt University in Berlin, Germany, 2023 | Images Courtesy of Mascha Creutz

documented for 8 years by an interdisciplinary group of researchers from the Technical University (TU) Berlin and the Humboldt University (HU) Berlin (Kg 2022). From April through October, lightweight planters are individually drained and collectively misted via a roof rainwater collection system. Due to urban soil contamination in the vicinity, stormwater evaporation was only permitted via phytoremediation. 950 climbing vines were planted, most of which have reached full building height. Compared to manual or automated textile shading systems, the building's deciduous plant screen, supported by a combination of wood, bamboo, steel cables, and rope, achieves significantly higher interior summer cooling of the glass facade (Kg 2022). Additionally, the VGS provides evaporative exterior cooling, establishes a vertical micro-habitat for birds and pollinators, filters dust, absorbs noise, mitigates glare both inside and out, and sequesters carbon. It also provides the delight of dappled sunlight patterns, the rustling of leaves, the scent of wisteria blossoms, and the chirping of birds (Fig. 5).

Nourishing the Seeds of Potential

Micro-Habitat Buildings Are Adaptive Investments in Public Health

> *......all organisms on our planet are fundamentally similar. It's kinship that makes our immune system work.*—Melanie Challenger, *How To Be Animal: A New History of What it Means to Be Human*

In the "Geography of Nowhere," James Howard Kunstler bemoans the results of a building and development ethos that "has nearly wrecked the human habitat in America." The term 'habitat' is of interest. Oxford Dictionary defines 'habitat' as "the natural home or environment of an animal, plant, or other organism." National Geographic states: "A habitat meets all the environmental conditions an organism needs to survive" (Urban Habitats—Green Roofs and Facades: A Habitat Template Approach n.d.). As the public health consequences of daily urban stressors indicate, the urban habitat, currently home to 50–80% of human populations depending on country, and projected to double by 2050 (Overview n.d.), needs salutogenic intervention. In habitat restoration interventions for other biological organisms, we seek to reestablish the living conditions that permit those organisms to flourish in alignment with their evolved physiological nature. The *economics of resilience* would suggest that we take the same approach to re-imagining human habitat as a place that fosters neurobiological energy optimization. In the context of built environments, the most comprehensive definition of 'habitat' I am currently aware of shall be quoted here in full:

> Habitat, in this context, encompasses the materials, structure, and program of the building; management of site metabolism, including energy needs and waste flows; concordance with the surrounding environment, both within the building and beyond the building façade; support for on-site biodiversity, from micro to macrofauna...; and perhaps most of all, a recognition that these features communicate habitability and community to human occupants through eons of evolutionary priming and that this appeal is both desirable, comfortable, and health-promoting" (Africa et al. 2019).

This deserves repeating: "on-site biodiversity, from micro to macrofauna, (…) communicates habitability and community" (Africa et al. 2019). While the need to significantly increase urban tree planting is self-evident, their placement in poorly ventilated streets may present challenges as densely foliated canopies can restrict airflow further, or there may not be sufficient space to sustainably support their root balls. The ideal way to integrate easy nature into tight and densely populated areas is to add *Ecotonal Building Envelopes* (EBEs) at scale to the underutilized real estate of acres of currently empty facades and roofs, transforming urban canyons into leafy bowers to improve the microclimate of each neighborhood and the city at large. Rather than isolated islands, a contiguous vegetated blanket can be folded across each city to cool, quiet, and delight, amplifying the individual contribution of each building towards a more healthful, more biodiverse urban environment where all may breathe more freely. *Terrapin Bright Green* uses the term *Biophilic Urban Acupuncture* to illuminate the salutary ripple effect even small-scale nature interventions generate: "Just as the practice of acupuncture is aimed at relieving stress in the human body, the goal of biophilic urban acupuncture is to relieve stress in the built environment" (Walker 2016). *Indirect/double-skin Green Façade Systems* are especially physiologically advantageous in this regard as they permit an enhanced refuge experience (Grahn & Stigsdotter 2009) sheltered by a transparent leafy veil that protects from prying eyes and the sensory overload of the street.

Let us explore the benefits of EBEs in response to the specific urban stressors discussed above. It should be noted that EBEs contribute on a multi-sensory level, and thus, the below categories all act in concert to support public health.

From Loneliness to Biophilia: A 2022 Columbia University study found that ecosystems, especially those experienced as children, serve as object relationships and that people are less able to safeguard these relationships for their own well-being when society does not recognize their significance (Bodnar et al. 2023). Through eons of evolutionary priming, *biophilia* represents our innate emotional connection to the biosphere that sustains our lives. As the Columbia study suggests, 'philia is family,' a community of relationships crucial for personality development and directly connected to well-being.

Shared suffering can open our hearts. During the isolation of the pandemic, many of us (re)discovered that nature offers ready companionship. We took to birdwatching, made friends with the equally lonely pigeon arriving daily on our apartment's fire escape, or struck up a conversation with the baby squirrel, whose playful curiosity charmed us to a smile. Perhaps our COVID loneliness persuaded us to reconnect with our innate biophilia (Salingaros 2015). Hopefully, it opened our eyes to the well-documented mental health benefits of leveling the playing field between 'us' and 'them,' to engage in up close and personal relationships with the buzzing, feathered, and furry fellow members of our urban community. "The intentionally cultivated ecotone promotes contact among people and between people and nature," Van der Ryn and Cowan wrote (Van Der Ryn and Cowan 1995). EBEs foster community, community with nature, and community among people. Building maintenance shifts towards stewardship, offering physiological benefits far beyond the reach of spreadsheets and bottom lines. People who reside in highly variable green environments have 37% lower odds of being hospitalized for cardiovascular disease (Yeager et al. 2020). As we know from our vacations, nature has a way of bringing people together with ease. From macro to micro, nature experiences restore as they delight, inspire as they awe. Awe gives our lives meaning by enticing us out of our lonely all-about-me-focus to connect with something larger than ourselves. We begin to belong again. Awe experiences make us more cooperative, more generous, and, so it turns out, profoundly strengthen our immune response. The groundbreaking work of UC Berkeley Prof. of Psychology Dacher Keltner must be mentioned here. Keltner found that awe experiences engage the parasympathetic nervous system, support the vagus nerve, and release oxytocin, thus reducing inflammatory cytokines (Keltner 2023). The homeostatic motivator of *pleasure/reward* kicks into full swing. Taking these significant health benefits out of the realm of the occasional into what Prof. Keltner refers to as "everyday wonder" falls squarely into the domain of every designer's responsibility. By applying a relatively simple recipe of ingredients, EBEs offer the possibility for multi-sensory, cortisol-reducing *everyday wonder* adaptable to a multitude of needy architectural settings. As Kate E. Lee's research of 'micro-breaks' suggests, mere 40-second views of nature have the power to improve cognitive function (Lee et al. 2015). Imagine the health benefits of frequent casual glances through seasonally dynamic leafy curtains that delight building occupants with gently swaying light patterns. Imagine the ability to partake in the lives of birds, praying mantises, and butterflies up close, on the 8th or 10th floor, and in the comfort of your own home. Awe experiences are homeostatically effective via *involuntary attention & soft fascination,* key components of *Attention Restoration Theory* (ART) (Kaplan 1995; Basu et al. 2018). As such, they contribute to overall stress-reduction and cognitive restoration. Richard Louv calls this effect "Nature's Ritalin" (Louv 2013). He writes: "Greenery in a child's everyday environment, even views of green through a window, specifically reduces attention deficit symptoms" (Louv 2013). We have entered the realm of *salutogenic design*, and EBEs make this possible and affordable daily and at scale.

The biodiversity of another community is in urgent need of support: our bodies' microbiotic community, which we know to be impaired in de-natured urban populations (Mills et al. 2019). Biodiversity interventions (Roslund et al. 2020) enhance immune regulation and can be viewed akin to preventative medicine:

> restoring urban biodiversity provides a low-risk and low-cost investment that is likely to rewild urban microbial processes and have potential to pay generational health dividends. We believe that these restoration investments are imperative for tackling the ongoing decline in urban green spaces, rapid growth in urban populations, and increases in microbially-mediated non-communicable diseases (Mills et al. 2019).

As an essential aspect of public health policy, all urban and suburban centers should follow the lead of biophilic cities such as San Francisco, Singapore, Los Angeles, and others to establish a city-wide biodiversity policy (San Francisco Biodiversity Program and Policy 2023; Handbook on the Singapore Index on Cities' Biodiversity

2008), comprehensive goals and guidelines integrating wild nature into urban infrastructure, not just on the ground but on all available surfaces in the form of vegetated building façades and roofs.

From Extreme Heat to Natural Airconditioning: On a hot late August day in Pennsylvania, I am gazing through the blooming vines of *Clematis Sweet Autumn,* filling the air with the scent of honey and the buzzing of bumblebees. We planted the prolific vine to shade the South and West facing glazing of our home, an effective and delightful curtain that, along with ivy-covered East and North facing walls, keeps our interiors naturally airconditioned. It also provides popular nesting and overnight accommodations for local birds. Experimental studies reveal that using plants as part of the building envelope significantly reduces ambient temperature close to the facade as well as interior temperatures, materially improving energy efficiency (Susorova et al. 2013). The temperature of building façades shaded by VGS was shown to be up to 7 °C (12.6°F) cooler than ambient air, thereby also positively affecting neighboring buildings by reducing heat reflection. Relative to shade sails, for example, evaporative cooling contributed 25–35% of the total cooling achieved by vegetated façades. The thermal advantages of vegetated façades suggest meaningful contribution to the sustainability of building designs and represent an *effective natural remedy for urban heat islands* (Bakhshoodeh et al. 2022; Shah et al. 2023). Even cities in humid climates benefit with a temperature decrease of 8.4 °C measured in Hong Kong, when both roofs and walls are vegetated (Alexandri and Jones 2008). The air space EBEs create between building wall and a double-skin VGS could be designed to generate a wind-assisted stack effect, pulling hot air out of poorly ventilated urban canyons. In all climates studied, the combination of vegetated roofs and vegetated walls leads to the highest mitigation of temperatures. Air masses enter urban canyons pre-cooled when roofs are also covered in vegetation (Alexandri and Jones 2008). It is of crucial importance to apply these techniques at scale:

> If applied to only one unit block, green roofs and green walls can create a small area of mitigated temperatures to the urban heat island effect, as has been shown in this microclimatic study. If applied to the whole city scale, they could mitigate raised urban temperatures, and, especially for hot climates, bring temperatures down to more "human-friendly" levels and achieve energy saving for cooling buildings from 32% to 100%. (Alexandri and Jones 2008).

The potential to measurably lower estimated levels of heat illness and death in anticipated electricity blackouts during prolonged extreme heat events lies in widespread interventions that improve evapotranspiration, shading, and solar reflection in built environments (Stone et al. 2023). Additionally, the *Living Architecture Performance Tool* recommends the integration of mechanical HVAC systems with living plant systems, such as strategic placement of HVAC units "to reduce the temperature of intake air" (Green Infrastructure Foundation n.d.; Wong et al. 2010a).

From Fine Particle Pollution to Phytoremediation: Through a variety of mechanisms, including accumulation, immobilization, and degradation, plants reduce the toxicity, mobility, and volume of aerosolized contaminants. Because only solar energy is required for these biological processes, phytoremediation is more affordable than engineering-based remediation techniques (Lee et al. 2021). Air pollution in urban canyons is affected by building surface materials, the ratio of building height to canyon width, and the resulting velocity of in-canyon air:

> Pollutant concentrations can be reduced by increasing dry deposition to surfaces. Compared to controlling emissions or enhancing dispersion, relatively little attention has been paid to deposition as a pollution control measure. An effective and accessible means of achieving an enhancement in pollutant deposition is to plant additional vegetation (Pugh et al. 2012).

Models show that vegetating street canyon walls offers significant potential for upgrading air quality values, particularly at low wind speeds when the effect of in-canyon vegetation is heightened. The model predicted reductions in NO_2 and PM_{10} concentrations of up to 40% and 60%, respectively, indicating that canyon wall vegetation forms a natural buffer against high-pollution episodes, frequently associated with low wind speeds and related acute impacts on human health (Pugh et al. 2012). Depending on species and size of leaves, VGS can serve as self-cleaning façades via autumnal leaf fall. Unsurprisingly, model results demonstrated that greening of urban in-canyon wall surfaces is more effective than greening of roofs in terms of reducing street-level pollutant concentrations because it acts directly upon the relatively small volume of air in the canyon (Pugh et al. 2012). In tight canyons: "Vegetated walls can improve canyon air quality more than street trees because better vertical airflow out of the canyon can be achieved" (Pugh et al. 2012). VGS also *mitigate urban dry island* (UDI) effects, characterized by drier atmospheres (less moisture and a greater vapor pressure deficit (VPD)), necessitating intensified urban greening efforts (Luo and Lau 2019). Additionally, VGS *sequester carbon, enhance canyon oxygen* levels, and *add negative air ionization*. Eva Selhub MD and Alan Logan ND write: "Researchers…linked the regional abundance of negative air ions to improved human health and longevity" (Selhub and Logan 2012). Increasing levels of negative air ions *positively affects Seasonal Affective Disorder and depression* (Terman and Terman 1995), as well as slowing *the aerosol transmission of Influenza and COVID viral particles* (Zhang et al. 2022).

From Light Pollution and Glare to Dynamic Dappled Light: To protect night-migrating birds, *three-dimensional Green Façade Systems* that add a double-skin-screen of vegetation can be effective because they reduce the overall glare of cities, diffusing the killer beacon effect of each illuminated building. Softened edges offer a softer landing in the buffering plant curtain that invites and then keeps its promise to shelter and even feed. As mentioned above, the maximum distance between building wall and plant curtain must be less than 3 feet to protect birds. From inside, our own circadian well-being can be shielded from the immediacy and glare of city lights. During daytime, the interior light quality becomes nuanced and softened, mitigating the uncomfortable glare-and stress-inducing condition of the single sun orientation so many urban rooms are limited to. Dappled sunlight patterns will move across our spaces, making breezes visible, thereby adding neurologically relevant circadian, seasonal, and weather-related information that supports our immune response and enlivens otherwise lifeless interior spaces. When all buildings adopt these adaptive mitigation strategies, we might even be able to see the stars again, affording ourselves the *soft fascination* of *every night wonder and awe experiences* we know to be associated with *Attention Restoration Theory* (ART) because they engage *ancient pleasure/reward motivators,* deeply meaningful because connected to physiological energy optimization.

From Noise Pollution to the Twitter of Birds: The first thing one notices when approaching a leafy building is the chirping of little birds who are busy making a life. There is a lot going on here. It's a community place that delights many of our senses. The key word is *life.* The building *feels alive* because it is. Our body-brain associates birdsong with well-being, with freshness and springtime, and with things generally being 'ok' since birds have a better view than we do and stop singing when peril invades the neighborhood, as every squirrel and chipmunk already knows. Anyone who has observed a cat or a hawk entering a bird-filled yard can testify to the immediate silence that ensues. (The 4 Ways Sound Affects Us - Julian Treasure, n.d.) As sound branding expert Julian Treasure puts it in one of his TED talks: "When the birds are singing, we feel alert. It's a good, body-relaxed, mind-alert state." He counts birdsong as one of our three favorite sounds (the other two being water and wind). Daily interactions with birds are linked to lasting mental health improvements (Hammoud et al. 2022). Whereas traffic noise was linked to depressive symptoms, birdsong was found to reduce anxiety and paranoia (Stobbe et al. 2022). We can conclude that *birdsong engages the homeostatically effective pleasure/reward motivator*. It is another of our 'vacation place ingredients,' fostering the *soft fascination* associated with *Attention Restoration Theory* (ART). Additionally, the combination of façade vegetation and the layer of air between plants and building wall can function as a sound-absorbing barrier. Depending on leaf size and density of vegetation, plants are effective at deflecting and absorbing sound waves, especially higher frequencies. A comprehensive 2010 study found that:

> The sound absorption coefficient of the vertical greenery system in the reverberation chamber has one of the highest values compared with other building materials and furnishings. The sound absorption coefficient increases with increased frequencies. In addition, it is observed that the sound absorption coefficient increases with greater greenery coverage (Wong et al. 2010b).

According to economists who examined health care costs and productivity losses due to heart disease and hypertension, a 5 dB reduction in noise levels could generate a $3.9 billion annual benefit (Baumgaertner et al. 2023).

Tending the Seeds of Potential

> *I like very much the idea of a "blooming building" and the notion that we might judge a building …*
> *by whether and how often it blooms.*—Timothy Beatley, *The Bird-Friendly City*

I have approached this topic from a primarily human-centric point of view because, for most of us, this is the normative narrative. Having subscribed to this narrative, however, we have become a lonely species, missed many opportunities, and created enormous challenges for ourselves and all other organisms on this planet. The narrative of human exceptionalism has quantifiably outlived its utility if ever it possessed any. It has been debunked by the extractive, neuro-biologically inefficient environments we have created and the current state of our planet in general (Richardson et al. 2023). As Melanie Challenger states: "If we want something more like a mutualistic approach to the life around us, we will have to see into the needs of other organisms" (Challenger 2021). Evolving into an adaptive, mutualistic approach permits progress with greater efficiency and, indeed, with the humility required to understand and implement biomimetic win-win solutions that honor and protect the dignity and well-being of all lives. As we stand at the dawn of disembodied intelligence, many of us discover that we yearn to define the best of human nature in biological and emotional kinship with other animals rather than with machines. Interestingly, this occurs at the exact moment we learn so much more about the language and consciousness of (other) animals. We have (re)emerged into a world that is humming with the conversations of fellow thinking beings, however enigmatic (Shah 2023). In *God, Human, Animal, Machine,* Megan O'Gieblyn observes:

> Today, as AI continues to blow past us in benchmark after benchmark of higher cognition, we quell our anxiety by insisting that what distinguishes true consciousness is emotions, perception, the ability to experience and feel: the qualities, in other words, that we share with animals. (O'Gieblyn 2022)

The extraordinary opportunity of this moment lies in the recognition that our evolutionarily-primed biophilia, our innate ability to recognize and value our shared ancestry with other living organisms, is the exact quality that illuminates the path we need to walk toward the survival imperative of reciprocity, placing our equally evolutionarily-primed self-interest into the service of stewardship. Biophilia opens a path of expansive affirmation, well-being, and affiliative beauty. As I listen to the first bird of morning, I am aware of pleasure, joy, and the extraordinary beauty with which this message is delivered. Bringing *sustainability into the range of homeostatic effectiveness by* engaging *ancient pleasure/reward motivators* is the most viable path toward the *economics of resilience*. As organisms who evolved to be warmed by the sun and cooled by evaporative transpiration, we require buildings that naturally maintain livable conditions when the next pandemic, heat wave, and power outage descends. We require buildings that restoratively engage our senses and, thereby, our nervous systems in support of a resilient immune response. Urgently needed is a paradigm shift towards designs that embrace *the duality of energy management optimization: quantitatively efficient designs* that foster climate-positive (carbon-negative) resilience of a building's carbon footprint (net-zero operational + low embodied + active sequestration) and *qualitatively efficient design* that fosters net-positive resilience of neuro-biological energy in support of public health. The qualitative can now be quantified in *body-nature-positive designs* that integrate biodiversity in *all* built environments. This cannot remain a luxury item available only to a select few. The solution to avoiding gentrification is to make this methodology commonplace as an essential aspect of our very humanity and the deeply intertwined twin goals of passive survivability and resilient public health. Evolutionary biology, biopsychology, neurobiology, chronobiology, sociobiology, and cognitive neuroscience provide comprehensive tools designers can use to deepen our work and curate well-being environments with clear public health-minded and science-informed intent. Equally important is remembering the wisdom of vernacular design, of working with patterns, spatial sequences, and proportions that are revered because they are neurologically compelling. Data suggest that places that restore and sustain us are the only value-adding investments both in terms of real estate value and the economic value of public health (The Economics of Biophilia 2023). These two goals directly overlap. Only places that are perceptually fluid and effortlessly cognitively legible permit emotional connection. As the pattern commonalities embedded in world heritage sites and other universally beloved buildings prove, we sustain what sustains us.

In a time of heightened focus on what divides us, let us find meaning in serving what connects us, namely the innate biological traits we share with one another and with our non-human kin. Designing and retrofitting buildings to shelter and nourish birds, bees, and butterflies offers a simple roadmap toward built environments that also shelter and nourish human life. Only places designed in alignment with the biology of all of life can be considered relevant in the context of the 21st-century's dire challenges. All things considered, I consider *climate/eco-positive micro-habitat buildings* to be the only truly equitable building typology that fully meets this moment, future-proofing our cities by addressing the needs of *all* urban residents. Edges matter! *Ecotonal Building Envelopes* (EBEs) must supersede the dead and deadening façade language of so many current environments. As an industry, we must embrace nature as our innate aesthetic language and, along the way, revise our outdated notions of 'tidy' nature, allowing for life to be tangled and prolific. Akin to *green immune cells, all buildings, new and existing, must adapt to become climate-resilient, restorative infrastructure investments in public health*. Local ecology must become intrinsic to each EBE's design, *actively sequestering carbon, mitigating air, noise, and light pollution, naturally cooling interiors and surrounding neighborhoods, and fostering urban micro-habitat biodiversity* in symbiotic realignment with the biology of all urban lives. The hour is late, and it is time for implementation at scale, "high time to reconnect approaches and perspectives on development to the biosphere foundation" (Folke et al. 2016), to realign urban habitats with salutogenesis. Let's clothe our buildings in living, cooling "superterrestrial gardens" (Alexandri and Jones 2008) and, while we're at it, let's infuse our ailing commercial districts (Badger 2023) with new life. EBEs create effective thigmotaxis, especially at street level, literally growing an upgraded sidewalk experience that will add palatable microclimatic benefits and, therefore, enhanced real estate value to the entire building above. Once the twitter of birds moves in, we know people will follow.

In his project, *The Flying Gardens of Maybe,* artist Andrew Yang collected seeds from the guts of birds who died colliding with buildings and planted them in beautiful ceramic jars. He tended those seeds, honoring the life and labor of each bird while mourning "the vast ecology of interruption" that killed them all (Andrew

Yang, Flying Gardens of Maybe | New-Empiricisms n.d.). Alternatively, imagine exhausted birds arriving in now verdant cities filled with blooming vertical gardens and rooftop prairies that offer safe landing, nourishment, and nesting opportunities. Let us commit to tending Andrew Yang's beautiful *seeds of potential* until their young shoots can scale the heights of our buildings and, once again, reach the sky.

References

Africa, J., Heerwagen, J., Loftness, V., & Balagtas, C. R. (2019). Biophilic Design and Climate Change: Performance Parameters for health. Frontiers in Built Environment, 5. https://doi.org/10.3389/fbuil.2019.00028

Air Pollution may Increase Risk for Dementia. (2023, April 5). News. https://www.hsph.harvard.edu/news/press-releases/air-pollution-may-increase-risk-for-dementia/

Alexandri E, Jones PJ (2008) Temperature decreases in an urban canyon due to green walls and green roofs in diverse climates. Build Environ 43(4):480–493. https://doi.org/10.1016/j.buildenv.2006.10.055

American Medical Association & American Medical Association. (2025, April 17). Trends in health care spending. American Medical Association. https://www.ama-assn.org/about/ama-research/trends-health-care-spending

An Ear for Nature. (2018). Terrapin Bright Green. https://www.terrapinbrightgreen.com/wp-content/uploads/2018/08/An-Ear-for-Nature_Terrapin_2018p.pdf

Andrews, K., & Andrews, K. (2022, February 2). London's largest living wall will "combat flooding". Dezeen. https://www.dezeen.com/2013/08/21/londons-largest-living-wall-will-combat-flooding/

Anthes, E. (2023, June 17). The new war on bad air. The New York Times. https://www.nytimes.com/2023/06/17/health/covid-ventilation-air-quality.html

Appleton J. (1975) The experience of landscape. John Wiley and Sons

Architecture, the urban environment and Severe Psychosis: Aetiology. (n.d.). Centre for Urban Design and Mental Health. https://www.urbandesignmentalhealth.com/journal2-psychosis.html

Association Between Particulate Matter (PM)2·5 Air Pollution and Clinical Antibiotic Resistance: A Global Analysis. (2023). Lancet. https://www.thelancet.com/journals/lanplh/article/PIIS2542-5196(23)00135-3/fulltext

Atlantic Council. (2021). Extreme heat: the economic and social consequences for the United States.

Badger, E. (2023, September 5). The ground-floor window into what's ailing downtowns. The New York Times. https://www.nytimes.com/2023/09/04/upshot/cities-downtowns-vacant-storefronts.html

Bakhshoodeh R, Ocampo C, Oldham C (2022) Exploring the evapotranspirative cooling effect of a green façade. Sustain Cities Soc 81:103822. https://doi.org/10.1016/j.scs.2022.103822

Basu A, Duvall J, Kaplan R (2018) Attention restoration theory: exploring the role of soft fascination and mental bandwidth. Environ Behav 51(9–10):1055–1081. https://doi.org/10.1177/0013916518774400

Baumgaertner, E., Kao, J., Lutz, E., Sedgwick, J., Taylor, R., Throop, N., & Williams, J. (2023, June 12). Noise could take years off your life. Here are the health impacts. The New York Times. https://www.nytimes.com/interactive/2023/06/09/health/noise-exposure-health-impacts.html

Beatley T (2016) Handbook of Biophilic city planning & design. Island Press

Bodnar S, Aljovin P, O'Neill P, Alavi S, Gamoran J, Liaqat A, Bitensky D, Bi H, Grella E, Kiefer M, Morenberg L, O'Leary C, Yadav P, Wasret A (2023) The environment as an object relationship: a two-part study. Ecopsychology 15(2):110–118. https://doi.org/10.1089/eco.2022.0070

Bustami RA, Belusko M, Ward J, Beecham S (2018) Vertical greenery systems: a systematic review of research trends. Build Environ 146:226–237. https://doi.org/10.1016/j.buildenv.2018.09.045
Challenger M (2021) How to be animal: a new history of what it means to be human. Penguin
Damasio A (2005) Descartes' error: emotion, reason, and the human brain. Penguin
Damasio A (2010) Self comes to mind: constructing the conscious brain. Vintage
Damasio A (2018) The strange order of things: life, feeling, and the making of cultures. Vintage
Ellard, C. (2024, June 14). Why boring streets make pedestrians stressed and unhappy | Aeon Essays. Aeon. https://aeon.co/essays/why-boring-streets-make-pedestrians-stressed-and-unhappy
Flying Gardens of Maybe | New-Empiricisms. (n.d.). New-empiricisms. https://www.andrewyang.net/maybe
Folke C, Biggs R, Norström AV, Reyers B, Rockström J (2016) Social-ecological resilience and biosphere-based sustainability science. Ecol Soc 21(3). https://doi.org/10.5751/es-08748-210341
Golembiewski JA (2015) The designed environment and how it affects brain morphology and mental health. Herd: Health Environ Res Des J 9(2):161–171. https://doi.org/10.1177/1937586715609562
Grahn P, Stigsdotter UK (2009) The relation between perceived sensory dimensions of urban green space and stress restoration. Landsc Urban Plan 94(3–4):264–275. https://doi.org/10.1016/j.landurbplan.2009.10.012
Green Infrastructure Foundation. (n.d.). Living architecture performance tool—Green Infrastructure Foundation. https://greeninfrastructurefoundation.org/lapt/#download-framework
Green Wall Basics | EFB. (n.d.). https://efb-greenroof.eu/green-wall-basics/
Greenscreen. (2023, July 11). Greenscreen—an innovative and unique modular trellising system. https://greenscreen.com/
Gunnar MR, Quevedo K (2007) The neurobiology of stress and development. Ann Rev Psychol 58(1):145–173. https://doi.org/10.1146/annurev.psych.58.110405.085605
Hammoud R, Tognin S, Burgess L, Bergou N, Smythe M, Gibbons J, Davidson N, Afifi A, Bakolis I, Mechelli A (2022) Smartphone-based ecological momentary assessment reveals mental health benefits of birdlife. Sci Rep 12(1). https://doi.org/10.1038/s41598-022-20207-6
Handbook on the Singapore Index on Cities' Biodiversity. (2008). https://www.cbd.int/doc/publications/cbd-ts-98-en.pdf
Hargraves, B. S. G. J. C. a. J. (2022, October 12). Spending on Preventive Services Represents a Small Fraction of Total Health Care Spending, but Costs to Individuals Could Be High without ACA Protection. HCCI. https://healthcostinstitute.org/hcci-originals-dropdown/all-hcci-reports/spending-on-preventive-services-represents-a-small-fraction-of-total-health-care-spending-but-costs-to-individuals-could-be-high-without-aca-protection
Hartig T (2017) Restorative environments ☆. Elsevier eBooks. https://doi.org/10.1016/b978-0-12-809324-5.05699-6
Herman JG, McKlveen JM, Ghosal S, Kopp BL, Wulsin AC, Makinson R, Scheimann JR, Myers B (2016) Regulation of the hypothalamic-pituitary-adrenocortical stress response. Comp Physiol:603–621. https://doi.org/10.1002/cphy.c150015
| Vertical Garden Patrick Blanc. (n.d.). https://www.verticalgardenpatrickblanc.com/medias
Kaplan S (1995) The restorative benefits of nature: toward an integrative framework. J Environ Psychol 15(3):169–182. https://doi.org/10.1016/0272-4944(95)90001-2
Kark S (2017) Effects of ecotones on biodiversity ☆. Elsevier eBooks. https://doi.org/10.1016/b978-0-12-809633-8.02290-1
Kelly BD, O'Callaghan E, Waddington JL, Feeney L, Browne S, Scully PJ et al (2010) Schizophrenia and the city: a review of literature and prospective study of psychosis and urbanicity in Ireland. Schizophrenia Res 116(1):75–89
Keltner D (2023) Awe: the new science of everyday wonder and how it can transform your life. National Geographic Books
Kennicott, P. (2023, September 21). How dream of air conditioning turned into dark future of climate change. Washington Post. https://www.washingtonpost.com/style/of-interest/interactive/2023/air-conditioning-climate-change/

Kg, P. V. G. (2022, August 23). Das Physikgebäude Berlin-Adlershof. NEUE LANDSCHAFT. https://neuelandschaft.de/artikel/das-physikgebaeude-berlin-adlershof-1504.html

Kimmerer RW (2020) Braiding sweetgrass: indigenous wisdom. Scientific Knowledge and the Teachings of Plants, Penguin UK

Kunstler JH (1994) Geography of nowhere: The rise and decline of America's man-made landscape. Simon and Schuster

Lakoff G (2012) Explaining embodied cognition results. Top Cogn Sci 4(4):773–785. https://doi.org/10.1111/j.1756-8765.2012.01222.x

Lee E, Williams K, Sargent LD, Williams NS, Johnson KA (2015) 40-second green roof views sustain attention: The role of micro-breaks in attention restoration. J Environ Psychol 42:182–189. https://doi.org/10.1016/j.jenvp.2015.04.003

Lee H, Jun Z, Zahra Z (2021) Phytoremediation: the sustainable strategy for improving indoor and outdoor air quality. Environments 8(11):118. https://doi.org/10.3390/environments8110118

Li Z, Yan H, Zhang X, Shah S, Yang G, Chen Q, Han S, Zhang D, Weinberger DR, Yue W, Tan H (2021) Air pollution interacts with genetic risk to influence cortical networks implicated in depression. Proc Natl Acad Sci U S A 118(46). https://doi.org/10.1073/pnas.2109310118

Lin H, Farkas ME (2018) Altered circadian rhythms and breast cancer: from the human to the molecular level. Front Endocrinol 9. https://doi.org/10.3389/fendo.2018.00219

Louv R (2013) Last child in the woods: Saving our children from nature-deficit disorder. Atlantic Books Ltd.

Lubell, S. (2020, November 3). How Milan's Bosco Verticale has changed the way designers think about sustainable design. Arch Digest. https://www.architecturaldigest.com/story/how-milans-bosco-verticale-has-changed-the-way-designers-think-about-sustainable-design

Luo M, Lau N (2019) Urban expansion and drying climate in an urban agglomeration of east China. Geophys Res Lett 46(12):6868–6877. https://doi.org/10.1029/2019gl082736

Madison AA (2019) Stress, depression, diet, and the gut microbiota: human–bacteria interactions at the core of psychoneuroimmunology and nutrition. Curr Opin Behav Sci 28:105–110. https://doi.org/10.1016/j.cobeha.2019.01.011

Mallgrave HF (2013) Architecture and embodiment: the Implications of the New Sciences and Humanities for Design. Routledge

McEwen, B. S. (2005, September 1). Stressed or stressed out: What is the difference? PubMed Central (PMC). https://www.ncbi.nlm.nih.gov/pmc/articles/PMC1197275/#:~:text=The%20term%20%E2%80%9Callostasis%E2%80%9D%20has%20been,cortisol%20and%20other%20chemical%20messengers

Medl A, Stangl R, Florineth F (2017) Vertical greening systems—a review on recent technologies and research advancement. Build Environ 125:227–239. https://doi.org/10.1016/j.buildenv.2017.08.054

Mills JG, Brookes JD, Gellie NJC, Liddicoat C, Lowe AJ, Sydnor HR, Thomas T, Weinstein P, Weyrich LS, Breed MF (2019) Relating urban biodiversity to human health with the 'Holobiont' concept. Front Microbiol 10. https://doi.org/10.3389/fmicb.2019.00550

Münzel T, Gori T, Babisch W, Basner M (2014) Cardiovascular effects of environmental noise exposure. Eur Heart J 35(13):829–836. https://doi.org/10.1093/eurheartj/ehu030

National Audubon Society (2023, February 21) Building collisions kill hundreds of millions of birds per year. Audubon. https://www.audubon.org/news/building-collisions-kill-hundreds-millions-birds-year

Noise as a Public Health Hazard. (2022, January 7). https://apha.org/Policies-and-Advocacy/Public-Health-Policy-Statements/Policy-Database/2022/01/07/Noise-as-a-Public-Health-Hazard

O'Gieblyn M (2022) God, human, animal, machine: technology, metaphor, and the search for meaning. Anchor

Office of the Assistant Secretary for Health (OASH). (2023, May 23). Surgeon general issues new advisory about effects social media use has on youth mental health. HHS.gov. https://www.

hhs.gov/about/news/2023/05/23/surgeon-general-issues-new-advisory-about-effects-social-media-use-has-youth-mental-health.html

Overview. (n.d.). World Bank. https://www.worldbank.org/en/topic/urbandevelopment/overview#

Perez G, Perini K (2018) Nature based strategies for urban and building sustainability. Butterworth-Heinemann

Pryor JT, Cowley LO, Simonds SE (2022) The physiological effects of air pollution: particulate matter, physiology and disease. Front Public Health 10. https://doi.org/10.3389/fpubh.2022.882569

Pugh TAM, Mackenzie AP, Whyatt D, Hewitt CN (2012) Effectiveness of green infrastructure for improvement of air quality in urban street canyons. Environ Sci Technol 46(14):7692–7699. https://doi.org/10.1021/es300826w

Richardson K, Steffen W, Lucht W, Bendtsen J, Cornell SE, Donges JF, Drüke M, Fetzer I, Bala G, Von Bloh W, Feulner G (2023) Earth beyond six of nine planetary boundaries. Sci Adv 9(37). https://doi.org/10.1126/sciadv.adh2458

Robinson S, Pallasmaa J (2015) Mind in architecture: neuroscience, embodiment, and the future of design. MIT Press

Roslund, M. I., Puhakka, R., Grönroos, M., Nurminen, N., Oikarinen, S., Gazali, A. M., Cinek, O., Kramná, L., Siter, N., Vari, H. K., Soininen, L., Parajuli, A., Rajaniemi, J., Kinnunen, T., Laitinen, O. H., Hyöty, H., & Sinkkonen, A. (2020). Biodiversity intervention enhances immune regulation and health-associated commensal microbiota among daycare children. Science Advances, 6(42). https://doi.org/10.1126/sciadv.aba2578

Salingaros, N. A. (2015). Biophilia & Healing Environments: Healthy principles for designing the built world (By Catherine O. Ryan). Terrapin Bright Green, LLC. https://www.terrapinbrightgreen.com/wp-content/uploads/2015/10/Biophilia-Healing-Environments-Salingaros-p.pdf. https://www.terrapinbrightgreen.com/wp-content/uploads/2015/10/Biophilia-Healing-Environments-Salingaros-p.pdf

San Francisco Biodiversity Program & Policy. (2023, August 14). sfenvironment.org—Our home. Our City. Our Planet. https://sfenvironment.org/article/the-biodiversity-program/biodiversity-program-summary

Selhub EM, Logan AC (2012) Your brain on nature: the science of nature's influence on your health, happiness and vitality. John Wiley & Sons

Shafi AA, Knudsen KE (2019) Cancer and the circadian clock. Cancer Res 79(15):3806–3814. https://doi.org/10.1158/0008-5472.can-19-0566

Shah, S. (2023, September 20). The animals are talking. What does it mean? The New York Times. https://www.nytimes.com/2023/09/20/magazine/animal-communication.html

Shah, I., Lau, S., Sekaran, V., & Ghahramani, A. (2023). Porous plant form-induced amplification of evapotranspiration for enhanced cooling in vertical greenery systems. Building and Environment, 245, 110904. https://doi.org/10.1016/j.buildenv.2023.110904

Stanton DE, Chávez JH, Villegas L, Benavides FV, Armesto JJ, Hedin LO, Horn HS (2014) Epiphytes improve host plant water use by microenvironment modification. Funct Ecol 28(5):1274–1283. https://doi.org/10.1111/1365-2435.12249

Sternberg EM (2001) The balance within: the science connecting health and emotions. Macmillan

Stobbe E, Sundermann J, Ascone L, Kühn S (2022) Birdsongs alleviate anxiety and paranoia in healthy participants. Scientific Rep 12(1). https://doi.org/10.1038/s41598-022-20841-0

Stone B, Gronlund CJ, Mallen E, Hondula D, O'Neill MS, Rajput M, Grijalva S, Lanza K, Harlan S, Larsen L, Augenbroe G, Krayenhoff ES, Broadbent A, Georgescu M (2023) How blackouts during heat waves amplify mortality and morbidity risk. Environ Sci Technol. https://doi.org/10.1021/acs.est.2c09588

Susorova I, Angulo M, Bahrami P, Stephens B (2013) A model of vegetated exterior facades for evaluation of wall thermal performance. Build Environ 67:1–13. https://doi.org/10.1016/j.buildenv.2013.04.027

Sussman A, Hollander JB (2021) Cognitive architecture: designing for how we respond to the built environment. Routledge

Tasnim N, Abulizi N, Pither J, Hart MM, Gibson DL (2017) Linking the gut microbial ecosystem with the environment: does gut health depend on where we live? Front Microbiol 8. https://doi.org/10.3389/fmicb.2017.01935

Terman M, Terman JS (1995) Treatment of seasonal affective disorder with a high-output negative ionizer. J Altern Complement Med 1(1):87–92. https://doi.org/10.1089/acm.1995.1.87

The 4 ways sound affects us - Julian Treasure. (n.d.). https://www.juliantreasure.com/blog/4-ways-sound-affects

The Economics of Biophilia. (2023) Terrapin Bright Green. http://www.terrapinbrightgreen.com/wp-content/uploads/2023/09/EOB2-Terrapin-2023-e.pdf

Turner, L. (2023, November 23). Harvesting the future: stormwater-powered living walls | Lily Turner. Living Arch Monitor. https://livingarchitecturemonitor.com/articles/harvesting-the-future-stormwater-powered-living-walls-fa23

Urban Habitats—Green Roofs and Facades: A Habitat Template Approach. (n.d.) Copyright (C) 2012 Urban Habitats. https://www.urbanhabitats.org/v04n01/habitat_full.html#cite64

Van der Kolk B (2014) The body keeps the score: brain, mind, and body in the healing of trauma. Penguin Books

Van Der Ryn, S., & Cowan, S. (1995). Ecological design.

van Os J, Kenis G, Rutten BPF (2010) The environment and schizophrenia. Nature 468(7321):203–212

Walker, J. (2016, May 12). Biophilic urban acupuncture: The importance of biophilia in urban places. Terrapin Bright Green. https://www.terrapinbrightgreen.com/blog/2015/10/biophilic-urban-acupuncture-biophilia-in-urban-places/

Westerink JH, Rajae-Joordens RJ, Ouwerkerk M, van Dooren M, Jelfs S, Denissen AJ, Penning de Vries E, van Ee R (2020) Deriving a cortisol-related stress indicator from wearable skin conductance measurements: quantitative model & experimental validation. Frontiers 2:39. https://doi.org/10.3389/fcomp.2020.00039

Wilson EO (2009) Biophilia. Harvard University Press

Wilson EO (2017) The origins of creativity. National Geographic Books

Wong NH, Tan AYK, Chen Y, Sekar K, Tan PY, Chan DJC, Chiang K, Wong NC (2010a) Thermal evaluation of vertical greenery systems for building walls. Build Environ 45(3):663–672. https://doi.org/10.1016/j.buildenv.2009.08.005

Wong NH, Tan AY, Tan PH, Chiang K, Wong NC (2010b) Acoustics evaluation of vertical greenery systems for building walls. Build Environ 45(2):411–420. https://doi.org/10.1016/j.buildenv.2009.06.017

Wood A, Bahrami P, Safarik D (2014) Green walls in high-rise buildings: an output of the CTBUH Sustainability Working Group. Images Publishing

Yeager R, Smith T, Bhatnagar A (2020) Green environments and cardiovascular health. Trends Cardiovasc Med 30(4):241–246. https://doi.org/10.1016/j.tcm.2019.06.005

Zhang C, Cui H, Zhang C, Chen Z, Jiang X, Liu J, Wan Z, Li J, Liu J, Yuwei G, Jin N, Guo Z (2022) Aerosol transmission of the pandemic SARS-CoV-2 and Influenza A virus was blocked by negative ions. Front Cell Infect Microbiol 12. https://doi.org/10.3389/fcimb.2022.897416

The Cosmogenesis of Building: Ancient (Eco)Logical Construction Practices

Eugenia Victoria Ellis

Abstract The impending threat of climate change on the health of our planet has awakened a new environmental awareness in our culture so that today we are reconsidering the constructed world with respect to nature and the position of the sun to create low-energy sustainable environments. This "new" approach to the design of the constructed world reconsiders nature-positive ancient traditions that have been forgotten due to the new technologies of the past century that have allowed us to defy nature. These ancient traditions were (eco)logical—the forces of nature were used to shape the constructed world to create comfortable dwellings that responded to prevailing environmental conditions. The ancient built world was auspicious because it was oriented with respect to the natural world: the positions of the sun, the stars and the planets. The cosmological origins of building will be demonstrated by considering the ancient practices of *Vāstu Śāstra* and *Feng Shui* as a way of reconsidering present-day nature-positive (eco)logical approaches to design.

Keywords Ecological design · Divination · Geomancy · Solar orientation

Introduction

The ancients modeled their constructed world in the image of the cosmos and in so doing imitated the work of the gods by considering all of celestial creation: the earth, the stars, the planets, the moon and the sun. Building was a gesture of cosmic harmony, of bringing the earthly body of the building in harmony with the cosmic body of nature. Today, new technological advances have allowed us to defy nature. Nowadays, it is possible to build anywhere regardless of climatic conditions. Even

E. V. Ellis (✉)
Department of Civil Architectural Environmental Engineering, Drexel University, Philadelphia, PA, USA
e-mail: eve22@drexel.edu

G. Iannaccone, E. V. Ellis (eds.), *Towards a Nature-Positive Built Environment*, Research for Development, https://doi.org/10.1007/978-3-031-95292-0_3

when the climate is desirable, we still construct hermetically sealed boxes that disregard the natural environment entirely.

The cosmological origins of building began with the human body as the first point of contact and genesis of the built world. In ancient times, in both China and India, human dwelling was considered to be a microcosm of the universe, and as such, in mirroring the cosmos earthly constructions were modeled after the operations of celestial creation. In mirroring the work of the gods, the act of building was considered to be a spiritual act; a procedure intended to communicate with the cosmos above to bring unity between the earthly and the divine, harmony between the built world and nature. In China, cities and dwellings on earth were modeled after the celestial nine-square grid where the Sky Emperor was believed to have lived at the center, venturing out periodically into the surrounding eight regions to inspect his kingdom. In India, the nine-square grid was also the model for dwelling, its center the belly of the *puruṣa* and the location where the earthly and the heavenly communicated.

Divining the Constructed World

> *From the trunk of a gum tree Numbakula fashioned the sacred pole (*kauwa-auwa*) and, after anointing it with blood, climbed it and disappeared into the sky. This pole represents a cosmic axis (*axis mundi*), for it is around the sacred pole that territory becomes habitable, hence is transformed into a world.* (Eliade 1957: 33)

Divining the constructed world was a talismanic operation that the Ancients used to orient their earthly creations to be "square with the world." The Ancients constructed according to divine co-ordinating principles to align their built works with the cardinal directions of the earth in deference to the cosmos.[1] This was an (eco)logical operation intended to embody the heavenly in an earthly construction that began with the human body at its center and origin. The body marked the beginning and the first point of contact with the heavens through its *axis mundi*, which in the body is the line of the spine in the erect human figure. In this way, the earthly microcosm of human built works could be brought into alignment with the macrocosm of nature: the sun, stars and cosmos.

Divination is a geomantic procedure. The word geomancy is derived from the Greek *geo*, literally meaning the earth, and *manteia*, meaning divination or coming from above. Geomancy is the act of projecting lines onto the earth from the sun above through marking the ground and encircling. This talismanic operation projects regulating lines upon the ground to provide auspicious conditions for the construction of the built environment and to protect the constructed world by being aligned with the natural environment.

[1] Vitruvius discusses this procedure in Book I, Chapter VI.6 of *De Architectura* (Granger, F., trans. 1931).

The divine resources for ancient geomantic procedures included the positions and the paths of the sun, the moon, the stars and the planets. The instrument the Ancients used to take their measurements was the gnomon, meaning literally, interpreter. The gnomon was a stick, often capped with the form of a human figure, used to help them interpret their position on earth with respect to the sun by being encircled: the intersection of the gnomon's cast shadow and the circle in the morning and the evening at the summer solstice located solar east and west from which north and south could be determined (Fig. 4). This (eco)logical procedure resulted in built works that considered the environment through solar orientation.

The following will be a discussion of two geomantic procedures that are used to this day to design buildings with respect to the environment: the Chinese method of *Feng Shui* and the Indian practice of *Vāstu Śāstra*.[2] These two practices have a common origin in the use of the gnomon for shadow casting and orienting with respect to the heavens; however, they have significant environmental and cosmological differences.

Polar Orientation: *Feng Shui*

> *By looking up in order to contemplate heavenly bodies, and by looking down to examine into the natural influences of the earth, man may acquire knowledge of the cause of darkness and light.* (*Yih King,* Chapter XIII, 1–12[3])

Chinese astronomers emphasized positional stellar astronomy in reference to the pole star and in India the motions of the sun, moon and planets had greater importance. This is most probably because the majority of China is north of the Tropic of Cancer, which runs central to India. In India, the position of the sun and its motion are much more critical in constructing the built world due to the environmental impact of the sun and resulting hot temperatures. In the northern latitudes of China, the motion of the stars about the pole star are much more visually appreciable, whereas the environmental concerns due to the position of the sun are less critical to their constructed world. In ancient times both systems relied on the gnomon to determine astronomical calculations.

It should be mentioned that the earth's axis is tilted with respect to the sun, which is why the celestial equator (the projection of the earth's equator onto the cosmic sphere) and the ecliptic (the zodiacal constellations and the apparent path of the sun

[2] Muslim geomantic practices are not discussed here because they are part of a great corpus of Islamic astronomy and astrology that places its emphasis on calculating the direction of the *qiblah* and locating the orientation toward Mecca. It should be noted that finding the direction of the *qiblah* is done with the aid of the "Indian circle"—this method places a gnomon at the center of a circle and locates the casting of the shadow to determine the north-south longitude and the east-west line perpendicular to it. As found in Nasr, S. H. 1976.

[3] As found in Carus, P. 1974: 55. This book is an abridged version of 1907 Open Court publication entitled *Chinese Thought*.

around the earth) do not align. In the northern hemisphere, because the earth's axis is tilted, the days are longer during the summer when the sun is high in the sky, the longest day and the sun's highest point is at the summer solstice; the days are shorter in the winter when the sun is lower in the sky, the shortest day and the sun's lowest point is at the winter solstice. The fall equinox and the spring equinox occur when the celestial equator and the ecliptic cross and are when the lengths of the day and the night are equal (Fig. 1).

The celestial pole was fundamental to ancient Chinese astronomy and formed the basis of their thinking regarding relationships between the microcosmic and the macrocosmic. According to myth, atop the celestial pole was the Imperial Star above the heavenly Central Palace where the sky emperor lived, central to his kingdom of eight surrounding "sky palaces." This location mirrored the position of the emperor on earth, around whom feudal land-holding practices were modeled as squares within squares based on a nine-square grid. Chinese maps were drawn with the north arrow pointing down. More than likely this was so that the map would be coordinated with the emperor's view while looking down on his kingdom. The Imperial Star was the pole star of the Great Bear constellation and is fixed, around which all the other stars revolve. According to Confucian authority, the ruler on earth should govern by the sky emperor's example: "He who exercises government by means of his virtue may be compared to the pole star, which keeps it place while all the stars turn around it."[4]

The fundamental basis of Chinese cosmology was the relationship between the pole star and the terrestrial horizon, which is divided by the cardinal directions into

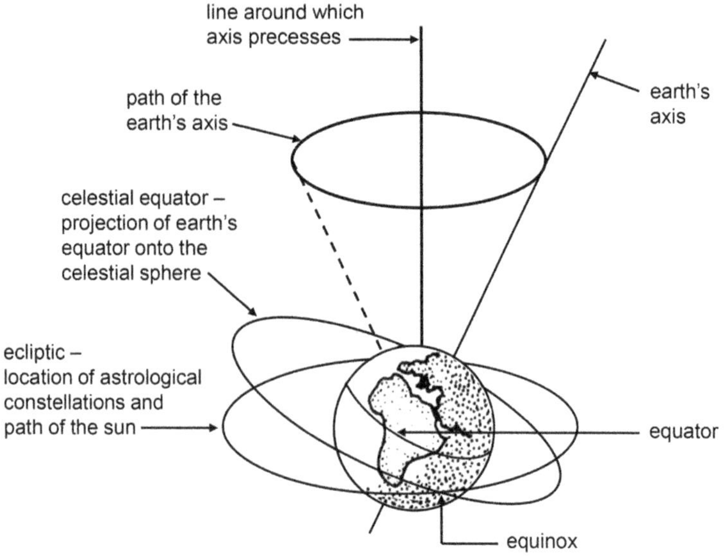

Fig. 1 Precession of the equinoxes (Ellis, after Ulansey 1989)

[4] *Lun Yü*, II, I, as found in Needham, J. 1959: 230.

the four directions of space (Fig. 2). Each direction of space corresponds to a seasonal palace: the Blue Dragon in the east (spring), the Vermilion Bird in the south (summer), the White Tiger in the west (fall), and the Black Tortoise in the north (winter). This designation dates to an antiquity of at least—14th century. There are seven *hsiu* grouped in each of these four equatorial palaces, for a total of twenty-eight houses. The twenty-eight divisions radiate from the pole star like the spokes of an umbrella that intersect the horizon line (Needham 1959).

The constancy of the Great Bear and the pole star was fundamental to ancient Chinese cosmology. However, when considered over the span of many millennia, the referent pole star the Chinese actually used changed over time due to the "wobble" of the earth's rotation on its axis. This is known as the precession of the equinoxes and makes one full cycle every 25,920 years. In short, the plane of the celestial equator is always perpendicular to the earth's axis (Fig. 1). The ecliptic, the location of the zodiacal constellations and the apparent path of the sun, always remains the same due to the earth's orbit around the sun. However, because of the earth's "wobble" on its axis, the two equinoctial points along the ecliptic, which is where the celestial equator and the ecliptic cross, precess through the constellations of the zodiac (Ulansey 1989). Although the constellation of the Great Bear always appears to circle in the general vicinity above the North Pole, the earth's axis or celestial pole does not always point to the same star. The effect of the precession on the position of the celestial pole is considerable: in effect, the celestial pole inscribes a large circle that has the pole of the ecliptic at its origin. This astronomical fact was fully recognized by the ancient Chinese as early as the +4th century (Needham 1959). The gnomon was the instrument that the Chinese used to calculate the precession of the equinoxes, to determine latitude in fixing provincial and other territorial boundaries, to construct their calendars, and to determine auspicious times (Needham 1959).

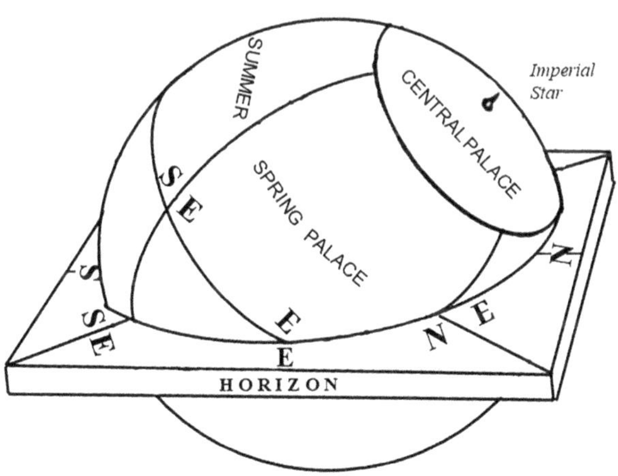

Fig. 2 Diagram of the ancient Chinese divisions of the celestial sphere and their relations with the horizon (Ellis, after de Saussure, as found in Needham 1959)

The sundial was also used to align the earthly with the universe. Sundials were related to the magnetic compass and the diviner's board (*shih*): the compass for preliminary orientation with the pole star and the diviner's board to determine auspices (Fig. 4). Magnetic north changes often with atmospheric conditions, is not exactly in the direction of the celestial pole and can vary from solar north by as much as 33°; however, its use was to preliminarily determine orientation until it could be verified using a gnomon. The Great Bear of this *shih* is on a central rotating round plate that signifies the heavens of the celestial pole, which sits on a square plate representing the earth. The square plate of the earth is marked with the eight trigrams of *Wen Wang* located at the cardinal and inter-cardinal directions, which are said to be in the "world of the senses" arrangement and relate to the real, phenomenal world in which we live; the world perceptible to the senses. Divining with the *shih*, then, relates the universe to the real world of persons, sense-perception and occurrences of daily living such as could discerned by the fortunes of the *I Ching* (*Yih King*).

Solar Orientation: *Vāstu Śāstra*

Sūtra 4: Who has the knowledge of circle and line is a sthāpaka. *Holding in hand a measuring rod of* khādira *wood and a cord of* darbha *grass fitted with a ring, this is his outer aspect. This knowledge is the knowledge of Art. From this knowledge arises divine knowledge, and such knowledge leads to liberation. This is verily the essence of the knowledge of Art.* (*Vāstusūtropaniṣad*, first chapter[5])

Geomancy in India is called *Vāstu Śāstra*: *vāstu* means to dwell and *śāstra* means science.[6] In the Indian tradition, *Vāstu Śāstra* is used to orient the built world to the environment through squaring the circle using a gnomon and the *vāstu puruṣa mandala*, requiring mastery of the circle and the line. This *mandala* is a magic square in the anthropomorphic form of a sacrificial victim lying face down in a *yantra.* A magic square is a gridded square that becomes magic if and when each of its squares when numbered sequentially add to the same number horizontally, vertically and on the diagonal.

[5] Boner, A., Sarma, S. R., and Bäumer, B. 1996. In India there is no principal distinction made between the Fine Arts and the Practical Arts. A *Sthāpaka* is one who has knowledge of the circle and the line and can be either a sculptor or an architect. *Śilpa* refers to sculpture and *vāstu* refers to architecture. Although this *upaniṣad* is directed toward both the sculptor and the architect, it is called a *vāstusūtra* instead of a *vāstuśāstra* most probably because it does not have to do with "modeling" the three-dimensional qualities of a work or sculpture but has to do with its layout and geometric composition, which in either discipline has to do with the organization of the work in plan and in elevation.

[6] *Vāstu Śāstra* is a geomantic art that was fully developed by +1st century. Its tenets were not written down in any one ancient manuscript but developed out of a corpus of over twenty texts referred to as *Vāstu Vidya* dating back to the +6th century, most of the previous material having been lost. The above information was found in: Padam, J. P. A. 1998 and Chakrabarti, V. 1999.

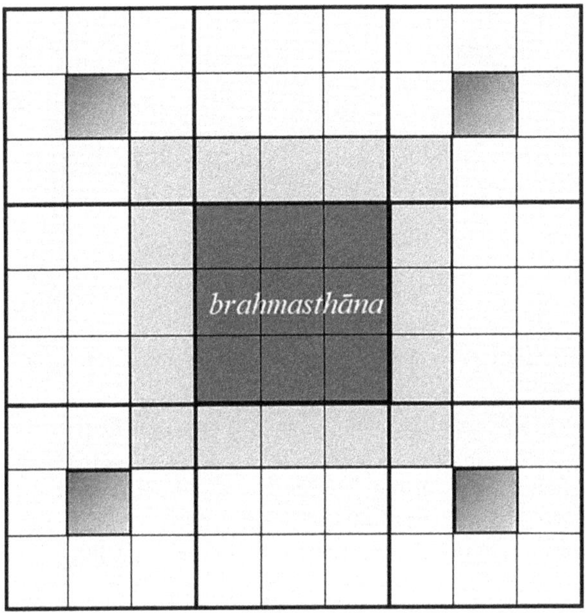

Fig. 3 *Vāstu Puruṣa Mandala.* Note central *brahmasthāna* (Ellis, after Kramrisch 1946)

A *yantra* is literally a visual and graphic symbol in the form of a square with four gates at the cardinal directions that is used for meditation purposes[7] and is characteristic of Hindu temple architecture. The temple's central space is called the *brahmasthāna* (Fig. 3) and is in the shape of a square. This central area of the temple is open to the sky and acts as a conduit for the exchange of communication between the earth and the heavens. This space also functions as a chimney to ventilate the temple and cool the building. The *brahmasthāna* is set within another square that is oriented "square" with the sun, which is "four-cornered" representing the earthly. The generic Hindu temple form is three by three squares, or a nine-square grid that can be further broken down. [8]

In its most simple form, a *mandala* is a *yantra* inscribed within a circle. The circle represents the motion of the revolution of the planets and symbolizes wholeness or totality. The magic square of the *puruṣa* is a *yantra* that becomes a *mandala* when it is encircled during the act of shadow casting to orient it with the sun. This geomantic art in India begins at the navel with the casting of a shadow using a gnomon, symbolic of the upright sacrificial person. In this way, the gnomon becomes a central axis pointing toward the sky that relates the earthly with the cosmos.

This vertical pole often takes the form of the human body and in India is synonymous with the sacrificial post *yūpa* and *axis mundi,* both of which rise from the

[7] Material on the meaning of the *yantra* was found in Zimmer, H. 1946 and Mookerjee, A. and Khanna, M. 1977.

[8] Corroborated in Bunce, F. W. 2002.

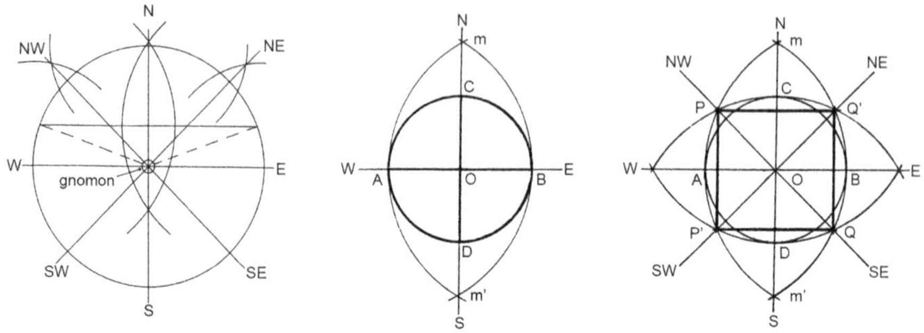

Fig. 4 Determining the cardinal points with a gnomon and using circles to define the original square

"altar at the navel of the earth" to intersect with the sky. The geomancer is a "sacrificer" who makes the gnomon "a ladder and a bridge to attain the world of heaven. Thus, having attained the world of heaven, the sacrificer finds support again in this world." The casting of the initial shadow to orient the *vāstu puruṣa mandala* to be "square with the world" initiates building orientation: "It is through the Sun that the Earth becomes the support for all beings." [9]

The individual "houses" (*pāda,* literally, feet) of the square correspond with parts of the human anatomy; the navel of the *puruṣa* is located at the square's origin in the center. In India, the human frame, the constructed temple, and the whole of the universe are analogical equivalents; therefore, the parts of the temple correspond both to the parts of the human anatomy and to the parts of the universe itself. All of these forms are "houses" and are dimensioned or measured-out (*māyā*). The Hindu noun *māyā* is formed from the root *mā*, which means "to measure or lay out" as in the ground plan of a building or the creation of the universe. As such, *māyā* is all of the created or material world.

Constructing the *yantra* is the first step in the design of the constructed world through the practice of *Vāstu Śāstra*. The eight tools of measurement are the *sutrashtaka*: scale, rope, cord, plumb line, tri-square, compass, level, and the eye. The scale and rope are of a prescribed length and are for measuring, while the rest are used for examining the site and for geometrical construction. The orientation of the site is established on an auspicious day when the sun is in the northern part of the sky. A "sunpillar" is erected and used to cast measured shadows. It is called a *yūpa,* a sacrificial person, or gnomon that is placed in a square area that is "as smooth as a mirror" and has been checked using a level (Pada 1998). A circle is inscribed with the gnomon as its center (Fig. 4), its radius the height of the gnomon (*r*). Two points are marked on the circle where it is intersected by the shadow at mid-morning and mid-afternoon (points A and B). These two points are joined by a straight line, thereby determining the "true" east-west axis. The closer the day is to

[9] The symbolism of the temple and the body in India was found in several writings of Coomaraswamy, A. K. 1997: "S*vayamātrnnā*: Janua Coeli," "An Indian Temple: The Kandarya Mahadeo," and "The Symbolism of the Dome".

the summer solstice, the "truer" the axis. Due to the tilt of the earth's axis, the closer the day is to either of the equinoxes, the greater the error: the line will be tilted south of east in the spring and north of east in the fall.

Once the true east-west axis is determined, the north-south axis is found by marking the arc of a circle with its origin at each of the east (B) and west (A) nodes, the radius of which equals the diameter of the original circle (*2r*). A line can be drawn where the two arcs intersect defining the north-south axis and it will be perpendicular to the east-west axis through the origin, the two together marking the cardinal directions. Using the points where the original circle intersect the north-south axis (C, D), construct two more arcs of radius *2r.* The points where the four arcs cross define the inter-cardinal directions with respect to the origin. By connecting these four points with lines, the earth has been "squared" through the construction of circles. Once the cardinal directions of the site have been determined on an auspicious day, the *vāstu puruṣa mandala* can be superimposed upon the site. The *pāda* (literally the feet) of the *puruṣa* then determine the function or use of the corresponding space to be constructed. For example, the "feet" can walk through the plan from the bedroom in the northeast, to the kitchen in the southeast and dining room to the west.

In India the "divine architect" is called Lord Viśvakarman: he who measures-out the cosmos. His counterpart on earth is called a *Sthāpaka*: he who is a master of the circle and the line. The *Sthāpaka* orients the built world by constructing a circle with a line from a gnomon at its *brahmabindu*, the center point of the tracing of the circle, which literally means "divine origin". The gnomon represents the *yūpa*, the sacrificial pillar or pole, which represents the *puruṣa*, the archetypal person. *Vāstu Śāstra* is a procedure used to orient the built world to the heavens above through the circle and the line using a gnomon and the *vāstu puruṣa mandala.* This is a procedure that "protects" the constructed world by encircling and by providing a "divine" framework to ensure that architecture will correspond to natural laws. This is an anthropomorphic procedure to guarantee unity between the body of the building and the body of the environment, because according to Hindu tradition: "He who is this (Brahman) in man, and he who is that (Brahman) in the sun, both are one".[10]

(Eco)Logical Design: Reconsidering *Feng Shui* and *Vāstu Śāstra*

While the nine-square grid as the generator of built form is common to both China and India and is reflected in the way they organized their cities and dwellings, their respective cosmological views of the square as a representation of the earth differ.

[10] This is a reference from Max Müller's *Sacred Books of the East, the Upaniṣads,* as found in *Bragdon Family Papers*, Department of Rare Books and Special Collections, University of Rochester Library, A.B81, Box 36, Folders 1-3, dated 04/20/1891.

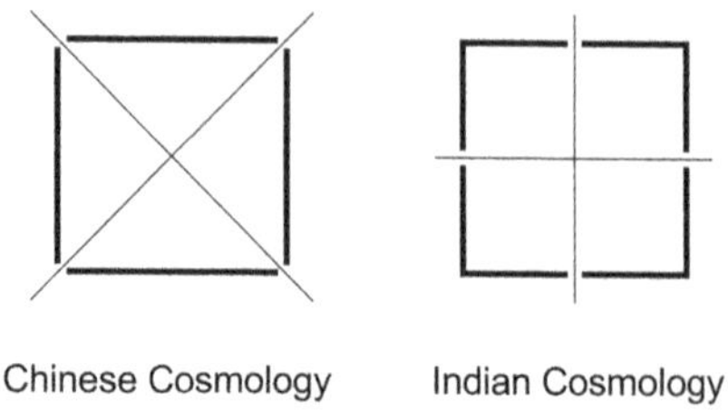

Fig. 5 Diagrammatic comparison of Chinese versus Indian cosmological views of the earth (Ellis)

India's square is a body-centered cosmology that emphasizes the crossing of the cardinal axes at the belly of the square, while China's emphasis is on the corners (Fig. 5). Ancient China looked to the cosmos in relationship to the "four corners of the earth" with respect to a central sacred mountain, an earthly cosmic axis relating the land to the heavens above. India's cosmos was reflected on earth as the *yantra*, or basic temple form with entrance gates at the cardinal directions. China's gateways were at the inter-cardinal directions, or the corners, dividing the earth into four seasonal palaces. To China, the center was the home of the Sky Emperor looking down from the Celestial Pole. On the other hand, ancient India looked to the cosmic in reference to a centered human body. Common to both is an emphasis on the center as a point of origin and place of communion on earth with the heavens above.

The reason the geomantic arts of China and India differ is due to fundamental differences in religious philosophy together with astronomical, astrological, climatic and geographic conditions.

Metaphysical Differences

The Chinese *Tao* emphasizes the balance of opposites (*yin* male and *yang* female) about a center to yield unity or one; whereas, Hindu philosophy is based on the active union of two opposites to equal a whole. In China, the *Tao* is a type of cosmic dance, a union of actives with passives, positives with negatives, in an unending interchange of one replacing the other in the play of life. In the Hindu practice of Tantra there are male and female forces; however, they are polarities that are aspects of one Reality, one indivisible whole or unity. On the creative plane the energy forces are called *Śiva* (male energy) and *Śakti* (female energy), two aspects of one creative force that are eternally conjoined. All phenomenal manifestation is based upon the earthly forces of Nature called *Puruṣa* (male principle) and *Prakriti* (female principle), which are essentially two aspects of the same principle. In keeping with Hindu tradition, the body of the building and the cosmic body were considered to be two aspects of the same reality—that the Brahman in the Sun and the Brahman on earth are One.

Astronomical/Astrological Comparisons

India uses the traditional zodiacal constellations attributed to having been discovered first by the Greeks; China has their own duodenary animal system. India refers to nine planets—the traditional seven of the ancients plus an additional two imaginary "planets," *rāhu* (dragon's head) and *ketu* (dragon's tail), which are the ascending and descending nodes of the moon with respect to the horizon (Mookerjee and Khanna 1977: 119). China has twenty-eight lunar mansions; in India each phase of the moon's orbit goes through one of the twenty-seven *nakshatras*,[11] which are star constellations distributed over the twelve *rashi* or zodiac signs at the ecliptic belt. As in China, *Vāstu Śāstra* relies on a complex system of astrological calculations to determine auspicious days for construction of the built environment. In *Feng Shui* geomancy begins with the pole star; however, in *Vāstu Śāstra* the position of the sun and the moon with respect to the earth is of prime importance, which may explain why there are two invisible "planets" that mark the point of passage of the moon with respect to the horizon as at appears and disappears in the sky.

Climatic/Geographic Influences

Favorable orientations for these two geomantic arts are quite distinct. First, the Chinese emphasize an axial relationship to the celestial pole, while in India the emphasis is on the crossing of the cardinal directions and an axial center. Second, in China the location of a building with respect to the pole star determines orientation, while in India the built world is oriented toward the path of the sun. Third, in China a building's main entrance looks out toward the south (as from the imperial position), whereas in India the main entrance should be either from the north or the east (looking toward creative origins as demonstrated by the daily birth of the sun).[12]

The auspices of the life force, or breath, as it moves through a building will organize the interior in a specific way to either China or India: In the more northern climate of China *Feng Shui* requires a spatial shift in moving through a building to restrict the movement of *chi* (cosmic breath) and to prevent it from going straight through, in one side and out the other; on the other hand, *Vāstu Śāstra* encourages the movement of *prāna* (vital breath) through the house as natural ventilation to cool the interior of buildings in a southern climate. Unique to a constructed world based on *Vāstu Śāstra* is an emphasis on a center that is open to the sky above to act as a cosmic pillar, chimney and *axis mundi*, and the requirement that the movement

[11] There is an occasional twenty-eighth constellation to average out the moon's orbital phases.

[12] India is reasonably close to the equator, which means that for a greater portion of the year the sun will rise high on the horizon line toward the northeast. For this reason, the preferred orientation in India is to the northeast or somewhere between north and east.

upward in a space as in a stairway should be in a clockwise direction to be consistent with the earth's rotation.

Conclusion

The distinctions between these two geomantic practices are cosmological and metaphysical, and geomancy itself is related to religious and spiritual practices; however, the founding principles of these practices are environmental due to the latitudinal relationships of China and India on the planet earth. These divination arts resulted from the scientific study of the forces of nature in an attempt to understand the surrounding world and to use this knowledge to shape the built environment. The resulting geomantic techniques were (eco)logical—the forces of nature were used to shape the constructed world to create comfortable dwellings that responded to prevailing environmental conditions. A practice that is even more important to consider with today's environmental concerns.

In China, *Feng Shui* developed in deference to the celestial pole because they are farther north and the stars in their sky appear to revolve around the pole star—a spectacular stellar phenomenon. Because of India's southern latitude and hot climate, *Vāstu Śāstra* considers the sun in determining auspices. In the more northern China, it would be beneficial to open a building toward the south for solar heat gain and to prevent *chi* from creating a draft in the cold winter months. On the other hand, in the more southern India, *prāna* and natural ventilation are necessary to cool down a building, and the *brahmasthāna* essential to act as a chimney to keep the breezes flowing.

The geomantic arts and the mythology of building have been at odds with post-Enlightenment rational, scientific thinking, which has taught us to forget about relationships between the natural and built environments. As a modern technological society, we are beginning to remember that daily living is always a communication between ourselves and the environment. The human species walks erect and through this posture points towards the heavens. Regardless of what science might tell us about the universe, our vertical bearing indicates we are always in communication with the cosmos while dwelling here on earth. Nowadays, more so than ever, it has become crucial for us to reconsider our relationship with the natural world and recognize that future constructions need to support a nature-positive built environment.

References

Boner A, Sarma SR, Bäumer B (1996) Vāstusūtra Upaniṣad: the essence of form in sacred art. Motilal Banarsidass Publishers (Original Publication 1982), Delhi

Bunce FW (2002) The iconography of architectural plans. D. K. Printworld (P) Ltd., New Dehli

Carus P (1974) Chinese astrology. Open Court Publishing Company, LaSalle, IL

Chakrabarti V (1999) Indian architectural theory: contemporary uses of Vastu Vidya. Oxford University Press, Delhi
Coomaraswamy AK (1997) The door in the sky. Princeton University Press, Princeton
Eliade M (1957) The sacred and the profane. Harper & Row, New York
Kramrisch S (1946) The Hindu temple. University of Calcutta, Calcutta
Mookerjee A, Khanna M (1977) The Tantric way. Thames and Hudson, London
Nasr SH (1976) Islamic science. Westerham Press Ltd, Great Britain
Needham J (1959) Science and civilisation in China Vol. 3 Mathematics and the science of the heavens and the earth. Cambridge University Press, Cambridge
Pada JPA (1998) Vãstu: reinventing the architecture of fulfillment. Management Publishing Co, Dehradun, India
Ulansey D (1989) The origins of the mithraic mysteries. Oxford University Press, New York
Vitruvius (1931) On architecture. Translated by Frank Granger. Harvard University Press, Cambridge
Zimmer H (1946) Myths and symbols in Indian art and civilization. Princeton University Press, Princeton

Ecology and the Urban Landscape

Urban Ecology and Biodiversity

Dane C. Ward and Heather L. Kostick

Abstract Urban ecology is a burgeoning field of study seeking to understand how the built environment and biotic communities interact. While urban ecology is a broad field of study, this chapter focuses on urban ecology at the landscape scale and seeks to understand the various ways landscape structure supports biodiversity. First, we will draw distinction between 'ecology in urbanized landscapes' and 'urban ecology', defining what it means to be 'urban' from an ecological lens, examine the processes of urbanization, and describe what ecological theories and principles help us to understand our urban landscapes. While modern contemporary urban landscapes vary as a result of each locality's geography, history, and culture we can draw comparisons by evaluating urban green and gray space availability and spatial dispersion. The biodiversity within these environments therefore is the result of many years of human interaction in the form of city planning, management, and socio-cultural community decision making. While each locality is unique, we generally observe predictable trends in urban biodiversity around the world, as urbanization creates a novel landscape of biophysical regimes. These biophysical conditions not only structure urban biodiversity but often lead to large systemic challenges for the human populations within our cities. This chapter closes with suggestions for partnering efforts to increase urban biodiversity, as well as ameliorate the ecological challenges of urbanization on human populations. Such efforts increase urban resilience, enhance ecological function, and increase benefits for public health in the face of climate change.

Keywords Urban ecology · Biodiversity · Urbanization · Air quality · Urban planning · Urban landscape

D. C. Ward (✉)
Department of History and Anthropology, Monmouth University, West, NJ, USA
e-mail: daward@monmouth.edu

H. L. Kostick
Department of Biodiversity, Earth & Environmental Science, Drexel University, Philadelphia, PA, USA
e-mail: hlk35@drexel.edu

G. Iannaccone, E. V. Ellis (eds.), *Towards a Nature-Positive Built Environment*, Research for Development, https://doi.org/10.1007/978-3-031-95292-0_4

Introduction

Urban landscapes offer a unique lens for testing ecological theories, understanding ecosystem function, and investigating the interactions of humans and the natural world. Urban landscapes are first and foremost built environments for dense human populations. As urban landscapes initially develop, they directly replace previously contiguous natural landscapes, reducing habitat availability, biodiversity, and ecosystem function. Historically, the built environment has directly replaced more natural environments while blue and green spaces comprise a more diverse suite of environments with increasingly more variable histories and disturbance regimes. In some instances, these influences facilitate the creation of new novel ecological systems which can contain unique suites of biodiversity (including non-native species) and ecological functions.

The variety of environments and their spatial properties (e.g. size, location, connectedness) lend to urban landscapes appearing markedly different from one another, however regardless of geographic location, population size, or cultural preferences all urbanized landscapes share some unique characteristics. Urban landscapes tend to have higher proportions of impervious surface, higher ambient temperatures, and greater concentrations of noise and air pollution when compared to the ecosystems they have replaced or their contemporary surrounding landscape. These characteristics which we might think of as unique to urban environments are simply 'amplified' from more natural counterparts, a convenience to urban ecologists who can employ many if not all of the same ecological theories, principles, and practices that were previously developed in more pristine study locations. One important change from classic ecological investigations is the treatment of humans as central to and within the ecosystem. In order to better understand how urban landscapes differ from more natural spaces, which are impacted by human activity in different ways, researchers have developed and rely upon the 'human ecosystem concept' (Machlis et al. 1997). The human ecosystem concept builds upon the long-held ecosystem model developed by Tansley (1935). The Tansleyan ecosystem concept examines interactions between biotic complexes and abiotic or physical complexes. Often excluding humans from the biotic complex and treating the effects of human activity as aspects of the physical complex. While this simplified model of ecosystems works well enough, it fails to integrate the most important aspects of human societies, the social complex and, for most societies, the built environment.

The incorporation of the social complex and the built environment help us to better understand how humans influence ecologies, and in some cases develop new ecologies because of the myriad interactions that occur from our urban tinkering. Human populations have long impacted natural systems, by diverting water, harvesting resources, and altering terrain. These same actions are observed in the process of urbanization and are often amplified. Where in the past a human population may have diverted water from a river for agriculture but maintained a dynamic floodplain, we now often observe urban populations diverting water from infrastructure, stabilizing riverbanks, and linearizing the course of the river. The human behaviors are similar, but we observe differential environmental impacts given the scale of the behavior. The differential impacts we observe in modern urban engineering reflect the broader social, economic, and cultural lenses of the population. Additionally, most urban landscapes are the result of multiple generation's ideals, preferences, and decision making in engineering over time. Those few urban landscapes which are not borne from changing human populations over time, are instead highly planned feats of engineering in the modern age. The level of planning and foresight needed to develop a 'planned urban landscape' is not possible for most populations as it requires stable human population size, a developed and stable economy, and a stable suitable location near provisioning resources and other urban centers. Because the conditions for planned urbanization are often unattainable, the majority of urban landscapes experience urban sprawl (unplanned or unpredictable expansion at the edges), infilling of undeveloped spaces, or replacement of previously developed spaces. The combination of unplanned growth in the built environment and replacement or reworking of previous built environments allows for the emergence of novel habitats and ecologies.

When we consider the dynamic nature of urban landscapes it becomes clear that urban landscapes amplify naturally occurring ecological processes and allow for the creation of novel ecological systems. In other words, most classic ecological paradigms are still held in urban landscapes and are worth studying as comparisons to more natural settings, but urban landscapes also lend to new paradigms and ecologies which lack comparison to more natural systems. As a result, we often discern between the study of ecology in urban landscapes and the study of urban ecology. For example, when studying ecology in urban landscapes researchers might study how biodiversity is affected by urbanization or how urbanization influences wildlife movement or behavior. On the other hand, when studying urban ecology researchers seek to elucidate the intersectionality of ecological and human systems with research topics including urban metabolism, green infrastructure, the role of urban green spaces in supporting biodiversity, or the social-ecological dynamics shaping urban sustainability.

Given the ecological amplification observed in urban systems and the lack of available habitat, many ecologists and land managers focus on promoting functional biodiversity instead of simply quantifying urban biodiversity (i.e. species richness) or broadly seeking to increase biodiversity without a function or service in mind. Fortunately, landscape architects, city planners and managers have a wealth of information and options for integrating functionally relevant species into the urban landscape. Functional diversity includes both species and ecosystems diversity which contributes to system level function, or emergent properties. These emergent properties or functions of ecosystems are often important contributors to ecosystem services which will be discussed later in this chapter.

Urbanization and How It Happens

With more than half of the world's population now living in cities (United Nations Department of Economic and Social Affairs, Population Division 2022), urbanization is the leading cause of global biodiversity loss, both directly through habitat replacement, and indirectly via the systems and networks required to provide resources for urban populations (Tilman et al. 2017). The United Nations estimates that at least 70% of all non-glaciated land mass has been altered by human activity, primarily for food provisioning (Benton et al. 2021). Climate change further compounds biodiversity loss because of globalization and a reliance on fossil fuels for transportation and energy provisioning for urban populations. Interestingly, biodiversity has a great potential to mitigate the impacts of climate change (Dobson et al. 2021). For the purpose of this chapter we will focus on the direct effects of urbanization on biodiversity.

Urbanization is the process of human populations gathering and residing in dense assemblages of built infrastructure. Urbanization is analogous with modern humanity itself, beginning with ancient urban centers such as Çatalhöyük or later, Mesopotamia. Since its inception, urbanization has occurred as the result of local environmental stability and stable or increasing human populations. Prior to the development of agriculture, small groups of humans would need to relocate in order to source enough food to sustain the group. With the advent of early agriculture, food provisioning became less of a limiting factor. Human groups grew in size with the ability to acquire surplus food resources allowing for the creation of small, localized settlements. These settlements would create socio-economic systems, leading to further expansion. Regional environmental stasis would allow urban populations to develop increasingly complex technology, specialization, and economics, which later in human history will come to be seen as primary drivers for urbanization. Early urban centers, though, enabled by environmental and socio-cultural stability, were reliant on nonurban landscapes for food and other resources, a pattern that continues in all urban systems to this day (Elmqvist et al. 2013).

While urbanization's origins date to ancient times, most of the global human population resided in more rural settings until the 19th century. Early modern urbanization in the latter 19th century, and modern urbanization through the 20th century were driven by industrialization and the emergence of economically important mega-cities (Sassen 2001; UN-Habitat 2016). Modern urban landscapes develop as a result of several interacting factors including geographic location, socio-cultural preferences, economy, the degree to which development is planned, and finally time or age. Planning of urban landscapes can include many of the aforementioned drivers, and can often be observed by analyzing the spatial configuration thereof. Highly planned urban landscapes seek to achieve a spatially heterogenous, equitable distribution of zoning types with efficient infrastructure networks. These highly planned urban landscapes are relatively rare, however, as they must occur in new areas without prior urbanization constraints.

As stated above, it is far more common that urban landscapes develop more spontaneously because of the cumulative impact of historical processes, urban growth dynamics, and evolving socio-economic trends over the age of the city (Alberti et al. 2003). Sprawl and infilling both have the potential to unravel previous urban planning by removing green space which supports urban biodiversity either directly as habitat or indirectly as connected habitat, serving as a source of biodiversity into the urban landscape. When multiple urban landscapes continue to expand outward from their core toward each other, as we see on the eastern seaboard of the United States, we observe a conurbation or megalopolis. These large spatial extents may include varying human population densities from city center-suburb-city center, but all of these spaces comprise an urban landscape. Within these urban landscapes, the built environment dominates land use types and more natural land use types (i.e. forest, grassland, and even agriculture) are greatly reduced in spatial extent and ecological function (more on these types of spaces in forthcoming sections).

To summarize, while urbanization is a long held phenomena of humanity, modern urbanization has relatively recently (< 200 years) become the dominant spatial model of human organization. Despite the differences between early and modern urbanization, several commonalities exist. First, urbanized landscapes directly replace previously occurring natural habitats, and subsequently urban populations are reliant upon additional non-urban altered landscapes for provisioning food, energy, and materials in order to sustain the urban population. Therefore, urban landscapes both directly and indirectly alter large portions of the terrestrial earth. Secondly, most urban landscapes are the result of long histories of iterative re-development which is driven by interactions between economics and socio-cultural norms, and is constrained by previous infrastructure or planning decisions. As urbanization continues, we observe urban centers sprawling into proximal undeveloped spaces and replacing or infilling relic green spaces that remained through prior decision making or by accident. In most instances the rate of urban sprawl and infilling results in limited naturally occurring habitat throughout. Finally, urban landscapes develop as large spatial mosaics of built environment (e.g. buildings, roadways, and infrastructure), 'gray spaces', slightly less altered aquatic environments referred to as 'blue spaces', and terrestrial vegetated areas or 'green spaces'. Despite urbanization's replacement of preexisting habitats, these urban landscape mosaics provide spaces for biodiversity to persist, new ecologies to emerge, and opportunities to jointly maintain ecosystem functions and services, and human populations, which is the goal of urban resilience.

Urban Biodiversity

Biodiversity is the sum of all species, genetic, and ecosystem diversity observed within a system. The colloquial use of the term, biodiversity, tends to describe species diversity, which is the number of unique species (i.e. species richness) and the relative abundance of each species (i.e. species evenness) within the area of interest. Generally speaking, having more species lends to greater species diversity. Additionally, having many species who are relatively even in abundance is indicative of even greater biodiversity as compared to a scenario where you have the same number of species, but one species makes up 95% of the observations made. To extend biodiversity to genetic diversity, we employ a similar model, having many genetically unique individuals within a population or species yields greater diversity than having many closely related individuals who share similar genetic composition. Finally, ecosystem diversity helps to explain the variety of unique habitats or living spaces available.

In naturally occurring systems, species are generally found within ecosystems that meet their physiological needs for survival, growth, and reproduction; therefore, identifying particular plant or animal species is likely to represent ecosystem diversity fairly well. Unfortunately, urban landscapes display significantly altered, and sometimes even novel, habitat types which may not contain the expected species assemblages that we observe in more natural areas. As a result, it is important to ascertain estimates of diversity at the species and ecosystems levels when studying urban biodiversity. When possible, gaining an understanding of genetic diversity would be valuable, but these data tend to be cost prohibitive and are typically only supported with research funding which seeks to examine how populations adapt and evolve under urban pressures. Because biodiversity includes genetic, species, and ecosystem diversity it plays a critical role in functioning and stability of ecosystems. Many of the functions of ecosystems (which will be discussed in greater detail later in this chapter) directly benefit humanity in the form of ecosystem services. Ecosystem services occur as a result of ecosystem structure and biodiversity and are subdivided into provisioning, regulating, supporting, and cultural services.

Biodiversity is often said to support sustainability; in ecology, we often discuss this sustainability as ecosystem resilience or resistance to change, but there are several means by which ecologists expect ecosystems to change over time. To resolve this potential contradiction, we will need to identify the drivers of biodiversity and elucidate how each driver is expected to behave at varying scales. Globally, there are three factors which define broadscale patterns in biodiversity: latitude, temperature, and precipitation. These three abiotic factors are used to define nine global biomes, which are the basis of our understanding and expectations for biodiversity. Because urban landscapes occur in at least 8 of the 9 biomes, excluding tundra, we often must adjust our expectations for what biodiversity can exist in each biome's urban landscapes. It would be inappropriate to expect that an urban landscape in the temperate regions could support as much or similar collections of biodiversity as an urban landscape in the tropics. The reason for this is not based solely from observed patterns of biodiversity, but instead the underlying mechanisms for how life on Earth works. Broadly speaking plant diversity (i.e. species richness) and abundance, measured by something called biomass, are driven by light and precipitation. Natural light from solar radiation is highly correlated with latitude, and as a result light determines temperatures. The light we are most concerned with is photosynthetically active radiation, visible light ranging from 400-700nm, which is utilized by plants to perform photosynthesis. Even more broadly, or rather by extension, the diversity and abundance of plants determines the diversity and abundance of animals through the consumption of these plants. Animals whose diets are almost solely derived from consuming other animals are also reliant upon plant productivity, and as such we must acknowledge that biodiversity alone does not maintain ecosystems but rather diversity and productivity form a positive feedback loop where high levels of productivity define how much biodiversity we can expect to observe. In turn, these higher levels of biodiversity help to maintain high levels of productivity through unique functional differences of species. This broadscale understanding of the drivers of biodiversity is the big picture.

At the landscape scale the drivers of biodiversity both in natural and built environments include habitat availability, habitat heterogeneity (i.e. connectivity and spatial configuration), ecological disturbance and species interactions, and time. Collectively these factors determine the rate of species accumulation, persistence, and extirpation. Our human expectations for biodiversity are defined by our cultural, social, and lived experiences and define how we view, value, interact with, and manage landscapes. Here we see more clearly the importance of the integrated human ecosystem model which we defined earlier in this chapter. While the ecological drivers of habitat availability, heterogeneity, disturbance, and time give us predictions for biodiversity in less human altered landscapes we can easily imagine how human preferences and decisions could easily outweigh the natural drivers of biodiversity.

It is necessary to understand how habitat availability, habitat heterogeneity (i.e. connectivity and spatial configuration), ecological disturbance and species interactions, and time build and maintain biodiversity. Remember that habitat, or ecosystem diversity, is a measure incorporated in biodiversity, and that various habitat types often exist as a patchwork mosaic across landscapes. Some plants and animals (called specialists) have strongly held habitat requirements while other species (called generalists) can be found across a wider range of habitats. The size and degree to which habitat patches are connected plays a large role in species presence or absence from a patch. Small and poorly connected habitat patches, like those which result from infilling and sprawl, tend to have lower species richness as a result of their limited size, and their reduced recruitment potential. Larger and better-connected habitat patches will tend to not only have greater biodiversity, but also have an increased potential to support more specialist species.

While the spatial arrangement of habitat patches within a landscape provides ecologists with some insights to their capacity to support biodiversity, these patches are not static. In fact, in most locations habitats and ecosystem evolve over time. This ecosystem 'evolution' or progression is called ecological succession and is the result of ecosystem accumulating energy and matter over time. A typical example of ecological succession in temperate forest biomes would see a grassland progressing into a mixed grass and woody shrubland, then onto a mixed forest-shrubland, until finally the forest is fully established and contains mostly dominate tree species. One can imagine that the energy within each of these ecological states is a bit different, and in fact grasslands tend to have less biomass and less productivity than forests.

In the temperate forest biome, progression from grassland to forest may take up to seventy-years.

One of the greatest factors impacting the timing of successional progression are ecological disturbances which are either naturally occurring or anthropogenic in origin. Ecological disturbances are characterized by the frequency of occurrence and the magnitude of the effects. Typical naturally occurring disturbances can include storms, floods, and wildfires while examples of anthropogenic disturbances can include logging, mining, habitat replacement (e.g. as seen through development), and climate change related disturbances. Because disturbance and succession are paired processes, and disturbances can often be cyclic in frequency, some ecosystems and species evolve disturbance-adapted species. A natural disturbance-adapted ecosystem can be found in the Pine Barrens of New Jersey, where repeated wildfires have selected for plant and animal species which are now fire-dependent. In this case, humans now try to manage the Pine Barrens with fire to maintain this globally rare ecosystem type, where without fire, the Pine Barrens might undergo ecological succession and become a more hardwood-dominated ecosystem. Similarly, some plants and animals, called urban-adapted species, have adapted to typical anthropogenic urban disturbance regimes and newly disturbed habitats where their natural predators might be absent. Even better adapted to urban landscapes are species termed urban-exploiters, species can often exploit human-associated resources like food waste and climate controlled buildings. Both urban-adapted and urban-exploiting species tend to be generalist species which can reproduce quickly and in abundance making use of newly available resources in the urban landscape.

Comparing urbanized landscapes to more intact proximal landscapes is a good way to understand how biodiversity is altered by human disturbances and the built environment. Although urban landscapes occur at all latitudes and in many different biomes, the structure and disturbance regimes within urbanized landscapes leads to a selection for generalist species, and globally, many ecologists suggest that urbanization is leading to a widespread biotic homogenization of biodiversity. Regardless of the difference in biodiversity between urban landscapes and their surrounding landscape, city planners and managers must consider facilitating biodiversity in the built environment in order to uplift or restore ecological function and productivity.

Ecological Function

Biodiversity plays an integral role in the functioning of ecosystems, both natural and urban alike. These ecosystem functions, in conjunction with urban biodiversity, provide essential ecosystem services that contribute to human well-being and urban sustainability. Ecosystem services are defined as those ecosystem functions which directly benefit people at no cost and are the result of natural systems' typical operation. The Millennium Ecosystem Assessment (Millennium Ecosystem Assessment 2005) defined ecosystem services as being grouped into four categories: (1)

provisioning services (e.g. food, medicine, timber, fiber, bioenergy); (2) supporting services (e.g. nutrient cycling, photosynthesis, primary production, soil formation); 3) regulating services (e.g. air quality regulation, climate regulation, water filtration, waste decomposition, crop pollination, human disease regulation, pest regulation, and natural hazard or storm regulation); and 4) cultural services (e.g. enrichment including knowledge systems and educational values, recreation and ecotourism, aesthetic, spiritual and religious values, sense of place and nature-relatedness). Urban forests, and remnant wetlands within the urban landscape are often cited examples of green spaces which provide important ecosystem services. Both urban forests and wetlands help mitigate urban flooding, regulate temperature, reduce air pollution, and provide habitat for wildlife (Nowak et al. 2008; Pickett et al. 2011).

Ecosystem services are not solely important because of their independent functions, but more holistically they allow for ecosystem-level stability. In ecology, ecosystem stability is often referred to as resistance and/or resilience, where resistance is an ecosystem's ability to remain unchanged in the face of disturbance or perturbation and resilience is the ability of an ecosystem to maintain or quickly return to desired function following disturbance or perturbation. When considering the human ecosystem model and that urban landscapes are designed to support large human populations, we quickly understand the importance of maintaining urban resilience. Meerow et al. (2016) offer a thoughtful definition of urban resilience following a meta-analysis of urban resilience literature from 1973 - 2013;

Urban resilience refers to the ability of an urban system-and all its constituent socio-ecological and socio-technical networks across temporal and spatial scales-to maintain or rapidly return to desired functions in the face of a disturbance, to adapt to change, and to quickly transform systems that limit current or future adaptive capacity.

In keeping with the context of this definition, urban landscapes must be planned and managed to maintain a diverse suite of ecosystem services in order to ensure the persistence of large urban populations. In the following sections we examine the four types of ecosystem services with examples in urban contexts.

Before exploring the ecosystem services, it is essential to understand terms used for land use within this field. Land-use definitions (e.g. developed, agriculture, forest, wetland, water) are used to analyze landscape structure and to make zoning or planning decisions. Within each land-use type greater descriptive resolution is available and can be much more helpful to spatial analysts in evaluating landscapes to resolve different questions. For instance, the 'developed' land-use type does not provide enough resolution for evaluating population density or residential proximity or adjacency to industry, as a result spatial analysts will further describe land-use within the developed type to elucidate for instance 'high density residential' from 'single family residential' from 'commercial' and so on. In urban landscapes, at the broad level land-use or land-cover types are typically binned into green, gray, and blue spaces. Gray spaces are those that include built infrastructure, deriving the name from concrete which is commonly utilized in buildings, sidewalks, and roadways. Blue spaces are those where land-cover is dominated by aquatic environments, rivers, streams, lakes, ponds, reservoirs. Green spaces are those that are vegetated. Again, we see that this level of classification is limited in resolution but provides an entry level for investigating urban landscapes (Daniels et al. 2018). One pervasive challenge in these simple analyses is how to treat land-use in the third (i.e. vertical) dimension.

Supporting Services

Supporting services generally encompass the ecological processes which are essential for ecosystem existence and sustainability. It might be difficult to think about supporting services when provisioning, cultural, and regulating services are much more relatable to our everyday lives, but supporting services are the foundation that enables ecosystems to exist. Without supporting services, the other ecosystem services are likely to be less effective or may be nonexistent. Imagine if our built environments and urban landscapes were designed to exist in a vacuum, instead of our buildings being well insulated from the natural world surrounding it, entire cities were encapsulated from the natural world. Could cities be sustainable in this thought experiment, if so, how? The answer, very simply, is no. Cities would require massive inputs in the form of food, water, energy and would need to export massive quantities of outputs in the form of waste and heat. While cities are spatially separated from required inputs to sustain large human populations, it has become increasingly clear that cities must better integrate ecological processes in order to be sustainable.

Some of the supporting services which can increase urban resilience and uplift other ecosystem services include: primary production, soil formation, nutrient cycling, climate regulation, and the maintenance of genetic diversity among urban wildlife. Together, primary production, soil formation, and nutrient cycling define and maintain ecosystems. Ecologists' expectations for primary productivity, measured as the grams or kilograms of carbon fixed per area per year through photosynthesis, are defined by latitude, precipitation, and temperature. These three variables are used to define major biomes on terrestrial earth and provide a sense of the functioning of ecosystems within these biomes. Urban landscapes are often significantly reduced in primary productivity from these biome-level predictions. In rare instances at local and regional scales where resource limitations (i.e. nitrogen or phosphorus limitation) decreases biome primary productivity, urban landscapes actually facilitate increases in primary productivity by increasing limited resource availability (Imhoff et al. 2004).

Reductions in primary productivity and in urban ecosystems, which have decreased species richness, are additionally impacted in their ability to maintain ecosystem functions such as carbon sequestration and nutrient cycling (Onandia et al. 2019; Schittko et al. 2022). In urban ecosystems that have increased biodiversity, including increased biodiversity attributable to non-native species, studies have found increased soil-related functionality, where soils with increased floral biodiversity increased the carbon storage of soils as well as increased soil faunal diversity

(Schittko et al. 2022). Interestingly, while increased biodiversity increases ecological function, increased native floral diversity had a greater positive effect than non-native floral diversity. Primary production and soil function play an integral role in the cycling of nutrients. Taken together, primary production, nutrient cycling, and the maintenance of soils in urban areas have the potential to play a significant role in mitigating climate change (Paustian et al. 2016).

Regulating Services

There are many regulating ecological functions, and here we will focus on those functions most germane to urban landscapes: stormwater management (natural hazard mitigation), urban heat (solar radiation), air quality, and water quality. Each of these components plays a vital role in managing urban environments. In a natural environment, these components are managed by natural resources. For example, in summer when solar radiation is at its peak, the land, trees, and other plant life help absorb the heat created by the sun and create shade to help regulate surface air temperatures. Plants, and in particular contiguous forest and other complex three-dimensional vegetation complexes, mitigate heat via evaporation and transpiration processes. In an urban environment, there is significantly less vegetated land cover, and a significant increase in built infrastructure with materials that have high heat retention properties. This creates urban heat islands (UHI) which are urbanized areas that experience higher temperatures than outlying rural areas or more natural environments due to the built environment (i.e., buildings, dark paved areas). UHIs cause several challenges for the urban communities: increased incidence of heat stress for human populations, increased energy consumption to cool buildings, higher emissions of air pollutants and greenhouse gases, and decreased water quality due to thermal pollution from stormwater runoff into blue spaces (Tan et al. 2010; Ulpiani 2021; Wang et al. 2021; EPA 2023). Urban heat islands also pose a problem for human health—they contribute to heat-related deaths and illnesses, respiratory difficulties due to poor air quality, heat exhaustion, and heat stroke (Shahmohamadi et al. 2011; Sagris and Sepp 2017; Mika et al. 2018; Piracha and Chaudhary 2022) (Tan et al. 2010; EPA 2023). Urban heat islands can be mitigated by adding or maintaining a combination of urban tree canopy and green spaces, blue spaces, and emerging technologies (Gunawardena et al. 2017; Targino et al. 2019; Peng et al. 2020; Jiang et al. 2021). Examples of UHI mitigation technologies include cool pavements (i.e., solar-reflective pavements that reduce solar heat absorption) (Pomerantz 2018) and using building materials that are more likely to dispel or release heat rather than absorb and contain heat (Gago et al. 2013; Qin 2015; Gunawardena et al. 2017; Leal Filho et al. 2018; Peng et al. 2020).

Air quality may be one issue when considering UHI, but summer or warmer seasons are not the only times when air quality is an issue. Generally, urban environments experience poor air quality due to motor vehicle emissions, UHI in warmer seasons, fumes from chemical production, natural gas and other fuel oils to heat homes, and by-products of manufacturing, power generation, and oil refineries (NIEHS 2024). These factors pollute the environment, in turn creating poor air quality which is connected to negative impacts like respiratory disease and illnesses, cancer, and heart disease (Manisalidis et al. 2020; Bikis 2023) (NIEHS 2024). These negative impacts are compounded indoors since urban buildings can have indoor pollutants negatively impacting air quality which is then intensified by the outdoor air quality (Tham 2016; Wolkoff 2018; Megahed and Ghoneim 2021). Although some European cities have banned or limited automobiles (Holman et al. 2015), it is unlikely that reducing vehicle emissions or manufacturing activities will happen on a larger scale without a major societal or global event, even though severely limited and reduced activities did show significant improvement of air quality in urban areas during partial or full lockdowns of the COVID-19 pandemic, because people were working remotely and manufacturing facilities were relying on limited personnel (Resmi et al. 2020; Nakada and Urban 2020; Kumari and Toshniwal 2020; Jephcote et al. 2021; Kumari et al. 2022). There are solutions, though, to improving air quality without reducing human activity including using green infrastructure that is specific to that urban environment (e.g., vegetation barriers, green roofs and walls, increasing green space and tree canopy, etc.), upgrading building indoor air filtration equipment, managing traffic differently, and reducing reliance on and use of fossil fuels (Hao and Wang 2005; Bigazzi and Rouleau 2017; Hewitt et al. 2020; Barwise and Kumar 2020; Viecco et al. 2021).

Natural hazard mitigation and water quality is another challenge in urban environments and comes in the form of stormwater management, flooding, and drinking water. In natural environments, rivers and creeks are surrounded by vegetation – typically trees and other plant life – that help absorb excess water. Additionally, natural flood plains (flat areas around rivers) help protect the area from the negative impacts of flooding, help provide erosion control, safeguard water quality, and provide wildlife habitat. In urban environments, these natural elements are heavily modified or built over, preventing them from mitigating the effects of heavy rainfall that causes flooding. Flooding in urban environments can not only damage or destroy infrastructure and personal and public property, but can also be a risk to human health by increasing exposure to infections and illnesses due to overflowing sewage, pollution to drinking water, or causing injury or death (Kaźmierczak and Cavan 2011; Ali and Hamidu 2014; Miller and Hutchins 2017; Han and He 2021; Mazzoleni et al. 2021).

Closely related to stormwater management is drinking water and water quality. Drinking water is integral to human health and is heavily impacted by the environment and the potential infrastructure that is used to clean, store, and transport it. Drinking water can come from rivers, reservoirs, lakes, and groundwater and urban areas tend to share centralized water sources to supply cities with enough water to support the large population (EPA 2023). Urban water sources take on a large amount of pollution from various sources including industrial, agriculture, residential and commercial waste, trash, pollutants, manufacturing, and automobiles (EPA 2023). Despite this, water quality tends to be better in urban areas compared to rural areas due to stronger environmental policies and management (Wang et al. 2008, 2021; Strosnider et al. 2017).

A common theme in managing these regulating functions is that many solutions are nature-based including increasing the quantity and quality of green space, using alternative materials or methods to absorb heat or water, increasing tree coverage, and using green infrastructure. All of these regulating functions are connected and can be impacted by one another especially during extreme weather events which are becoming increasingly common due to climate change (Mann et al. 2017, 2018). The impacts of climate change cannot be understated or overlooked in the functions of urban landscape especially as the WHO projects that nearly 70% of all humans will live in urban areas by 2050 (World Health Organization 2018).

Provisioning Services

While provisioning services are among the most directly valuable to human populations they most often do not occur within cities, but instead help to define the ecological footprint of urban populations. Provisioning of food, timber, fibers, etc. is now linked to the global movement of goods, and often greatly impacts the surrounding land-use composition of the landscapes adjacent to dense urbanization. One exception to this is urban agriculture. Urban agriculture can be done at a variety of scales from the private home garden and backyard or hobby farming to community gardens to growing produce and raising livestock in urban areas. Growing and producing food directly in urban areas increases food security and access, creates urban resiliency by reducing reliance on importing food from rural areas, and increases functionality and optimization of green space with regards to human needs. Further, it can help with nutrient and water recycling in the environment (Ferreira et al. 2018; Rufí-Salís et al. 2020; Lal 2020; Langemeyer et al. 2021; Ouyang et al. 2022).

In addition to providing food within the urban matrix, urban agriculture can be incorporated as a form of green space where primary productivity, nutrient cycling, and soil function can be maintained. As such, urban agriculture and urban gardens are a mix of native and non-native foodstuffs which support increases in faunal diversity via the introduction of new food species as well as providing foraging opportunity for other organisms (e.g. pollinators, and other subsidized wildlife) (De Zeeuw et al. 2011; Orsini et al. 2014). In addition to green urban agriculture, hydroponic operations in gray infrastructure can provide provisioning services and, in some cases, can been adapted for use in wastewater treatment (Magwaza et al. 2020).

Cultural Services

Cities and urbanized landscapes often contain diverse natural habitats which provide cultural ecosystem services that positively contribute to the well-being and quality of life of human populations. While not the most biodiverse land-use type, recreational opportunities can be found within urban green spaces such as parks, gardens, and waterfronts. These spaces serve as popular areas for both passive and active exposure to nature including walking, picnicking, jogging, cycling, gardening, birdwatching, and religious and cultural practices and gathering. Urban green spaces offer opportunities for exercise and relaxation as well as social interaction (Schipperijn et al. 2010). Research highlights the positive impact of urban green spaces on mental health and stress reduction (Shanahan et al. 2015, 2016) (Ulrich et al. 2010), dietary quality and diversity (Milliron et al. 2022), sense of place and belonging (Adevi and Grahn 2011), and social cohesion (Peters et al. 2010; Jennings and Bamkole 2019) indicating the crucial role natural spaces play in promoting psychological well-being among urban dwellers.

Cultural ecosystem services in urban areas encompass aesthetic values derived from visual and sensory experiences of nature within the built environment. Urban green spaces, street trees, and green infrastructure elements enhance the visual appeal of cities, providing scenic beauty and aesthetic pleasure to residents and visitors (Nassauer and Opdam 2008). Studies have shown that exposure to green spaces and natural landscapes in urban settings can evoke positive emotional responses, foster a sense of place attachment, and contribute to a sense of belonging within the community (Kaplan and Kaplan 1989; Carrus et al. 2015). These aesthetic benefits of urban nature play a crucial role in shaping the identity and character of cities. Additionally, the aesthetic value derived from natural or green spaces within cities can play a large role in the valuation of residential housing, resulting in gentrification and inequity for the spaces without as much green space (Dooling 2009; Rigolon and Németh 2020). This has led many city planners and scholars to seek methods for greening cities to maintain ecological function without further creating environmental injustice in the form of green space accessibility, often discussed as 'just green enough' (Wolch et al. 2014).

Measuring Functional Diversity

In the field of biology, scientists have often used the motto 'structure begets function'. In ecology, as an integrative sister discipline of biology, the same motto is held as truth. While all ecosystems have function(s), the degree or magnitude to which any particular ecosystem function works is related to the spatial scale of the ecosystem and its biological structure, in the form of biodiversity. When managing or planning for ecosystem function it is often important to think about biodiversity in terms of species-specific function, referred to as functional diversity. As a result of the need to understand species functional roles towards ecosystem function researchers have developed several measures for quantifying the distribution of functional and phenotypic differences within and between ecological communities. These functional diversity measures attempt to, in the simplest form, quantitatively elucidate the variation within a single trait in a community (Cadotte et al. 2009). Because most ecosystem functions will be influenced by multiple functional traits, more complicated functional diversity measures seek to incorporate multivariate measurements - see (Cadotte et al. 2011) for a nice summary of functional diversity measures. Additionally, some of the multivariate functional diversity measurements seek to consider species abundances to give weight to species' trait contribution to ecosystem function. Unfortunately for planners, managers, and ecologists, the relationship between species richness and functional diversity is not always a 1:1 positive correlation (Wright et al. 2006; Cadotte et al. 2011). Among the reasons for these less predictable positive relationships are 1) most ecological communities contain functionally redundant species, 2) species do not have the same magnitude of importance towards a single function, and 3) species richness and functional traits can change relative to environmental factors such as light, temperature, and precipitation (Petchey et al. 2004; Cadotte et al. 2011).

Establishing functional diversity measures requires significant quantitative study of the species (and their traits) which commonly co-occur in community assemblages. Many studies have confirmed that functional diversity in both plants and animals better predicts ecosystem functioning than species richness or species-based indices (Gagic et al. 2015). Even more interestingly, a recent meta-analysis of 15 peer reviewed articles (from 2009-2015) found that there is a positive correlation between species functional redundancy (i.e. species richness within functional groups) and increased ecological stability or resilience, while species richness alone was not correlated with effect size (Biggs et al. 2020). At this point, readers might be questioning how they should move forward with incorporating ecological function into urban planning and management, given that the current state in this field of research suggests 1) biodiversity is important for ecosystem function, but at the same time we see that 2) species functional diversity and redundancy are far more important in maintaining ecosystem function than species richness alone. It is imperative to note that there is a large data gap in our understanding of how species functional traits occur generally, and even more so in urban landscapes where abiotic factors might alter species functional attributes. To further complicate this urban management challenge, planners must acknowledge that urban systems tend to be less species-rich than neighboring ecosystems, have different environmental stressors, and therefore might limit planners' suite of species to include in planning urban green spaces. Guidance for urban planners with regards to managing urban ecosystem services is notably sparse and lacking (Woodruff and BenDor 2016). Equally important are the socio-cultural considerations of people in the urban landscape being managed, and our collective understanding of and willingness to work with non-native species, which are abundant in most urban ecosystems (Schwarz et al. 2017).

Understanding and planning for the enhancement of urban ecosystem function is critical for building resilient cities. As demonstrated throughout this chapter, urban ecosystems provide a myriad of services in the form of ecological functions. However, ensuring the persistence and enhancement of these functions necessitates comprehensive strategies informed by scientific research. One approach involves the development and implementation of ecosystem-based management practices that prioritize ecological integrity and biodiversity conservation within urban landscapes (McPhearson et al. 2016). This includes promoting green infrastructure initiatives, such as urban forests, green roofs, and permeable pavements, which have been shown to enhance ecosystem services, mitigate urban heat island effects, and improve air and water quality (Pataki et al. 2011; Escobedo et al. 2011). Additionally, integrating principles of landscape ecology and urban planning can help identify and protect ecological corridors and connectivity, facilitating species movement and enhancing habitat quality within urban environments (Ahern 2007). By employing these interdisciplinary approaches and leveraging scientific knowledge, policymakers, planners, and stakeholders can work collaboratively to foster ecologically functional and resilient cities for current and future generations.

Furthermore, measuring and monitoring urban ecosystem function is essential for evaluating the effectiveness of management strategies and guiding decision-making processes. Utilizing advanced techniques such as remote sensing, geographic information systems (GIS), and ecological modeling can provide valuable insights into the spatial distribution, dynamics, and interactions of urban ecosystems and their services (Seto et al. 2012). Long-term ecological monitoring programs, such as the National Ecological Observatory Network (NEON), offer comprehensive data collection on urban biodiversity, land cover change, and ecosystem processes, enabling researchers to assess ecosystem health and resilience over time (Avolio et al. 2018). Additionally, engaging community scientists through participatory monitoring initiatives can enhance data collection efforts and foster community stewardship of urban ecosystems (Pocock et al. 2018). By integrating these diverse approaches and fostering interdisciplinary collaboration, we can advance our understanding of urban ecosystem function and develop evidence-based strategies to promote ecological sustainability and human well-being in urban areas.

Urbanization's Influence on Biodiversity and Ecological Function: Observed Patterns

Urbanization and the provisioning of resources for dense human populations plays a significant role in altering biodiversity both at the global scale as well as within each urban landscape (McDonald et al. 2019). Urbanization has been implicated in the global decline of biodiversity through the direct replacement and fragmentation of habitat (McDonnell and Hahs 2015). In fact, urbanization has been identified to cause the greatest rates of local extinction through the elimination of native species (McKinney 2002). Urbanization, unlike other forms of habitat destruction, is long-lasting and is difficult to reverse. The expansion of infrastructure, roads, and buildings reduces available habitat for biodiversity, fragmenting natural landscapes and isolating populations (Fahrig 2003). Isolated populations whose geographic distributions are reduced in spatial extent have higher extinction rates than more naturally occurring or contiguous populations. Globally, climate change poses additional challenges for biodiversity (both urban and non-urban) altering temperature and precipitation patterns, and affecting species distributions, phenology, and ecosystem processes (McDonnell & Hahs, 2015).

Within urban landscapes human activities further accelerate declines in biodiversity via habitat degradation such as soil alterations, pollution, anthropogenic disturbance, the introduction of invasive species, and overexploitation of natural resources. Where these activities do not replace natural habitat, they further reduce habitat quality for wildlife populations (McKinney 2008). Across all spatial scales and all taxonomic groups, urbanization tends to decrease native biodiversity in the form of species richness, alter populations relative abundances, and exerts evolutionary selective pressures on urban wildlife populations (Faeth et al. 2011; Johnson and Munshi-South 2017; Piano et al. 2020; Lambert and Donihue 2020). More specifically, animal diversity tends to decrease with urbanization while plant diversity across urban landscapes can increase due to human introduction of non-native species. Rega-Brodsky et al. (2022) conducted a meta-analysis of 1,209 studies and identified that 56.2% of all studies were conducted on urban plants (37.6%) and urban birds (18.6%). While this meta-analysis included studies conducted from 1990 onward, only 5% of the studies attempted to link urban biodiversity to ecological function or ecosystem service. An earlier study of urban birds and plants (Aronson et al. 2014) found that while abundances of both urban plants and birds have decreased, that most of the species observed were native to urban landscapes but saw reductions in representation in urban landscapes when compared to neighboring more intact natural ecosystems; only 8% of native birds and only 25% of native plants were found in studied cities.

The processes of urbanization and human behaviors lead to several landscape-level patterns that are common observations among most cities and city surroundings. Primarily, we observe that urban landscapes are heterogenous mosaics of land-use and land-cover types. It is of particular importance in landscape ecology that land-use types be large and contiguous in nature. Imagine the last time you were in a forest, you likely noticed that the humidity increased, and the temperature decreased as you walked into the forest from forest edge. This observation is size dependent, where if you were to walk under a singular tree you would be far less likely to observe significant changes in humidity, but you might still observe a slight decrease in temperature due to shading from the tree. Many of the functions of ecosystems behave similarly and lead researchers, specifically spatial or landscape ecologists, to examine the spatial distribution of land-use or land-cover types. The response variables often examined include land-use patch size, shape, connectivity, and spatial autocorrelation.

While keeping in mind that many important ecosystem functions are dependent on spatial scale, we can examine urban landscapes spatial composition to elucidate areas which can be leveraged to uplift or further promote valuable ecosystem functions, to identify communities experiencing environmental injustice and identify opportunities to resolve inequity or lacking ecological function. Some of the most common examples come from the United States where racist red-lining practices segregated urban landscapes (Schell et al. 2020). Funding was differentially provided to maintaining ecosystem services, economic stability or growth, and increased health outcome in affluent communities. In recent years, several studies have examined that historically red-lined, marginalized or primarily Black and Brown communities are found in portions of cities experiencing higher levels of urban heat and lower urban tree coverage or access to green spaces. More broadly, in urbanized landscapes we generally observe a large degree of heterogeneity, reduced patch size, and poor ecological connectivity. Because of reduced patch size, and poor patch connectivity, urban landscapes are often absent of core habitat and instead display higher edge to core ratios among urban green spaces and habitats. These edge habitats are often far more variable biophysically and provide challenges to habitat specialist species.

While urban landscapes exhibit spatial heterogeneity and reduced wildlife habitat availability as compared to non-urbanized landscapes, studies have shown that urban green spaces, including parks, gardens, and natural habitat remnants, act as local biodiversity hotspots supporting higher levels of biodiversity compared to built gray areas (Hope et al. 2003; McKinney 2008). Biodiversity patterns vary across cities depending on land-use, proximity to natural areas, habitat fragmentation, socio-economic factors (Phillips and Lindquist 2021) and historical land use legacies (Aronson et al. 2017). There are instances where legacy green spaces within cities act as refuges or local diversity hotspots for native plants and wildlife (Kantsa et al. 2013). The drivers of urban diversity in these urban hotspots include the presence of diverse vegetation types, microhabitats, and structural complexity within urban green spaces (Goddard et al. 2010; Fabián et al. 2021; Gonsalves et al. 2022). In addition to patches of urban green spaces, studies are increasingly examining the role that individual street trees, private gardens, and engineered green infrastructure play in contributing to green space connectivity within urban landscapes (Filazzola et al. 2019; Mimet et al. 2020; Cadavid-Florez et al. 2020) (Cadavid-Florez et al., 2020; Mimet et al., 2020; Filazzola et al., 2019).

Moving Forward

Despite the challenges posed by urbanization, cities also offer opportunities for conservation and restoration of biodiversity through green infrastructure, urban planning, and community engagement (McPhearson et al. 2016). Research suggests that city planning should incorporate green space wherever possible and planners should promote green spaces and the positive benefits to community members (Barbosa et al. 2007) (Green et al. 2016; Kondo et al. 2020). Built infrastructure that includes green design elements such as green roofs and green walls contribute to biodiversity conservation by creating habitat for a variety of plants, insects, and birds, particularly in densely populated urban areas (Getter and Rowe 2006). Additionally, biodiversity influences the perceptions of urban green space and people tend to value spaces with high biodiversity (Gunnarsson et al. 2017; Fischer et al. 2018). Increasing biodiversity through well-planned implementation of urban green space can improve ecosystem function, further benefiting the urban-dwelling humans.

Conserving and enhancing urban biodiversity can draw upon some classic conservation biology and ecological know how. Maintaining current green space availability, while finding opportunities to increase green space availability and increasing urban habitat connectivity will be imperative. City planners, landscape architects, engineers, and ecologists should additionally invest in both restoring urban habitats when possible, but equally become more comfortable working in novel urban ecosystems (Lundholm 2015). Novel urban ecosystems can arise from a human activities, urbanization, and iterative land-use change and may include the introduction of non-native species (Hobbs et al. 2006). For example, vacant lots, brownfields, and other disturbed urban landscapes may be colonized by pioneer species adapted to harsh environmental conditions (Pataki et al. 2011) . These novel ecosystems may provide habitat for wildlife, promote biodiversity conservation, and contribute to urban resilience by enhancing ecosystem functions such as carbon sequestration, soil stabilization, and water infiltration (Pickett et al. 2011). The management of these spaces however must not create environmental inequities or further burden communities which have been classically marginalized.

As a result of the need to increase ecological function and urban biodiversity decision makers must make every effort to engage local communities through participatory planning processes, shared decision making, and community science initiatives (Lyytimäki and Sipilä 2009). Effective conservation and management of urban biodiversity also require collaboration and partnerships between government agencies, non-profit organizations, academic institutions, and local communities (Colding et al. 2006). Community science initiatives, such as urban biodiversity monitoring programs and community-based conservation projects, can help collect data, raise awareness, and mobilize public support for biodiversity conservation (Dickinson et al. 2012). Moreover, integrating biodiversity considerations into urban planning, design, and management practices can help minimize negative impacts on biodiversity and maximize opportunities for habitat creation and restoration (Lundholm 2015).

In addition to the management challenges and the need for community stakeholder involvement, scientists will need to continue to partner biodiversity surveys with functional trait analyses and evaluation of urban biodiversity redundancy. When taking a functional approach to managing urban biodiversity, decision makers will need to decide between managing for native species or for more cosmopolitan species which might perform better in novel urban ecosystems. Several challenges will continue to hinder urban biodiversity conservation efforts, including climate change, invasive species, habitat loss, and inadequate funding for conservation programs (McKinney 2008). Addressing these challenges requires interdisciplinary research, collaborative partnerships, and innovative approaches that integrate biodiversity and ecological function with urban planning, design, and management (Andersson and Colding 2014) in order to increase urban resilience and support human populations.

References

Adevi AA, Grahn P (2011) Attachment to certain natural environments: a basis for choice of recreational settings, activities and restoration from stress? ENRR 1:36. https://doi.org/10.5539/enrr.v1n1p36

Ahern J (2007) Green Infrastructure For Cities The Spatial Dimension. Cities of the future towards integrated sustainable water and landscape management. IWA Publishing, London, In

Alberti M, Marzluff JM, Shulenberger E et al (2003) Integrating humans into ecology: opportunities and challenges for studying urban ecosystems. BioScience 53:1169. https://doi.org/10.1641/0006-3568(2003)053[1169:IHIEOA]2.0.CO;2

Ali D, Hamidu S (2014) Environmental hazard: climate change and flooding, the impact on the built environment in Nigeria. Journal of Environmental Sciences and Resources Management, 6:136-144.

Andersson E, Colding J (2014) Understanding how built urban form influences biodiversity. Urban Forest Urban Green 13:221–226. https://doi.org/10.1016/j.ufug.2013.11.002

Aronson MFJ, La Sorte FA, Nilon CH et al (2014) A global analysis of the impacts of urbanization on bird and plant diversity reveals key anthropogenic drivers. Proc R Soc B 281:20133330. https://doi.org/10.1098/rspb.2013.3330

Aronson MF, Lepczyk CA, Evans KL et al (2017) Biodiversity in the city: key challenges for urban green space management. Front Ecol Environ 15:189–196. https://doi.org/10.1002/fee.1480

Avolio ML, Pataki DE, Trammell TLE, Endter-Wada J (2018) Biodiverse cities: the nursery industry, homeowners, and neighborhood differences drive urban tree composition. Ecological monographs 88:259–276. https://doi.org/10.1002/ecm.1290

Barbosa O, Tratalos JA, Armsworth PR et al (2007) Who benefits from access to green space? A case study from Sheffield, UK. Landsc Urban Plan 83:187–195. https://doi.org/10.1016/j.landurbplan.2007.04.004

Barwise Y, Kumar P (2020) Designing vegetation barriers for urban air pollution abatement: a practical review for appropriate plant species selection. NPJ Clim Atmos Sci 3:12. https://doi.org/10.1038/s41612-020-0115-3

Benton TG, Bieg C, Harwatt H, Pudasaini R, Wellesley L (2021) Food system impacts on biodiversity loss three levers for food system transformation in support of nature. Chatham House.

Bigazzi AY, Rouleau M (2017) Can traffic management strategies improve urban air quality? A review of the evidence. J Transp Health 7:111–124. https://doi.org/10.1016/j.jth.2017.08.001

Biggs CR, Yeager LA, Bolser DG et al (2020) Does functional redundancy affect ecological stability and resilience? A review and meta-analysis Ecosphere 11:e03184. https://doi.org/10.1002/ecs2.3184

Bikis A (2023) Urban air pollution and greenness in relation to public health. J Environ Public Health 2023:1–18. https://doi.org/10.1155/2023/8516622

Cadavid-Florez L, Laborde J, Mclean DJ (2020) Isolated trees and small woody patches greatly contribute to connectivity in highly fragmented tropical landscapes. Landsc Urban Plan 196:103745. https://doi.org/10.1016/j.landurbplan.2020.103745

Cadotte MW, Cavender-Bares J, Tilman D, Oakley TH (2009) Using phylogenetic, functional and trait diversity to understand patterns of plant community productivity. PLoS One 4:e5695. https://doi.org/10.1371/journal.pone.0005695

Cadotte MW, Carscadden K, Mirotchnick N (2011) Beyond species: functional diversity and the maintenance of ecological processes and services. J Appl Ecol 48:1079–1087. https://doi.org/10.1111/j.1365-2664.2011.02048.x

Carrus G, Scopelliti M, Lafortezza R et al (2015) Go greener, feel better? The positive effects of biodiversity on the well-being of individuals visiting urban and peri-urban green areas. Landsc Urban Plan 134:221–228. https://doi.org/10.1016/j.landurbplan.2014.10.022

Colding J, Lundberg J, Folke C (2006) Incorporating green-area user groups in urban ecosystem management. AMBIO J Hum Environ 35:237–244. https://doi.org/10.1579/05-A-098R.1

Daniels B, Zaunbrecher BS, Paas B et al (2018) Assessment of urban green space structures and their quality from a multidimensional perspective. Sci Total Environ 615:1364–1378. https://doi.org/10.1016/j.scitotenv.2017.09.167

De Zeeuw H, Van Veenhuizen R, Dubbeling M (2011) The role of urban agriculture in building resilient cities in developing countries. J Agric Sci 149:153–163. https://doi.org/10.1017/S0021859610001279

Dickinson JL, Shirk J, Bonter D et al (2012) The current state of citizen science as a tool for ecological research and public engagement. Front Ecol Environ 10:291–297. https://doi.org/10.1890/110236

Dobson A, Rowe Z, Berger J et al (2021) Biodiversity loss due to more than climate change. Science 374:699–700. https://doi.org/10.1126/science.abm6216

Dooling S (2009) Ecological gentrification: a research agenda exploring justice in the city. Int J Urban Reg Res 33:621–639. https://doi.org/10.1111/j.1468-2427.2009.00860.x

Elmqvist T, Redman CL, Barthel S, Costanza R (2013) History of urbanization and the missing ecology. In: Elmqvist T, Fragkias M, Goodness J et al (eds) Urbanization, biodiversity and ecosystem services: challenges and opportunities. Springer, Netherlands, Dordrecht, pp 13–30

Environmental Protection Agency. (2023). Why urban waters?. https://www.epa.gov/urbanwaterspartners/why-urban-waters

Escobedo FJ, Kroeger T, Wagner JE (2011) Urban forests and pollution mitigation: analyzing ecosystem services and disservices. Environ Pollut 159:2078–2087. https://doi.org/10.1016/j.envpol.2011.01.010

Fabián D, González E, Sánchez Domínguez MV et al (2021) Towards the design of biodiverse green roofs in Argentina: assessing key elements for different functional groups of arthropods. Urban Forest Urban Green 61:127107. https://doi.org/10.1016/j.ufug.2021.127107

Faeth SH, Bang C, Saari S (2011) Urban biodiversity: patterns and mechanisms. Ann N Y Acad Sci 1223:69–81. https://doi.org/10.1111/j.1749-6632.2010.05925.x

Fahrig L (2003) Effects of habitat fragmentation on biodiversity. Annu Rev Ecol Evol Syst 34:487–515. https://doi.org/10.1146/annurev.ecolsys.34.011802.132419

Ferreira AJD, Guilherme RIMM, Ferreira CSS, Oliveira MDFMLD (2018) Urban agriculture, a tool towards more resilient urban communities? Curr Opin Environ Sci Health 5:93–97. https://doi.org/10.1016/j.coesh.2018.06.004

Filazzola A, Shrestha N, MacIvor JS (2019) The contribution of constructed green infrastructure to urban biodiversity: a synthesis and meta-analysis. J Appl Ecol 56:2131–2143. https://doi.org/10.1111/1365-2664.13475

Fischer LK, Honold J, Cvejić R et al (2018) Beyond green: broad support for biodiversity in multicultural European cities. Glob Environ Chang 49:35–45. https://doi.org/10.1016/j.gloenvcha.2018.02.001

Gagic V, Bartomeus I, Jonsson T et al (2015) Functional identity and diversity of animals predict ecosystem functioning better than species-based indices. Proc R Soc B 282:20142620. https://doi.org/10.1098/rspb.2014.2620

Gago EJ, Roldan J, Pacheco-Torres R, Ordóñez J (2013) The city and urban heat islands: a review of strategies to mitigate adverse effects. Renew Sustain Energy Rev 25:749–758. https://doi.org/10.1016/j.rser.2013.05.057

Getter KL, Rowe DB (2006) The role of extensive green roofs in sustainable development. HortSci 41:1276–1285. https://doi.org/10.21273/HORTSCI.41.5.1276

Goddard MA, Dougill AJ, Benton TG (2010) Scaling up from gardens: biodiversity conservation in urban environments. Trends Ecol Evol 25:90–98. https://doi.org/10.1016/j.tree.2009.07.016

Gonsalves S, Starry O, Szallies A, Brenneisen S (2022) The effect of urban green roof design on beetle biodiversity. Urban Ecosyst 25:205–219. https://doi.org/10.1007/s11252-021-01145-z

Green OO, Garmestani AS, Albro S, Ban NC, Berland A, Burkman CE, Gardiner MM, Gunderson L, Hopton ME, Schoon ML, Shuster WD (2016) Adaptive governance to promote ecosystem services in urban green spaces. Urban Ecosyst 19:77–93. https://doi.org/10.1007/s11252-015-0476-2

Gunawardena KR, Wells MJ, Kershaw T (2017) Utilising green and bluespace to mitigate urban heat island intensity. Sci Total Environ 584–585:1040–1055. https://doi.org/10.1016/j.scitotenv.2017.01.158

Gunnarsson B, Knez I, Hedblom M, Sang ÅO (2017) Effects of biodiversity and environment-related attitude on perception of urban green space. Urban Ecosyst 20:37–49. https://doi.org/10.1007/s11252-016-0581-x

Han J, He S (2021) Urban flooding events pose risks of virus spread during the novel coronavirus (COVID-19) pandemic. Sci Total Environ 755:142491. https://doi.org/10.1016/j.scitotenv.2020.142491

Hao J, Wang L (2005) Improving urban air quality in China: Beijing case study. J Air Waste Manage Assoc 55:1298–1305. https://doi.org/10.1080/10473289.2005.10464726

Hewitt CN, Ashworth K, MacKenzie AR (2020) Using green infrastructure to improve urban air quality (GI4AQ). Ambio 49:62–73. https://doi.org/10.1007/s13280-019-01164-3

Hobbs RJ, Arico S, Aronson J et al (2006) Novel ecosystems: theoretical and management aspects of the new ecological world order. Glob Ecol Biogeogr 15:1–7. https://doi.org/10.1111/j.1466-822X.2006.00212.x

Holman C, Harrison R, Querol X (2015) Review of the efficacy of low emission zones to improve urban air quality in European cities. Atmos Environ 111:161–169. https://doi.org/10.1016/j.atmosenv.2015.04.009

Hope D, Gries C, Zhu W et al (2003) Socioeconomics drive urban plant diversity. Proc Natl Acad Sci U S A 100:8788–8792. https://doi.org/10.1073/pnas.1537557100

Imhoff ML, Bounoua L, Ricketts T et al (2004) Global patterns in human consumption of net primary production. Nature 429:870–873. https://doi.org/10.1038/nature02619

Jennings V, Bamkole O (2019) The relationship between social cohesion and urban green space: an avenue for health promotion. IJERPH 16:452. https://doi.org/10.3390/ijerph16030452

Jephcote C, Hansell AL, Adams K, Gulliver J (2021) Changes in air quality during COVID-19 'lockdown' in the United Kingdom. Environ Pollut 272:116011. https://doi.org/10.1016/j.envpol.2020.116011

Jiang Y, Huang J, Shi T, Li X (2021) Cooling island effect of Blue-Green corridors: quantitative comparison of morphological impacts. IJERPH 18:11917. https://doi.org/10.3390/ijerph182211917

Johnson MTJ, Munshi-South J (2017) Evolution of life in urban environments. Science 358:eaam8327. https://doi.org/10.1126/science.aam8327

Kantsa A, Tscheulin T, Junker RR et al (2013) Urban biodiversity hotspots wait to get discovered: The example of the city of Ioannina, NW Greece. Landsc Urban Plan 120:129–137. https://doi.org/10.1016/j.landurbplan.2013.08.013

Kaplan R, Kaplan S (1989) The experience of nature: a psychological perspective. Cambridge University Press, Cambridge

Kaźmierczak A, Cavan G (2011) Surface water flooding risk to urban communities: analysis of vulnerability, hazard and exposure. Landsc Urban Plan 103:185–197. https://doi.org/10.1016/j.landurbplan.2011.07.008

Kondo MC, Oyekanmi KO, Gibson A, South EC, Bocarro J, Hipp JA (2020) Nature prescriptions for health: a review of evidence and research opportunities. Int J Environ Res Public Health 17(12):4213. https://doi.org/10.3390/ijerph17124213

Kumari P, Toshniwal D (2020) Impact of lockdown on air quality over major cities across the globe during COVID-19 pandemic. Urban Clim 34:100719. https://doi.org/10.1016/j.uclim.2020.100719

Kumari S, Yadav AC, Saharia M, Dev S (2022) Spatio-temporal analysis of air quality and its relationship with COVID-19 lockdown over Dublin. Remote Sensing Appl: Soc Environ 28:100835. https://doi.org/10.1016/j.rsase.2022.100835

Lal R (2020) Home gardening and urban agriculture for advancing food and nutritional security in response to the COVID-19 pandemic. Food Sec 12:871–876. https://doi.org/10.1007/s12571-020-01058-3

Lambert MR, Donihue CM (2020) Urban biodiversity management using evolutionary tools. Nat Ecol Evol 4:903–910. https://doi.org/10.1038/s41559-020-1193-7

Langemeyer J, Madrid-Lopez C, Mendoza Beltran A, Villalba Mendez G (2021) Urban agriculture —A necessary pathway towards urban resilience and global sustainability? Landsc Urban Plan 210:104055. https://doi.org/10.1016/j.landurbplan.2021.104055

Leal Filho W, Echevarria Icaza L, Neht A et al (2018) Coping with the impacts of urban heat islands. A literature based study on understanding urban heat vulnerability and the need for resilience in cities in a global climate change context. J Clean Prod 171:1140–1149. https://doi.org/10.1016/j.jclepro.2017.10.086

Lundholm JT (2015) Green roof plant species diversity improves ecosystem multifunctionality. J Appl Ecol 52:726–734. https://doi.org/10.1111/1365-2664.12425

Lyytimäki J, Sipilä M (2009) Hopping on one leg—the challenge of ecosystem disservices for urban green management. Urban Forest Urban Green 8:309–315. https://doi.org/10.1016/j.ufug.2009.09.003

Machlis GE, Force JE, Burch WR (1997) The human ecosystem Part I: the human ecosystem as an organizing concept in ecosystem management. Soc Nal Resour 10:347–367. https://doi.org/10.1080/08941929709381034

Magwaza ST, Magwaza LS, Odindo AO et al (2020) Partially treated domestic wastewater as a nutrient source for tomatoes (Lycopersicum solanum) grown in a hydroponic system: effect on nutrient absorption and yield. Heliyon 6:e05745. https://doi.org/10.1016/j.heliyon.2020.e05745

Manisalidis I, Stavropoulou E, Stavropoulos A, Bezirtzoglou E (2020) Environmental and health impacts of air pollution: a review. Front Public Health 8:14. https://doi.org/10.3389/fpubh.2020.00014

Mann ME, Rahmstorf S, Kornhuber K et al (2017) Influence of anthropogenic climate change on planetary wave resonance and extreme weather events. Sci Rep 7:45242. https://doi.org/10.1038/srep45242

Mann ME, Rahmstorf S, Kornhuber K et al (2018) Projected changes in persistent extreme summer weather events: the role of quasi-resonant amplification. Sci Adv 4:eaat3272. https://doi.org/10.1126/sciadv.aat3272

Mazzoleni M, Mård J, Rusca M et al (2021) Floodplains in the anthropocene: a global analysis of the interplay between human population, built environment, and flood severity. Water Resour Res 57:e2020WR027744. https://doi.org/10.1029/2020WR027744

McDonald RI, Mansur AV, Ascensão F et al (2019) Research gaps in knowledge of the impact of urban growth on biodiversity. Nat Sustain 3:16–24. https://doi.org/10.1038/s41893-019-0436-6

McDonnell MJ, Hahs AK (2015) Adaptation and adaptedness of organisms to urban environments. Annu Rev Ecol Evol Syst 46:261–280. https://doi.org/10.1146/annurev-ecolsys-112414-054258

McKinney ML (2002) Urbanization, biodiversity, and conservation. BioScience 52:883. https://doi.org/10.1641/0006-3568(2002)052[0883:UBAC]2.0.CO;2

McKinney ML (2008) Effects of urbanization on species richness: a review of plants and animals. Urban Ecosyst 11:161–176. https://doi.org/10.1007/s11252-007-0045-4

McPhearson T, Pickett STA, Grimm NB et al (2016) Advancing Urban Ecology toward a Science of Cities. BioScience 66:198–212. https://doi.org/10.1093/biosci/biw002

Meerow S, Newell JP, Stults M (2016) Defining urban resilience: a review. Landsc Urban Plan 147:38–49. https://doi.org/10.1016/j.landurbplan.2015.11.011

Megahed NA, Ghoneim EM (2021) Indoor Air Quality: Rethinking rules of building design strategies in post-pandemic architecture. Environ Res 193:110471. https://doi.org/10.1016/j.envres.2020.110471

Mika J, Forgo P, Lakatos L et al (2018) Impact of 1.5 K global warming on urban air pollution and heat island with outlook on human health effects. Curr Opin Environ Sustain 30:151–159. https://doi.org/10.1016/j.cosust.2018.05.013

Millennium Ecosystem Assessment (2005) Ecosystems and human well-being: our human planet—summary for decision-makers

Miller JD, Hutchins M (2017) The impacts of urbanisation and climate change on urban flooding and urban water quality: a review of the evidence concerning the United Kingdom. J Hydrol: Regional Studies 12:345–362. https://doi.org/10.1016/j.ejrh.2017.06.006

Milliron B-J, Ward D, Granche J et al (2022) Nature relatedness is positively associated with dietary diversity and fruit and vegetable intake in an urban population. Am J Health Promot 36:1019–1024. https://doi.org/10.1177/08901171221086941

Mimet A, Kerbiriou C, Simon L et al (2020) Contribution of private gardens to habitat availability, connectivity and conservation of the common pipistrelle in Paris. Landsc Urban Plan 193:103671. https://doi.org/10.1016/j.landurbplan.2019.103671

Nakada LYK, Urban RC (2020) COVID-19 pandemic: impacts on the air quality during the partial lockdown in São Paulo state. Brazil Sci Total Environ 730:139087. https://doi.org/10.1016/j.scitotenv.2020.139087

Nassauer JI, Opdam P (2008) Design in science: extending the landscape ecology paradigm. Landsc Ecol 23:633–644. https://doi.org/10.1007/s10980-008-9226-7

National Institute of Environmental Health Sciences (NIEHS). (2024) Air pollution and your health. https://www.niehs.nih.gov/health/topics/agents/air-pollution. mnj

Nowak D, Crane D, Stevens J et al (2008) A ground-based method of assessing urban forest structure and ecosystem services. AUF 34:347–358. https://doi.org/10.48044/jauf.2008.048

Onandia G, Schittko C, Ryo M et al (2019) Ecosystem functioning in urban grasslands: the role of biodiversity, plant invasions and urbanization. PLoS One 14:e0225438. https://doi.org/10.1371/journal.pone.0225438

Orsini F, Gasperi D, Marchetti L et al (2014) Exploring the production capacity of rooftop gardens (RTGs) in urban agriculture: the potential impact on food and nutrition security, biodiversity and other ecosystem services in the city of Bologna. Food Sec 6:781–792. https://doi.org/10.1007/s12571-014-0389-6

Ouyang X, Xu J, Li J et al (2022) Land space optimization of urban-agriculture-ecological functions in the Changsha-Zhuzhou-Xiangtan Urban Agglomeration. China Land Use Policy 117:106112. https://doi.org/10.1016/j.landusepol.2022.106112

Pataki DE, Carreiro MM, Cherrier J et al (2011) Coupling biogeochemical cycles in urban environments: ecosystem services, green solutions, and misconceptions. Front Ecol Environ 9:27–36. https://doi.org/10.1890/090220

Paustian K, Lehmann J, Ogle S et al (2016) Climate-smart soils. Nature 532:49–57. https://doi.org/10.1038/nature17174

Peng J, Liu Q, Xu Z et al (2020) How to effectively mitigate urban heat island effect? A perspective of waterbody patch size threshold. Landsc Urban Plan 202:103873. https://doi.org/10.1016/j.landurbplan.2020.103873

Petchey OL, Hector A, Gaston KJ (2004) How do different measures of functional diversity perform? Ecology 85:847–857. https://doi.org/10.1890/03-0226

Peters K, Elands B, Buijs A (2010) Social interactions in urban parks: stimulating social cohesion? Urban Forest Urban Green 9:93–100. https://doi.org/10.1016/j.ufug.2009.11.003

Phillips D, Lindquist M (2021) Just weeds? Comparing assessed and perceived biodiversity of urban spontaneous vegetation in informal greenspaces in the context of two American legacy cities. Urban Forest Urban Green 62:127151. https://doi.org/10.1016/j.ufug.2021.127151

Piano E, Souffreau C, Merckx T et al (2020) Urbanization drives cross-taxon declines in abundance and diversity at multiple spatial scales. Glob Chang Biol 26:1196–1211. https://doi.org/10.1111/gcb.14934

Pickett STA, Cadenasso ML, Grove JM et al (2011) Urban ecological systems: scientific foundations and a decade of progress. J Environ Manage 92:331–362. https://doi.org/10.1016/j.jenvman.2010.08.022

Piracha A, Chaudhary MT (2022) Urban Air pollution, urban heat island and human health: a review of the literature. Sustainability 14:9234. https://doi.org/10.3390/su14159234

Pocock MJO, Chandler M, Bonney R et al (2018) A vision for global biodiversity monitoring with citizen science. Advances in ecological research. Elsevier, In, pp 169–223

Pomerantz M (2018) Are cooler surfaces a cost-effect mitigation of urban heat islands? Urban Clim 24:393–397. https://doi.org/10.1016/j.uclim.2017.04.009

Qin Y (2015) A review on the development of cool pavements to mitigate urban heat island effect. Renew Sustain Energy Rev 52:445–459. https://doi.org/10.1016/j.rser.2015.07.177

Rega-Brodsky CC, Aronson MFJ, Piana MR et al (2022) Urban biodiversity: state of the science and future directions. Urban Ecosyst 25:1083–1096. https://doi.org/10.1007/s11252-022-01207-w

Resmi CT, Nishanth T, Satheesh Kumar MK et al (2020) Air quality improvement during triple-lockdown in the coastal city of Kannur, Kerala to combat Covid-19 transmission. PeerJ 8:e9642. https://doi.org/10.7717/peerj.9642

Rigolon A, Németh J (2020) Green gentrification or 'just green enough': do park location, size and function affect whether a place gentrifies or not? Urban Studies 57:402–420. https://doi.org/10.1177/0042098019849380

Rufí-Salís M, Petit-Boix A, Villalba G et al (2020) Recirculating water and nutrients in urban agriculture: An opportunity towards environmental sustainability and water use efficiency? J Clean Prod 261:121213. https://doi.org/10.1016/j.jclepro.2020.121213

Sagris V, Sepp M (2017) Landsat-8 TIRS data for assessing urban heat island effect and its impact on human health. IEEE Geosci Remote Sensing Lett 14:2385–2389. https://doi.org/10.1109/LGRS.2017.2765703

Sassen S (2001) The global city: New York, London, Tokyo. Princeton University Press

Schell CJ, Dyson K, Fuentes TL et al (2020) The ecological and evolutionary consequences of systemic racism in urban environments. Science 369:eaay4497. https://doi.org/10.1126/science.aay4497

Schipperijn J, Stigsdotter UK, Randrup TB, Troelsen J (2010) Influences on the use of urban green space—a case study in Odense, Denmark. Urban Forest Urban Green 9:25–32. https://doi.org/10.1016/j.ufug.2009.09.002

Schittko C, Onandia G, Bernard-Verdier M et al (2022) Biodiversity maintains soil multifunctionality and soil organic carbon in novel urban ecosystems. J Ecol 110:916–934. https://doi.org/10.1111/1365-2745.13852

Schwarz N, Moretti M, Bugalho MN et al (2017) Understanding biodiversity-ecosystem service relationships in urban areas: a comprehensive literature review. Ecosyst Serv 27:161–171. https://doi.org/10.1016/j.ecoser.2017.08.014

Seto KC, Güneralp B, Hutyra LR (2012) Global forecasts of urban expansion to 2030 and direct impacts on biodiversity and carbon pools. Proc Natl Acad Sci U S A 109:16083–16088. https://doi.org/10.1073/pnas.1211658109

Shahmohamadi P, Che-Ani AI, Etessam I et al (2011) Healthy environment: the need to mitigate urban heat island effects on human health. Procedia Eng 20:61–70. https://doi.org/10.1016/j.proeng.2011.11.139

Shanahan DF, Fuller RA, Bush R et al (2015) The health benefits of urban nature: how much do we need? BioScience 65:476–485. https://doi.org/10.1093/biosci/biv032

Shanahan DF, Bush R, Gaston KJ et al (2016) Health benefits from nature experiences depend on dose. Sci Rep 6:28551. https://doi.org/10.1038/srep28551

Strosnider H, Kennedy C, Monti M, Yip F (2017) Rural and urban differences in air quality, 2008–2012, and community drinking water quality, 2010–2015—United States. MMWR Surveill Summ 66:1–10. https://doi.org/10.15585/mmwr.ss6613a1

Tan J, Zheng Y, Tang X, Guo C, Li L, Song G, Zhen X, Yuan D, Kalkstein AJ, Li F, Chen H (2010) The urban heat island and its impact on heat waves and human health in Shanghai. Int J Biometeorol 54:75–84. https://doi.org/10.1007/s00484-009-0256-x

Tansley AG (1935) The use and abuse of vegetational concepts and terms. Ecology 16:284–307. https://doi.org/10.2307/1930070

Targino AC, Coraiola GC, Krecl P (2019) Green or blue spaces? Assessment of the effectiveness and costs to mitigate the urban heat island in a Latin American city. Theor Appl Climatol 136:971–984. https://doi.org/10.1007/s00704-018-2534-1

Tham KW (2016) Indoor air quality and its effects on humans—a review of challenges and developments in the last 30 years. Energ Buildings 130:637–650. https://doi.org/10.1016/j.enbuild.2016.08.071

Tilman D, Clark M, Williams DR et al (2017) Future threats to biodiversity and pathways to their prevention. Nature 546:73–81. https://doi.org/10.1038/nature22900

Ulpiani G (2021) On the linkage between urban heat island and urban pollution island: three-decade literature review towards a conceptual framework. Sci Total Environ 751:141727. https://doi.org/10.1016/j.scitotenv.2020.141727

Ulrich RS, Berry LL, Quan X, Parish JT (2010) A conceptual framework for the domain of evidence-based design. HERD: Health Environments Research & Design Journal 4(1):95–114. https://doi.org/10.1177/193758671000400107

United Nations Human Settlements Programme (UN-Habitat) (2016) Urbanization and development: emerging futures. UNHabitat, Nairobi, Kenya

United Nations Department of Economic and Social Affairs, Population Division (2022) World population prospects 2022: summary of results

Viecco M, Jorquera H, Sharma A et al (2021) Green roofs and green walls layouts for improved urban air quality by mitigating particulate matter. Build Environ 204:108120. https://doi.org/10.1016/j.buildenv.2021.108120

Wang J, Da L, Song K, Li B-L (2008) Temporal variations of surface water quality in urban, suburban and rural areas during rapid urbanization in Shanghai, China. Environ Pollut 152:387–393. https://doi.org/10.1016/j.envpol.2007.06.050

Wang F, Wang Y, Zhang K et al (2021) Spatial heterogeneity modeling of water quality based on random forest regression and model interpretation. Environ Res 202:111660. https://doi.org/10.1016/j.envres.2021.111660

Wolch JR, Byrne J, Newell JP (2014) Urban green space, public health, and environmental justice: The challenge of making cities 'just green enough.' Landsc Urban Plan 125:234–244. https://doi.org/10.1016/j.landurbplan.2014.01.017

Wolkoff P (2018) The mystery of dry indoor air—an overview. Environ Int 121:1058–1065. https://doi.org/10.1016/j.envint.2018.10.053

Woodruff SC, BenDor TK (2016) Ecosystem services in urban planning: Comparative paradigms and guidelines for high quality plans. Landsc Urban Plan 152:90–100. https://doi.org/10.1016/j.landurbplan.2016.04.003

World Health Organization (2018) COP24 special report: health and climate change. World Health Organization. https://iris.who.int/handle/10665/276405

Wright JP, Naeem S, Hector A et al (2006) Conventional functional classification schemes underestimate the relationship with ecosystem functioning. Ecol Lett 9:111–120. https://doi.org/10.1111/j.1461-0248.2005.00850.x

Partnering with Spontaneous Nature in the City and the Suburbs

Scott Quitel

Abstract Spontaneous nature refers to natural ecosystems and habitats that arise on their own - seemingly from out of nowhere. In cities, spontaneous nature sprouts up on abandoned lots, buildings, and infrastructure. In suburbs, it springs up along roadsides, in abandoned fields, and in lawns that go unmowed. Urban and suburban dwellers often view unplanned vegetation and uninvited wildlife negatively, labeling them with terms such as 'weed,' 'pest,' 'non-native,' 'invasive,' or even 'noxious.' On a planet maligned by massive habitat destruction and impacted by rapidly rising temperatures, especially in urban environments, spontaneous nature ought to be viewed not negatively, but, rather as perhaps the most valuable free benefit to ever be bestowed upon a city or suburb. Spontaneous nature provides wildlife habitat, increases local biodiversity, spurs pollination and seed dispersal, and sequesters carbon. In cities, where extreme conditions often impair quality of life and health, spontaneous nature filters soil, replenishes groundwater, purifies air, manages stormwater, reduces runoff and erosion, and moderates ambient temperature.... Near downtown Philadelphia, hidden in plain view from most pedestrians, just 20 feet overhead, spans a century-plus-old railroad structure: the Reading Viaduct, which carried trains into the city. In 1984, the Viaduct was replaced by a tunnel, but the structure was left standing. In the 40 years since, a diverse habitat of forest and meadow has colonized the entire structure. In 2011, a plant survey revealed over 50 plant species growing atop the Viaduct. In 2016, another survey tallied 78 species. In 2019, the city of Philadelphia closed a 150-acre golf course in FDR park, in South Philadelphia, after 80 years of operation. In just a few years, the fairways and greens have been reclaimed by robust meadow and emergent wetland and upland forests—all in an area that was once a vital part of Philadelphia's estuary. Alarmingly, despite the nearly incalculable value of these two new urban natural areas, which have cost Philadelphia citizens zero, both of these novel ecosystems are currently threatened by planning visions that look to severely tame the wildness of the viaduct and install soccer fields and tennis courts in the estuary.... Our conceptions and

S. Quitel (✉)
LandHealth Institute, School of Entrepreneurship, Drexel University, Philadelphia, PA, USA
e-mail: scott@landhealthinstitute.org

G. Iannaccone, E. V. Ellis (eds.), *Towards a Nature-Positive Built Environment*, Research for Development, https://doi.org/10.1007/978-3-031-95292-0_5

views of nature need to change—fast. And then, informed action—or purposeful inaction—must follow. We must directly acknowledge: The original invasive species is Homo sapiens. We are the 'weeds' that scattered all over the land and supplanted so much of the native habitat. The most direct, impactful way to improve the ecological health of Earth is to partner with nature. In the city and the suburbs, this means embracing spontaneous nature, protecting it, nurturing it, educating others about its wonders and values, and enabling it to flourish.

Keywords Spontaneous nature · Biodiversity · Pollinator corridors · Natural ecosystems

Prologue. The Incredible Hulk

The asphalt is severely buckled and broken. A thick tree trunk emerges powerfully through and above the hardscape. Sizable surface roots ripple through and along the pavement, supporting the massive tree trunk. The orange bark of the surface roots, striated with living, breathing, lenticels, contrasts markedly with the grey and black of the asphalt. The scene evokes memories from my youth, watching the television series, "The Incredible Hulk". In the show, the character Bruce Banner, played by Bill Bixby, transforms from an average-sized male into a super-human, exaggeratedly muscular entity called the Hulk whenever he gets upset. I remember next to nothing of the TV show but for the lasting image of a normal-sized human upper arm suddenly rippling, bulging, and tearing through a shirt sleeve, transformed into the most massive bicep imaginable. Quite similar to the emerging orange surface roots.

"The Incredible Hulk", based on a Marvel comic book character, is pure fiction; the scene described above is not. In the present-day scenario, the starring role is played by a white mulberry tree (*Morus alba*), which somehow germinated from an intact parking lot, parking stripes still visible, fenced to keep out cars several decades ago. Today, the white mulberry towers above an 'asphalt garden', joined in a thriving urban plant community by a diverse array of other species. No human being planted any of these plants. The plants did it all. All by themselves.

The white mulberry, and each of its urban neighbors, have an invaluable message to share with us. They are communicating with us, directly and quite clearly. They have been for a long time. Why are we still not listening to the mulberry?

The Context: Urban Spontaneous Nature

Nature, by its own nature, is opportunistic. All living organisms sense opportunity—to survive, to live, to thrive, to establish, to colonize, to live on. Opportunism is genetic. All organisms, from the tiniest microbes to the largest animals and the tallest trees, are wired to pounce on opportunity (Bozorgmehr 2012). Sometimes, ecosystems experience disturbance—for example, fire, flood, high wind, ice, drought, and so on; and, for example, mowing, pruning, harvesting, clearing, channelizing, herbiciding, fertilizing, and so on (Pickett and White 1985; Seidl et al. 2017; Burton et al. 2020). Typically, following ecosystem disturbance, many and varied organisms seize upon new opportunities created by disturbance. A volcano erupts, lava flows downhill, and much of a meadow or forest is wiped out. However, soon after the lava cools, new plants and animals are often quick to recolonize the scorched landscape. A river floods with extreme force and turbulence and washes away an established floodplain forest. However, seeds from plants a hundred miles upstream find themselves deposited upon and within freshly laid sediments on the denuded riverbank.

Some natural ecosystems are characterized by the occurrence of periodic disturbances . For example, in the New Jersey Pine Barrens, named after the pitch pine (*Pinus rigida*), pinecones stay tightly closed for years, waiting for a wildfire to clear the accumulated, slow-to-biodegrade organic matter lying upon the forest floor and simultaneously to force open the cones and allow the extremely light seeds to flutter down to the ground and make direct contact with the newly exposed soil (Schäfer 2011). In addition to enabling new pitch pine seedlings to germinate from the sandy forest floor, the passing fire sparks vibrant resprouting of various understory and groundcover plants, such as mountain laurel (*Kalmia latifolia*), sheep laurel (*Kalmia angustifolia*), lowbush blueberry (*Vaccinium angustifolium*), eastern teaberry (*Gaultheria procumbens*), bearberry (*Arctostaphylos uva-ursi*), beach heather (*Hudsonia tomentosa*), and the curiously occurring broom crowberry (*Corema conradii*), which is endemic much further north, to northeastern New England, and Canada.

In mostly or largely natural ecosystems, where human interference is minimal or limited, disturbance tends to promote opportunities for species movement and new establishment, enhanced biodiversity, and even new speciation. However, such is typically not the case in urbanized and suburbanized areas. For purposes of this discussion, "urban" and "suburban" are to be regarded as endpoints of a spectrum characterizing level of human disturbance of land. More specifically, on one end of the spectrum is extremely densely developed urban land, typical of city centers comprised mostly of buildings and hardscape. On the opposite end of the spectrum is mildly, densely developed land, marked by detached residences with substantive yards and sprawling commercial areas distinguished by abundant horizontal consumption of land. Throughout the spectrum, human disturbance predominates, with opportunities for the establishment of healthy natural ecosystems extremely limited, except in protected natural areas (Olmos-Martínez et al. 2022). In all areas within the urban-suburban spectrum, 'spontaneous nature' refers to the pockets of habitat that are able to opportunistically sprout up within patches of land or infrastructure that somehow evade human usage or maintenance in otherwise highly developed, urbanized areas. Examples of places in cities and suburbs where spontaneous habitats may arise include vacant lots, abandoned or poorly maintained buildings, edges of railways, roadsides, median strips, abandoned infrastructure, and utility rights-of-way.

The difference between opportunistic emergence of nature in natural or mostly natural settings and opportunistic spontaneous emergence of nature in urban and suburban settings lies within the difference of disturbance type between the two settings. Disturbance in a natural area tends to be event-based: an ice storm, a fire, a flood, for example. An event happens, an existing ecosystem is materially impacted, the event passes, an opportunity for a new ecosystem to establish itself opens up. In urban and suburban areas, disturbance tends to be permanent or long-lasting or insidious. In dense sections of cities, land is blanketed with hardscape, streets, and buildings, all designed to last indefinitely. Rivers and creeks, which naturally tend to meander laterally over time, are locked into hard-armored channels and stripped of all or most of their riparian vegetation. Sometimes, streams are buried underground and diverted into sewer pipes. In the suburbs, fields and residential yards are mowed regularly. Fallen leaves and woody debris are quickly cleared from the ground. Many yards and fields received regular applications of herbicide. Neighborhoods are sprayed with insecticide. Golf courses are over-fertilized. Within the upper soil layers of many suburban landscapes lie light-weight seeds deposited by wind, heavier seeds dropped by birds and other animals, and acorns and other nuts buried by squirrels. Repeatedly, but unnoticed by most, diverse dormant seedbanks lying beneath suburban lawns and fields yield nascent stems that emerge above the ground plane, only to be cut back by the blades of mowers or the cords of string trimmers.

Spontaneous Nature in Action: The Reading Viaduct

In a densely developed section of Philadelphia, a richly diverse ecosystem of plants and wildlife resides on an elevated railway, hidden in plain view from most who traverse the city streets, twenty feet or so below. The Reading Viaduct was formerly a strategic section of a commuter railway system, owned and operated by the Reading Railroad Company, which transported passengers into and out of Center City Philadelphia. In 1984, after about a century of operation, the Reading Viaduct ceased carrying trains into the Reading Terminal, replaced by a commuter tunnel, which presently transports train passengers below ground into Center City. Today, a surviving section of the Viaduct runs for just over a half-mile from Fairmount Avenue and North Ninth Street to near Vine Street and North Eleventh Street. Since 1984, the Viaduct structure has lain relatively undisturbed.

Almost as soon as trains ceased running atop the Reading Viaduct, plant life began to rise from the silent railbed. Over the past 40 years, a diverse mix of plant communities have colonized the entire surface of the abandoned structure. Each different plant community type appears to have become established in response to the specific surrounding site conditions. For instance, sections of the Viaduct that are relatively more sheltered by nearby buildings tend to have a richer diversity of tree species. More exposed areas tend to be comprised primarily of meadow or grassland species. A walk along the transect reveals an array of plant communities, which based upon predominant species, could aptly be characterized as 'empress tree forest', 'little bluestem - dogbane - mugwort meadow', 'staghorn sumac - little bluestem - woolly mullein savannah', and 'mixed oak forest' (Figs. 1 and 2).

In September 2016, LandHealth Insitute was commissioned by the Pennsylvania Horticultural Society to conduct a plant survey of the Reading Viaduct. Five years earlier, in December 2011, a different organization repeatedly conducted a plant survey and identified over fifty species growing upon the Viaduct. LandHealth, in its later survey, identified seventy-eight species - a strong indication of increasing biodiversity.

Many of the catalogued species were those that typically grow well in urban conditions, such as tree of heaven (*Ailanthus altissima*), white mulberry (*Morus alba)*, empress tree (*Paulownia tomentosa)*, Japanese honeysuckle (*Lonicera japonica*), common ragweed (*Ambrosia artemisifolia)*, common dogbane (*Apocynum cannabinum*), mugwort (*Artemisia vulgaris*), Queen Anne's lace (*Daucus carota*), Japanese knotweed (*Fallopia japonica*), purple loosestrife (*Lythrum salicaria*), white sweet clover (*Melilotus alba*), broadleaf plantain (*Plantago major*), and narrow-leafed plantain (*Plantago lanceolata*). In addition to these mostly exotic species, other native species, with relatively high wildlife value, were identified. Examples include common juniper (*Juniperus communis*), black cherry (*Prunus serotina*), willow oak (*Quercus phellos*), red oak (*Quercus rubra*), white oak (*Quercus alba*), and Pennsylvania blackberry (*Rubus pensilvanicus*).

Connected to the Reading Viaduct, until recently, was the abandoned Spring Garden Street train station. Except for the roof, which had been pierced in places by the elements, along with a few empress trees and trees of heaven, the building

Figs. 1 and 2 Mossy life growing upon Viaduct wall (on the left); spontaneously emerging woolly mullein (on the right)

structure was mostly intact. Dappled sunlight penetrated through holes in the roof and the open windows, helping to create a unique urban micro-climate. The inside of the relatively damp, partially lighted station served uniquely as a sort of 'urban fernery', which included four different fern species: maidenhair spleenwort (*Asplenium trichomanes*), common fragile fern (*Cystopteris fragilis*), marginal woodfern (*Dryopteris marginalis*), and sensitive fern (*Onoclea sensibilis*). In addition to plants, the abandoned historic train station was also used by homeless individuals. Viewing such use of the structure as undesired, the absentee owner of the Reading Viaduct, Reading International (no relation to the defunct Reading Railroad Company), demolished the Spring Garden Street station in 2021. As is the case with the old building, the ferns are no more (Fig. 3).

Not only does the former railbed surface of the Reading Viaduct serve as a platform for spontaneous nature; other spaces connected to the structure do as well. One such space is the parking lot described at the beginning of this chapter. The vacant lot sits beneath the Viaduct, near Hamilton and North Tenth Street. It is owned by Reading International. In addition to the previously referenced white mulberry (*Morus alba*), four other tree species vigorously grow out of the asphalt: empress tree (*Paulownia tomentosa*), London planetree (*Platanus x acerifolia*), black willow (*Salix nigra*), and eastern cottonwood (*Populus deltoides*). In addition to the trees, the parking lot houses several established populations of herbaceous species,

Fig. 3 Reading Viaduct flanked by Spring Garden Street Station (house of 'urban fernery'), just prior to its demolition in 2021

including common reed (*Phragmites australis*), common milkweed (*Asclepias syriaca*), Canada goldenrod (*Solidago canadensis*), and plume poppy (*Macleaya cordata*).

Less than a block from the parking lot, the stone-faced wall of a buttress supporting the Viaduct serves as beneficial habitat for a large population of house sparrows (*Passer domesticus*). The wall appears to be a rock cliff face, as many crevices have formed within the facade, due to eroding mortar and missing stone blocks, which have fallen from the wall over time. During springtime, the stone wall is alive with sparrow activity, as adult birds continuously fly out and back from the wall, carrying food and nest materials into the multiple holes and crevices, which house nests for chicks. In places, adult sparrows peer out from the holes, apparently on guard. The chipping and chirping of the sparrows is louder than the surrounding din of the traffic. House sparrows tend to fare well in cities because their high-pitched chirping enables them to communicate effectively in noisy settings. The present condition of this one small component of the Viaduct makes for ideal habitat for sparrows (Figs. 4 and 5).

Carlton Street and Wood Street are two very narrow streets that pass beneath the Viaduct, perpendicular to the structure. Each street is overhung by an archway, composed in part of limestone and concrete. Each dark, narrow passageway truly feels like an actual cave in the city. Acidic rainwater, draining through the Viaduct over many years, has created an intricate variety of limestone formations inside the tunneled portions of Carlton Street and Wood Street. Hanging down from the arched ceilings of each passageway are hundreds of stalactites of varying lengths, actively growing, as evidenced by small droplets of water suspended precariously at their sharp tips. Along the street and walkway beneath the arched ceiling are numerous smooth patches of limestone, representing would-be stalagmites, were it not for

Figs. 4 and 5 Dogbane seeds dispersing in winter. Urban bird nest aloft in empress tree

regularly passing pedestrians and cars. Several ridged, often glistening wet limestone columns line parts of the walls. Water seepage is active and very perceptible, and, at times, audible. Geologically, the tunnel sections of Carson Street and Wood Street function quite similarly to natural active caves.

Barrier Island Spontaneity

Atlantic City, New Jersey, is a densely developed city, encompassing the northern half of the barrier island Absecon. Since its founding, Atlantic City has served as a popular resort destination for millions of people. Because of its popularity, essentially all of Atlantic City is covered by human development. There is no designated preserved natural area in Atlantic City, nor on the entire occupiable (by humans) whole of Absecon Island. Viable opportunities for spontaneous nature to arise and survive are extremely rare.

A barrier island, when left alone by humans, is a naturally moving landform, which responds to prevailing ocean currents (McNamara and Lazarus 2018). In the case of Absecon, the island is naturally inclined to erode in the north and accrete in the south. This means that without human intervention, the island would continually lose sandy land in northern Atlantic City and gain land at the southern end, where Longport sits. A barrier island is also an extremely dynamic ecosystem, directly exposed to recurring high winds and violent wave activity. Not surprisingly, Atlantic City has had to repeatedly rebuild substantial portions of itself in the aftermaths of powerful storms over the years.

In the Inlet section of northern Atlantic City, storms over time destroyed much of the Boardwalk, which Atlantic City is famous for, and negatively impacted residential buildings and houses in the area. After Hurricane Sandy in 2012, a very large seawall was constructed along the ocean-facing side of the Inlet, which is the area of Absecon Island most susceptible to storm intensity. In addition to the seawall, a series of seven large jetties was constructed to extend several hundred feet out into the sea, positioned perpendicular to the shore.

The newly built jetties immediately attracted anglers to the area. Today, the Inlet is a very popular fishing destination. Apparently, the artificial additions to the Inlet have served to offer suitable habitat for a wide variety of fish species that typically reside further out in the ocean. This outcome is predictable, as rocky substrate frequently makes for beneficial marine habitat. What I have found to be highly noteworthy, however, are the increasing numbers and varieties of waterfowl that have been visiting the jetties over the past few winters.

For many years, I have regularly frequented the Inlet while traversing the Boardwalk on foot or on bicycle. Generally, over the years, I commonly encountered various species of gulls and terns. Perhaps, some mallards or a few cormorants, as well. However, at some point in the recent past, I began to notice loons and bufflehead and other migrating waterfowl in places where I had not noted them before. And then I began to notice that both bird numbers and species diversity were increasing each year. In the weeks before this writing, I spotted as many as one hundred loons in an area, where just a few years ago, I may have spotted a few, if any.

In the winter of 2023–24, species that I have observed in the vicinity of the Inlet jetties include: common loon (*Gavia immer*), double-crested cormorant (*Nannopterum auritum*), bufflehead (*Bucephala albeola*), long-tailed duck (*Clangula hyemalis*), American black duck (*Anas rubripes*), mallard (*Anas platyrhynchos*), black scoter (*Melanitta americana*), lesser scaup (*Aythya affinis*), greater scaup (*Aythya marila*), brant (*Branta bernicla*), common merganser (*Mergus merganser*). This list is not an official survey. Based upon reported observations from other birders, I believe it is far from complete. My list is based merely off of a few random opportunistic viewings. What is important to note is this: to keep its northern end from washing away, Atlantic City hard-armored its Inlet section with a series of jetties. Almost immediately, the fish came. Soon after, the winter birds arrived. And, as of today, they seem to be coming more and more, both in numbers and in terms of species richness. On a recent visit to the Inlet, I encountered two women with binoculars, and we shared our observations with one another. One of them exclaimed, "With this, who needs to go to Forsythe!" She was referring to the nearby Edwin B. Forsythe National Wildlife Refuge. While her statement may have been hyperbolic, a few years ago, it would have sounded outrageous.

The southern end of Atlantic City is mostly residential. Up and down the ocean-facing beaches of New Jersey, Atlantic City included, the State requires the construction of a primary dune, sown with plugs of beach grass (*Ammophila breviligulata*), to serve as a buffer against storms and tidal surges. Running along the inland-facing side of the Boardwalk, roughly between Dover Avenue and Jackson Avenue (the southern border of Atlantic City), is an unmanaged, generally ignored, narrow line of sand, lying between the Boardwalk and the connected property lines of the beachfront homes and apartment buildings built along the beach and Boardwalk. The width of this sandy area averages around ten feet. Sheltered by the primary dune and Boardwalk on one side and the low seawalls on the other side, this neglected, relatively miniscule strip of sand may contain the richest wild species diversity on Absecon Island. Trough-like in form, this area has more moisture-holding capacity than the nearby dune. Lying slightly lower than adjacent areas, the sandy strip acts as a catchment for seeds blown from plants on the dune and from plants on residential landscapes, as well as seeds deposited by birds. An incomplete plant species inventory of the strip includes: eastern red cedar (*Juniperus virginiana)*, Japanese black pine (*Pinus thunbergii*), tree of heaven (*Ailanthus altissima*), empress tree (*Paulownia tomentosa*), black cherry (*Prunus serotina*), willow (*Salix sp.*), grey birch (*Betula populifolia*), northern bayberry (*Myrica pensylvanica*), staghorn sumac (*Rhus typhina*), winged sumac (*Rhus copallina*), beach rose (*Rosa rugosa*), American holly (*Ilex opaca*), Virgina creeper (*Parthenocissus quinquefolia*), virgin's bower (*Clematis virginiana*), Japanese honeysuckle (*Lonicera japonica*), common reed (*Phragmites australis*), cattail (*Typha latifolia*), beach grass (*Ammophila breviligulata*), little bluestem (*Schizachyrium scoparium*), seaside goldenrod (*Solidago sempervirens*), evening primrose (*Oenothera biennis*), prickly pear (*Opuntia humifusa*), common milkweed (*Asclepias syriaca*), hawkweed

Figs. 6 and 7 Staghorn sumac in autumn (on the left); little bluestem meadow giving way to mixed deciduous forest (on the right)

(*Pilosella officinarum*), rose mallow (*Hibiscus moscheutos*), purple loosestrife (*Lythrum salicaria*), tickseed (*Coreopsis sp.*), blue balloon flower (*Platycodon grandiflorus*), and sensitive fern (*Onoclea sensibilis*).

The thin sandy strip acts as a showcase of spontaneous vegetation. It also presents a powerful message that nature is tenacious and will take anything it can get its roots into (Figs. 6 and 7).

A Rewilded Golf Course

In the 1920s, the City of Philadelphia filled in many acres of estuarine wetlands and adjacent natural area to lay the grounds for a world's fair to celebrate the United States Sesqui-Centennial in 1926. In subsequent years, much of the grounds was used to create FDR Park. Approximately 150 acres of former fairground was used to create a municipal golf course, which operated for about eighty years. FDR Park and the adjacent golf course site are located in South Philadelphia, not far from the confluence of the tidal Schuylkill and Delaware Rivers. Remnants of tidal creeks still run through parts of this area. Not surprisingly, the area frequently floods.

In 2019, the City closed the golf course. Almost immediately, spontaneous nature arrived and made its presence felt. Early successional grasses and forbs quickly replaced many of the greens and fairways. Wetland plants began to colonize the wetter areas, and mesic species began to colonize the upland areas. During the second growing season following the golf course closure, thousands of tree seedlings could be spotted emerging amidst the vigorously colonizing herbaceous species. Sweetgum (*Liquidambar styraciflua*), silver maple (*Acer saccharinum*), red maple

(*Acer rubrum*), green ash (*Fraxinus pennsylvanica*), black cherry (*Prunus serotina*), river birch (*Betula nigra*), black willow (*Salix nigra*), northern catalpa (*Catalpa speciosa*), red oak (*Quercus rubra*), black oak (*Quercus velutina*), swamp white oak (*Quercus bicolor*), pin oak (*Quercus palustris*), white oak (*Quercus alba*), tulip poplar (*Liriodendron tulipifera*), eastern white pine (*Pinus strobus*), bald cypress (*Taxodium distichum*), and other tree seedlings began making their way toward the sky. Early forest succession was clearly underway.

Local residents generally rejoiced at the nature-infused transformation of the formerly off-limits area. They readily made use of the existing pathways, which were formerly used by golfers to make their way throughout the golf course. Local residents gave the area a name: the Meadows—in response to the sudden appearance of tall, reedy grasses and forbs, swaying in the wind, reclaiming the previously manicured landscape.

Wildlife abounds throughout the former golf course, birds especially. Located very close to the Delaware River Flyway, the area is a magnet for migrating birds. Amphibian life is also robust. On many late winter and spring evenings, the chorus of spring peepers (*Pseudacris crucifer*) drowns out the noise of the surrounding city. Equally musically deafening is the cacophony of American toads (*Anaxyrus americanus*). Also prevalent are green frogs (*Lithobates clamitans*) and American bullfrogs (*Lithobates catesbeianus*).

When spontaneous nature is given the opportunity to colonize a large previously managed landscape, like a golf course, the visual impact of the transformation is dramatic. While a ten-foot-wide sandy strip of land next to a boardwalk can go essentially unnoticed, a 150-acre former golf course gone wild cannot. Equally dramatic is the ecological impact of giving spontaneous nature free rein over a formerly managed landscape. Regardless of the species type of newly arriving plants, native or non-native, any new vegetation offers new structure, new habitat, new food sources—new opportunities for continually emerging plants and wildlife. New vegetation also means substantially more roots in the ground, thereby dramatically improving the well-being of the soil profile. The predominant plant type of a golf course is turf grass, which has very shallow roots. In contrast, meadow grasses and forbs typically have roots that penetrate deeply into the soil layers. Tree roots tend to spread out laterally, widely. The combination of herbaceous and woody root growth serves to promote natural stormwater management that continuously improves drastically over time.

The story of the rewilding of the old FDR Park golf course would have had a happy ending, had the City simply accepted the invaluable gift of nature, including the thousands of free plants and animals that came with it. However, the City, operating through the Fairmount Park Conservancy, unveiled a land plan for the former golf course that includes various amenities, including a newly installed ecological core and athletic fields made of artificial turf. As of this writing, in preparation for reworking the site, hundreds of trees have been cut down and removed from the site. More trees, including many classified as ‘heritage trees,’ are scheduled for removal. A large mound of excavated soil, slated to become a grassy hill, sits atop an area that recently was in the early stages of becoming a mixed oak forest.

The FDR Park plan is ambitious. Regardless of its merits, the plan is extremely expensive, both in dollar terms ($250+ million) and in terms of opportunity cost. A century ago, a vast area of the Delaware Estuary was converted into a vast recreational area, which came to include a golf course. A few years ago, the golf course closed. Immediately, at no cost to the City, nature—as an interconnected system—stepped in and began, in its own natural way, to 'reimagine' and to 're-engineer' the site. Within just a couple of years, a walker traversing the newly developing ecosystems could easily perceive that a new 'vision' for the site, comprised of an interconnected mix of different ecological communities, was in the early stages of being realized. The City and the Fairmount Park Conservancy have, in effect, decided that 'nature's vision' is not good enough. And so, they have created their own vision instead.

Spontaneous Nature in the Suburbs

About twenty years ago, my family and I moved from Philadelphia to a suburban neighborhood just outside the city. We bought a house on an acre of land. At the time, most of that acre was lawn. Our newly acquired property did have a handful of trees growing in spots. It also had several well-maintained plant beds, generally located next to or close to our house, on all four sides of our residence. Overall, the overwhelming majority of our property was covered by turf grass, subject to periodic mowings throughout the growing season.

Soon after moving into our new house in the suburbs, I replaced several of our inherited plant beds with native gardens, including a large prairie garden and a woodland garden. I then turned a river-rock-lined swale that quickly conveyed stormwater to the street into a densely planted ephemeral streambed, loaded with deep-rooted plants to hold the banks in place and minimize erosion. Also, over the years, I have purchased and planted a wide variety of species throughout the property.

After living with my suburban landscape for a few years, I remained frustrated by how much lawn still comprised my land. I then decided to team up with spontaneous nature, in a concerted effort to add more plant coverage to my property.

The core of my plan centered around designating several areas of my land as 'no-mow zones'. In these areas, I allowed any plant that germinated to grow without interference. I kept a watchful eye out for emerging seedlings that had evaded mowing. Sometimes, when I spotted a desirable spontaneous plant growing in an undesirable place, such as right next to the driveway or too close to the house, I transplanted the specimen to a more favorable spot. I dug a pond out back and maintained it minimally. I took the excavated soil from the hole and built a mound next to a retaining wall. I then left the mound alone to be colonized spontaneously. To this day, I leave the leaves, unlike my neighbors, who annually rake or blow their leaves to the curb, where a township truck sucks them up and takes them to a composting yard (so that residents can order leaf mulch from the township and pay to have it delivered back to suburban properties). I also leave fallen woody debris where it falls or move it to wooded areas of my yard.

Today, in 2024, I count over fifty different plant species growing on my property that arrived on their own, not purveyed by me. In terms of quantity, I estimate the total number of desirable spontaneously derived plants on my property to be well in the thousands. I have nearly fifty trees at least twenty feet tall that arrived on their own. My ephemeral stream bioswale is almost entirely forested, dominated by red maples and pin oaks that seeded themselves and shaded out most of the prairie grasses and forbs that I originally planted in and along the swale. My pond serves as habitat for over-wintering bullfrogs and green frogs and occasionally hosts breeding American toads. The mound I constructed from the soil I excavated to build my pond is covered by staghorn sumac (*Rhus typhina*), and river oats (*Chasmanthium latifolium*), among other species, all of which seeded themselves. Eastern red-backed salamanders (*Plethodon cinereus*) live beneath the woody debris and fallen leaves that I have allowed to accumulate in shaded areas. The debris also serves as habitat for eastern garter snakes (*Thamnophis sirtalis sirtalis*) and DeKay's brown snakes (*Storeria dekayi*). Bird activity throughout my property is generally robust.

Following is a species list of plants that have seeded themselves on my land. Note that several of the prairie species originated from plants that I planted in my meadow garden.

Woody plants	Herbacious plants
Red maple (*Acer rubrum*)	Common milkweed (*Asclepias syriaca*)
Silver maple (*Acer saccharinum*)	Vernonia noveboracensis (*New York ironweed*)
Box elder maple (*Acer negundo*)	Sneezeweed (*Helenium autumnale*)
Pin oak (*Quercus palustris*)	Foxglove beardtongue (*Penstemon digitalis*)
Red oak (*Quercus rubra*)	Celandine poppy (*Stylophorum diphyllum*)
American sweetgum (*Liquidambar styraciflua*)	Grass-leaved goldenrod (*Euthamia graminifolia*)
Northern catalpa (*Catalpa speciosa*)	Canada goldenrod (*Solidago canadensis*)
River birch (*Betula nigra*)	Gray goldenrod (*Solidago nemoralis*)
Gray birch (*Betula populifolia*)	Common fleabane (*Erigeron philadelphicus*)
Tulip poplar (*Liriodendron tulipifera*)	Aster (*Aster spp.*)
Green ash (*Fraxinus pennsylvanicus*)	Spotted joe-pye weed (*Eutrochium maculatum*)
White ash (*Fraxinus americana*)	Hollow joe-pye weed (*Eutrochium fistulosum*)
Black cherry (*Prunus serotina*)	Tall thoroughwort (*Eupatorium altissimum*)
Black walnut (*Juglans nigra*)	Ageratina altissima (*White snakeroot*)
American sycamore (*Platanus occidentalis*)	Rose mallow (*Hibiscus moscheutos*)
American pussy willow (*Salix discolor*)	Rattlesnake master (*Eryngium yuccifolium*)
Eastern redbud (*Cercis canadensis*)	Virginia bluebells (*Mertensia virginica*)
American holly (*Ilex opaca*)	Lesser celandine (*Ficaria verna*)
Ginkgo (*Ginkgo biloba*)	Dandelion (*Taraxacum officinale*)
White mulberry (*Morus alba*)	River oats (*Chasmanthium latifolium*)
Tree of heaven (*Ailanthus altissima*)	Tussock sedge (*Carex stricta*)
European hornbeam (*Carpinus betulus*)	Seersucker sedge (*Carex plantaginea*)
Callery pear (*Pyrus calleryana*)	Soft rush (*Juncus effusus*)
Staghorn sumac (*Rhus typhina*)	Rough horsetail (*Equisetum hyemale*)
Pennsylvania blackberry (*Rubus pensilvanicus*)	Field horsetail (*Equisetum arvense*)
Wineberry (*Rubus phoenicolasius*)	Sensitive fern (*Onoclea sensibilis*)
Mock orange (*Philadelphus coronarius*)	Christmas fern (*Polystichum acrostichoides*)
	Interrupted fern (*Osmunda claytoniana*)

Why the Embrace of Spontaneous Nature Is Essential in Cities and Beyond

Cities are often harsh environments. They are typically hotter than surrounding areas. They are often more polluted than outlying areas. They are characterized by an over-abundance of hard surfaces and structures and a corresponding paucity of living plant cover. Accordingly, cities tend to be more prone to occurrences of dangerous temperatures and dangerous air quality levels than surrounding areas. One of the greatest issues facing nearly all cities is the need for much more living plant mass.

For purposes of discussion, let us invent a figurative, generic plant species: *Anytree anyplace*. *Anytree anyplace* possesses various traits, each of which conveys value to humans and all other living organisms. *Anytree anyplace,* or any real plant species, regardless of its land of origin:

- Manages water
- Sequesters carbon
- Manufactures oxygen
- Provides food for wildlife and often for humans
- Provides cooling and often shade
- Filters air, water, and soil
- Aerates soil
- Reduces erosion
- Provides habitat
- Offers wonder
- Beautifies in the eyes of many
- Supports healthy recreation
- May reduce stress, anxiety, and depression

Any plant, planned or spontaneous, purchased or free, native or non-native, displays the above traits.

Today, many cities, such as New York and Philadelphia, recognize the values offered by plants, and they are implementing projects to increase urban tree canopy cover. These initiatives are very costly, calling for purchases of thousands, or even millions, of trees. A spontaneous tree is a free tree. Each year, many abandoned urban spaces—vacant lots, untended medians, border edges of fields and yards, the Reading Viaduct, the former FDR Park golf course, the untended sandy strip in Atlantic City - serve as hosts to countless germinations of seeds, already wired to consume carbon and grow. Spontaneous nature presents cities with immense untapped opportunity, if they can simply learn to embrace it.

In Philadelphia, the empress tree (*Paulownia tomentosa*), is considered by the Parks and Recreation Department to be an unwanted tree, an invasive, exotic weed tree. Each year, city crews clear hundreds of empress trees that have sprouted up on their own throughout various urban spaces. Yet, in Logan Square Park, the Philadelphia Parks and Recreation Department plants and maintains a border of empress trees, which surround the famous Swann Fountain. Each spring, loads of people stop by to admire the stunning periwinkle flowers that adorn the trees. Both the trees lining the fountain and the trees popping up in vacant lots throughout the metropolis are the exact same species. Whether intentional or spontaneous, each empress tree offers the same array of valuable ecosystem services. Each tree cools its surroundings, sucks in carbon, stems erosion, and manages stormwater. Each, regardless of its location, blooms in brilliant periwinkle. So why does the City choose to pay to plant empress trees in one area and pay to remove empress trees from other areas?

Not only cities, but the entire world needs to learn to embrace the benefits of spontaneous nature. The most pressing global problems of the moment include natural habitat destruction, mass extinction of species, and global warming. Spontaneous nature directly combats all three (Tredici 2010). In spaces and places where it is allowed to flourish, spontaneous nature creates new habitat where none or little existed before. Whether in vacant fields and lots, on top of abandoned infrastructure, or within eaves and crevices, spontaneous nature puts back new habitat. By creating new habitat, spontaneous nature fosters biodiversity and genetic diversity (Deparis et al. 2023). Left to its own devices, it can spark new speciation. Each spontaneously derived plant does its own small, but real, part to directly reduce global warming. Each plant pulls in gaseous carbon from its surrounding air and stores it as solid carbon in its tissue, thereby bringing about a continual net loss of carbon dioxide as long as it lives. If all humans disappeared from Earth, spontaneous nature would take over in an instant. And in that instant, global warming would begin to slow down, and biodiversity would start to rebound. We can learn from this.

How to Partner with Spontaneous Nature

Spontaneous nature is our ally. If we truly accepted this concept and lived accordingly, our cities and suburbs would look dramatically different. They would be much greener. They would be vibrant. They would be teeming with wildlife. They would be cooler, healthier, more beautiful, safer, more alluring. There are multiple ways in which we can go about forming and building a mutually beneficial relationship with spontaneous nature. Some ways are attitude-based. Others are action-based. We need to practice both ways. Following are a dozen steps we can take to establish a true partnership with spontaneous nature.

1. *Acknowledge our genetic relationship with spontaneous nature*

   Spontaneous nature, all nature, is our relative. We evolved from nature. Under a microscope, all of our cells are more similar than they are different. Humans invented the idea of 'conquering nature' and separating ourselves from nature. Yet we are all related, and we are all members of the same community, be it a human-created city or suburb or the planet as a whole. And, while nature would survive and thrive without us, we cannot survive without nature.

2. *Express plant gratitude*

    Plants are the source of all life. Photosynthesis is the miraculous biochemical reaction of life. Plants give us our food. They give us our oxygen. The next time you take a walk, think of *Anytree anyplace*. Select a random tree. Focus on it. Feel it. Smell it. Admire its structure. Connect. Empathize. Breath in the deep understanding that that tree, by its very existence, helps to give you life.
3. *Sense the wonder of spontaneous nature*

    After you express gratitude for the presence and presents of plants, sense and enjoy the wonder of spontaneous urban ecosystems. Become aware of and present with whole plant communities that sprout up on roadsides and median strips. Observe closely how a tree manages to grow outward and upward from a wall in the city. Marvel at the diverse life that grows upon and out of curbsides. Take note of the dripping stalactites that hang down from urban overpasses. Under these same overpasses, admire the craft of birds, of spiders, of mud daubers, which take advantage of our urban infrastructure to construct dwellings and places to rear their young. Live spontaneous nature in the city is as enthralling as any nature show on television. We just need to take it in.
4. *Confront our take on 'invasives'*

    Take a look at yourself in the mirror. Ask yourself, "Am I native to the country I reside in? Am I non-native. Am I invasive? A weed?" If we regard certain plants and animals as 'invasives', then we are the original invasives. The plants and animals did not invade. We brought them here, either by accident or on purpose. They just learned how to live in a new place. When our ancestors came to North America, not only did they bring non-native plants and animals; they resculpted the land. They cleared forests. They dammed streams. They filled wetlands. They cut hills and filled valleys. Today, most North American native soil has long since washed out to sea. Given that so much of our original, native soils, land topography, land types, and hydrology is so utterly different from how it all originally laid out, what actually does constitute a 'native species' on our markedly human-altered urban and suburban landscapes of today?

    We humans are so quick to label. When we find a plant or animal to our dislike, we label it 'invasive'. Sometimes we label it 'noxious'. We call plants that we do not like 'weeds'. We sometimes reward children for crushing scores of spotted lanternflies, as if the bugs are 'bad'. Why? What message are we sending to our kids? At a minimum, we need to be mindful of our actions, and we need to have thoughtful reasoning behind our actions. Human intervention in natural processes occurring in urbanized environments is very arbitrary. Regardless of any merits of human intervention in urban ecosystems, we do need to acknowledge that each crushed lanternfly, each euthanized red-eared slider, and each weeded plant represents a human decision to end a life. If we cannot connect with life at the individual species member level, we are not likely to connect at the ecosystem level.
5. *Flip our attitude toward phragmites*

    Certain species are here to stay, at least for the foreseeable future. In addition, we encounter them nearly everywhere, and their presence is palpable. One such species is common reed or phragmites (*Phragmites australis*). Common reed is ubiquitous in urban and suburban environs. And it is dramatically conspicuous on the landscape. It grows in dense monocultures. It reaches heights of twenty feet. It dominates many urban estuaries, such as the New Jersey Meadowlands and much of the Delaware Estuary. Common reed is typically regarded as a pervasive, undesired invasive. Many express deep disdain for the plant. Once, at a meeting held at the John Heinz National Wildlife Refuge at Tinicum, Refuge leaders expressed their pride at having successfully removed large swaths of common reed encroaching on the wetlands through the application of glyphosate herbicide, sprayed from a helicopter. In addition to the Heinz Refuge, many people view species like common reed as common enemies.

Phragmites is a star in the spontaneous nature world, adept at colonizing new spaces vegetatively or by seed (Meyerson et al. 2016). It serves as wildlife habitat for various marsh dwellers. Its roots bind soil very effectively. It sequesters lots of carbon. Its particular ability to grow upon itself, atop its stiff dead biomass, suggests that the plant may serve as a buffer against sea level rise in back bays and tidal marshes (Chambers et al. 2012; Milke et al. 2020). And, common reed may be regarded as majestic or beautiful. The tall reeds sway rhythmically in the wind. Mature inflorescences are deep burgundy. In the film "American Fiction", there is a scene in which the main character and his romantic partner walk along a sandy path toward a New England beach. They are flanked on both sides by phragmites in bloom, taller than they are. It is a lovely image.

Flipping our long-held views on prevalent species like phragmites will point us constructively in the direction of forming alliances, not battles, with spontaneous nature. When we fight assertive and persistent species like phragmites, we ultimately lose, even when we employ helicopters. When we focus our efforts on finding beneficial ways to live in community with nature, including cutting picturesque pathways through the reeds, all living members of an urban or suburban community experience improved quality of life.

6. *Dandelion love*

   Many suburban residents go to great lengths to rid their lawns of all dandelions (*Taraxacum officinale*). They often pay landscape companies significant sums of money to apply herbicide that kills all plants but the grass. Why do so many people hate dandelions? Dandelion flowers are festively yellow. The plant's taproot systems extend into the soil well beneath the level reached by turf grass roots, thereby creating micro-tunnels that promote infiltration. All plant parts of the dandelion are edible (Kania-Dobrowolska and Baraniak 2022). And, dandelions encourage interaction between plants and humans. Who can resist pulling the stem of a mature, fluffy seedhead and blowing the seeds up into the sky? Engaging with 'weeds' like dandelions may spark engagement with other plants, like spotted jewel weed (*Impatiens capensis*), which features a spring-like mechanism that catapults seeds to new germination spots. What constitutes a weed is in the eye of the beholder. Regarding plants like dandelions as 'serendipitous species' or as 'fun flowers,' instead of as 'weeds,' might spur a suburban dweller to better understand, appreciate, and enjoy their landscape as a living ecosystem, as opposed to a staid, tamed patch of turf. Once a connection with a dandelion or other living organism is made, another connection is more likely to follow.

7. *Leave the leaves*

   There is no reason, other than for aesthetics, to remove leaves or other organic matter from a natural or naturalized landscape. Leaves break down and add organic content to the soil. They also aid in moisture retention. Further, they buffer both small plants and animals from temperature extremes. Fallen woody debris provides shelter and habitat for wildlife. Both fallen leaves and woody debris promote growth of beneficial fungi in the upper soil layers. Post a sign on your land informing others that you proudly leave your leaves. Perhaps they will be moved to do the same.

8. *Cease the use of all chemicals*

   There is no such thing as a benign herbicide or pesticide. Most urban soils contain measurable levels of legacy chemical contamination from industrial land usage. Each time it rains in the city, runoff from roads transports pollutants onto fields and planted landscapes and into streams. While a specific chemical may succeed in ridding a landscape of a targeted pest, no net gain in ecological health ever accrues to an ecosystem exposed to herbicide or pesticide. We already have more than enough chemical pollution pervading our ecosystems and our backyards. Let us now stop putting new chemicals into circulation simply for getting rid of living pests we do not like. Any relationship between humans and nature, spontaneous or otherwise, begins with the cessation of all chemical applications.

9. *Establish no-mow zones*

   Designate areas on your land where mowing is avoided and nature is invited to drop in and get settled. Many suburban yards contain diverse dormant seedbanks. Simply refraining from mowing a select patch of lawn and observing what new growth emerges can be thrilling. When a suburban neighborhood is designed and developed, it typically has a planting plan that calls for the planting of a handful of street tree species, such as red maple, red oak, sweet gum, eastern redbud, flowering dogwood, and others. If residents in a neighborhood curtailed all mowing, the green areas of the neighborhood would likely succeed into a forest comprised of the surrounding prevalent street trees. Curtailing mowing on chosen patches is likely to yield free trees, among other spontaneous species (Lerman et al. 2018).

10. *Accept the gift and further it*

    The experience of receiving an unexpected gift is often packed with excitement and intrigue. When we get one, we rush to open it and check it out. Spontaneous nature is just such a gift. Our first inclination ought to be to keep it. Certainly not to pull it or cut it or spray it. Lack of or loss of habitat continues to be of epidemic proportions, both locally and globally. We need all the new nature we can get. When some tree or ecosystem gifts itself to us, seemingly from out of the blue, we should run to it with excitement and check it out. We should then welcome it and simply accept our gift.

Spontaneous nature is a true gift, as long as you regard it as such. Accept the gift of newly arriving free plants and animals with pride and gratitude. Protect newly emerging ecosystems. Encourage further spontaneous emergence by planting a garden in one section of your property and allowing seeds from the garden to be transported to and to germinate in other areas of your yard. Take pride in knowing that any spontaneous ecosystem that you welcome or initiate on your land goes a material way toward adding biodiversity, reducing habitat loss, and combatting global warming (Newbold et al. 2019).

11. *Design with spontaneous nature*

    To truly partner with spontaneous nature, we can consciously design with it. When a new tree arrives on its own in your yard, it could inspire you to plant a woodland garden around it. When bird activity is robust in a field or a lot near your home, consider planting a garden to attract some of those birds to your land. Actively seek inspiration from spontaneous nature by exploring passive urban areas to learn how spontaneous vegetation arranges itself (Bonthoux et al. 2019). In some sections of the Reading Viaduct, little bluestem grows in rows next to patches of dogbane, while tall flower stalks of woolly mullein tower over the landscape sporadically. The scene looks as if it could have been designed by a human. Nature is an excellent designer.

    Opportunities for humans to co-design with spontaneous nature in the city abound. For example, Philadelphia is peppered with approximately forty thousand vacant lots. The City pays substantially to have about thirteen thousand of these lots cleaned up and managed. The typical approach on these lots is to clear the area of trash, plant turf grass, border the lot with a split-rail fence, and install a few trees or planting beds. In most cases, around 90% or more of the lot is maintained as lawn and receives periodic mowings and cleanups. Over the past two decades, the City's approach to managing these lots has made positive impacts on several social issues plaguing blighted neighborhoods. Perhaps now is a fine time to design and install some wild areas on the parcels.

    LandHealth Insititute is a nonprofit organization that takes a different approach to vacant land. On a two-acre patch of former abandoned railyard in West Philadelphia, LandHealth operates a native plant nursery. In addition to the nursery, LandHealth has allowed various sections of its property to transform themselves into urban meadows, comprised of various spontaneously sown species, such as several clovers, mugwort, dogbane, common milkweed, chickory, Queen Anne's lace, and others. In some spots, LandHealth has added some of its own propagated plants. Throughout the site, LandHealth has woven a walking trail and installed rustic amenities, such as picnic tables, benches, a fire pit, a meditation circle with tree-stump seating, and colorfully hand-painted bird boxes.

    In other parts of Philadelphia, LandHealth has turned abandoned spaces into designed urban wild areas. Rustic seating, viewing spots, bird or bat houses, and informational graphics are typically part of the design. In most cases, spontaneous nature is allowed to flourish, often in tandem with intentionally planted native species. In all cases, people are invited into these landscapes to immerse themselves in some urban nature. In practice, LandHealth regards spontaneous nature as a design partner.

12. *Educate and advocate*

    Perhaps the most effective way to partner with spontaneous nature is to educate about its merits and advocate for its promotion and protection. One effective way to accomplish this is to simply start spreading the word. Most people have very little connection to local natural ecosystems. Let your friends and relatives know what you know about the wonders and values of nature that establishes itself on its own in urban and suburban settings. Post a small sign on your land that informs: 'No-Mow Zone' or 'I Proudly Leave the Leaves.' A curious passerby is bound to delve into the scenario and then follow suit on their own property.

Individual or group advocacy efforts at the local government level can lead to profound changes on the land. Cities and towns have various boards or commissions that advise on or decide on land use issues and policies. Examples include planning commissions, zoning boards, environmental advisory boards, shade tree commissions, open space committees, and the like. You can join one of these entities. Or you can attend their regular meetings, which are open to the public. At these meetings, citizens at large are able to share their views or ideas. It is generally very feasible for informed citizens at large to materially impact decisions or ordinances or policies on the management of public land, use of chemical herbicides, protection of sensitive areas, and stormwater management approaches. Occasionally, a municipality receives an opportunity to acquire a large tract of land. A quarry closes. A large estate goes on the market. A golf course closes. Such instances offer significant opportunities to invite nature back to where it used to reside. An informed, involved advocate may have a lot of impact in such instances.

Leaders of municipal governments and NGOs often do not share an intimate connection with nature and natural systems. Most organizational leaders do not possess a comprehensive understanding of the inner workings of spontaneous ecosystems, nor of the immense health and well-being benefits that are realized when spontaneous nature is invited back to spaces it once inhabited and is welcomed as an ally and fellow community member by humans. Organizational heads need to understand. And those of us already informed need to educate them.

The most open-minded audience to educate about the wonders and rewards of spontaneous nature is children. In my own extensive experience, I have led many children of all ages on exploratory walks and hikes through urban ecosystems in places ranging from vacant lots to abandoned piers to the Reading Viaduct to the FDR Park old golf course site to large preserved city parks. When kids are allowed to immerse themselves in the urban and suburban outdoors, they typically dive in with reckless abandon. They explore and they play and they learn through their senses. They blow dandelion seeds and milkweed seeds into the sky and frolic as they do so. They relish eating ripe mulberries from the urban-tough trees and sipping nectar from honeysuckle growing on fences. They play tag amongst the shrubs and tall grasses. When they come to a tree - any tree, native or non-native - they do not label it; they climb it.

It is cliché and patently unfair to say, “The kids are our future.” Or, “It’s up to the young to fix the environment.” It is irresponsible to even imply that one age group of individuals is burdened with dealing with our global environmental problems, especially when they had nothing to do with putting humans in our current pridacament. When it comes to the various environmental crises confronting our cities and suburbs and Earth, every single one of us, re/gardless of age, is responsible for responding. However, in general, the younger the child, the less judgmental, jaded, myopic, and pessimistic the response; and the more ambitious, idealistic, playful, and creative the response. When we introduce young students to spontaneous nature, the kids are more likely than adults to fall in love at first sight. Enmeshing our young students within spontaneous ecosystems can be a powerful catalyst for long-term ecological uplift in cities and beyond, with our kids starring as the change agents.

Epilogue

Spring 2024. The incredible white mulberry hulk stirs to life. Another growing season begins. New leaves appearing. Flower buds thickening. Juicy light and dark berries should be ripe by June. Sinewy surface roots imperceptibly bulge a bit, breaking up a little more asphalt. Twenty feet above the incredible hulk, and the surrounding urban neighborhood, the incredible Reading Viaduct ecosystem also awakens for spring. It is now forty years since the last train ran. The fuzzy, camel hair-like buds of the empress tree already look ready to give way to bursts of fragrant periwinkle. A red-tailed hawk looks down from the catenary remains. Perched in wait for a white-footed mouse to scurry across an opening in the meadow growing atop the old railbed. A squealing woodcock flushes, making a rare appearance. A mockingbird perched on the tip of a staghorn sumac branch sings an intricately beautiful melody. The squirrels, who years ago sowed acorns of three different species of oak, wonder: “Is this the year the tallest willow oak first bears fruit?”

Will this growing season be the last for the incredible hulk? It certainly could be. Just a few years ago, seemingly on a whim, Reading International saw fit to demolish the historic, architecturally significant Spring Garden Street train station. No real warning was issued. The old parking lot that serves as home for the hulk has a lot of economic value, as real estate prices in the Viaduct neighborhood continue to soar. Reading International could fell all the trees in the asphalt garden in a day. The days for the Viaduct urban wild ecosystem may be numbered as well. The nonprofit Rail Park friends group has visionary renderings that reimagine the Viaduct as a greenway of sorts—one that looks much tamer, more manicured, more corporate than the present Eden of urban wilderness. Clearly missing from the conceptual design is any preserved existing habitat or any hint of wildness or serendipity.

In 1984, when the City of Philadelphia shut off the electricity powering the trains that rode above the Viaduct for a century, the City was unknowingly granted the most valuable natural gift it had ever received. But not many view the re-wilding of the Viaduct as a gift to be preserved largely in its current state. Certainly no person in a local leadership role does. Just as the wonders of the Reading Viaduct hide in plain view from most who live or work or walk in its shadow, most leaders, both local and global, cannot seem to realize that a simple yet immensely powerful way to confront issues like habitat loss and global warming is to warmly accept the hand of a spontaneously incoming ecosystem, and the hands of all the plants and animals that comprise it.

We begin to heal our ailing planet Earth once we reconnect ourselves and our souls with her and with her living, life-giving systems, including spontaneous. Spontaneous nature surely wishes to connect with us.

References

Bonthoux S, Voisin L, Bouché-Pillon S, Chollet S (2019) More than weeds: spontaneous vegetation in streets as a neglected element of urban biodiversity. Landsc Urban Plan 185:163–172. https://doi.org/10.1016/j.landurbplan.2019.02.009

Bozorgmehr JEH (2012) Natural selection as a paradigm of opportunism in biology. J Bioecon 14:61–75. https://doi.org/10.1007/s10818-010-9094-5

Burton PJ, Jentsch A, Walker LR (2020) The ecology of disturbance interactions. BioScience 70:854–870. https://doi.org/10.1093/biosci/biaa088

Chambers RM, Meyerson LA, Dibble KL (2012) Ecology of Phragmites australis and responses to tidal restoration. In: Roman CT, Burdick DM (eds) Tidal Marsh restoration. Island Press/ Center for Resource Economics, Washington, DC, pp 81–96

Deparis M, Legay N, Isselin-Nondedeu F, Bonthoux S (2023) How managers and city dwellers relate to spontaneous vegetation in cities: Towards an integrative approach. Urban Forest Urban Green 82:127876. https://doi.org/10.1016/j.ufug.2023.127876

Kania-Dobrowolska M, Baraniak J (2022) Dandelion (Taraxacum officinale L.) as a source of biologically active compounds supporting the therapy of co-existing diseases in metabolic syndrome. Foods 11:2858. https://doi.org/10.3390/foods11182858

Lerman SB, Contosta AR, Milam J, Bang C (2018) To mow or to mow less: Lawn mowing frequency affects bee abundance and diversity in suburban yards. Biol Conserv 221:160–174. https://doi.org/10.1016/j.biocon.2018.01.025

McNamara DE, Lazarus ED (2018) Barrier Islands as coupled human–landscape systems. In: Moore LJ, Murray AB (eds) Barrier dynamics and response to changing climate. Springer International Publishing, Cham, pp 363–383

Meyerson LA, Cronin JT, Pyšek P (2016) Phragmites australis as a model organism for studying plant invasions. Biol Invasions 18:2421–2431. https://doi.org/10.1007/s10530-016-1132-3

Milke J, Gałczyńska M, Wróbel J (2020) The importance of biological and ecological properties of phragmites Australis (Cav.) Trin. Ex Steud., in phytoremendiation of aquatic ecosystems—the review. Water 12:1770. https://doi.org/10.3390/w12061770

Newbold T, Adams GL, Albaladejo Robles G et al (2019) Climate and land-use change homogenise terrestrial biodiversity, with consequences for ecosystem functioning and human wellbeing. Emerg Top Life Sci 3:207–219. https://doi.org/10.1042/ETLS20180135

Olmos-Martínez E, Romero-Schmidt HL, Blázquez MDC et al (2022) Human communities in protected natural areas and biodiversity conservation. Diversity 14:441. https://doi.org/10.3390/d14060441

Pickett STA, White PS (1985) The ecology of natural disturbance and patch dynamics. Academic Press, San Diego, New York, Berkeley

Schäfer KVR (2011) Canopy stomatal conductance following drought, disturbance, and death in an upland Oak/Pine Forest of the New Jersey Pine Barrens, USA. Front Plant Sci 2. https://doi.org/10.3389/fpls.2011.00015

Seidl R, Thom D, Kautz M et al (2017) Forest disturbances under climate change. Nat Clim Change 7:395–402. https://doi.org/10.1038/nclimate3303

Tredici PD (2010) Spontaneous urban vegetation: reflections of change in a globalized world. Nat Cult 5:299–315. https://doi.org/10.3167/nc.2010.050305

Park in a Truck: A Community Led Green Initiative

Kimberlee Douglas and Drew Harris

Abstract Park in a Truck (PiaT) is an initiative of Thomas Jefferson University's Landscape Architecture Program and Laboratory for Urban and Social Innovation (LUSI). PiaT brings nature back into everyday city life by linking adjacent available lots to create block-by-block green networks and corridors of high-quality, outdoor spaces. It has been well documented that clean air, clean water, and the protection of limited green spaces is critical to healthy urban neighborhoods. However, in Philadelphia, underserved neighborhoods have historically suffered disproportionately when it comes to environmental, social and economic justice. PiaT is a community-led initiative to create a citizen-operated green network throughout Philadelphia established through low-cost, fast-turnaround renovations of vacant lots by a community's residents. The DIY Open source PiaT Toolkit is a roadmap communities can follow to improve their neighborhoods by creating public green space into a network of green healthy spaces. Building parks can be an expensive and complicated undertaking – the Toolkit enables a community to do all phases of the planning and design by providing its citizens with all the information necessary to design, build and maintain their new park.

Keywords Green public spaces · Community gardens · Biodiversity · Social innovation · Urban parks

K. Douglas (✉) · D. Harris
College of Architecture & The Built Environment, Thomas Jefferson University, Philadelphia, PA, USA
e-mail: kimberlee.douglas@jefferson.edu; drew.harris@jefferson.edu

G. Iannaccone, E. V. Ellis (eds.), *Towards a Nature-Positive Built Environment*, Research for Development, https://doi.org/10.1007/978-3-031-95292-0_6

Philadelphia History: Environmental, Social and Economic Background

Trees provide shade, which cools cities and towns in hot weather. They improve air quality and contribute to storm water management. They increase real estate values, boosting municipal revenues. They provide venues for outdoor recreation, socializing and relaxing (Frumkin 2013).

Many events have contributed to the socio-cultural-environmental condition of today's Philadelphia. However, two planning decisions stand out: initial land planning by William Penn in the 17th century and the redlining that occurred in the 20th century.

In 1683, the City of Philadelphia was laid out by the surveyor Thomas Holme for William Penn (Fig. 1). It is evident that the connection to Nature was important to the city's founder: marked by four green public spaces in each of its quadrants and punctuated with land for its city hall at the center, the city was laid out spanning an area bounded by the Delaware River to the east and the Schuylkill River to the west—both rivers flow to the Delaware River Watershed (Fig. 2). The Delaware River became the city's conduit to Europe facilitating large ships for international trade. Named the "hidden river" by the Dutch, the Schuylkill River became the locus for industrial production.

As the city grew and developed, the two rivers began to be used for their own specific purposes. The Delaware River could accommodate large sailing ships, called galleons. These ships were the mainstay of maritime commerce until the early 19th century but much too large with too deep of a draft to ever sail on the

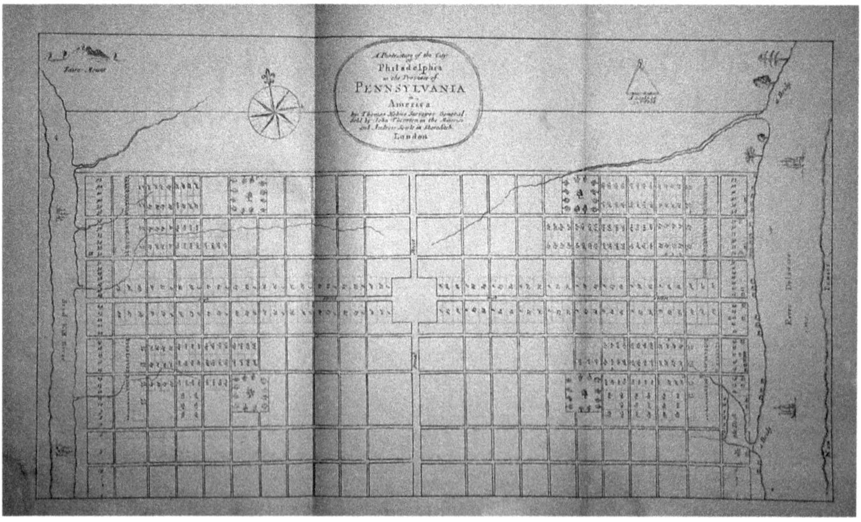

Fig 1 1683 Thomas Holme, "A Portraiture of the City of Philadelphia" (Historic Maps of Pennsylvania)

Fig. 2 1752 N. Scull and G. Heap, "A Map of Philadelphia and Parts Adjacent" (Historic Maps of Pennsylvania)

Schuylkill River. As such, the two rivers developed into locations for two very different, but compatible industries. The Delaware River became the site of commerce for Philadelphia with one pier after another located along its riverbank. These piers were wharves that housed grains, sugar, and other products necessary to support Philadelphia's economy.

On the other hand, the Schuylkill River became the site of industry and the Fairmount Waterworks located along its banks provided drinking water for the city

(Fig. 3). The Fairmount Dam was constructed in 1822 to provide higher levels of water for pumping at the waterworks; to the north of the dam was the source of the City's drinking water which still provides 42% of the city's drinking water today (PWD 2023). By 1854, downstream of the dam at Point Breeze two industries were thriving: the Atlantic Refining Company which shipped nearly one-half of the world's illuminating fluid and the Point Breeze Gas Works which was then considered the largest in the world (Weigley 1982). The delicate ecosystem of this hidden river had become severely compromised by the end of the 19th century due to the petroleum industry (Kitchel et al. 2002). The hope is for future planning to include restoring natural wetlands along its shoreline, beautifying the river, and protecting the city against the encroaching floodwaters of climate change.

In 1937, the Homeowners Loan Corporation (HOLC) created an appraisal map for Philadelphia that strikingly illustrates discriminatory lending practices for this city: the appraisal map "redlined" the city into four distinct zones from the "best" to "hazardous." HOLC refinanced home mortgages in default or at risk of foreclosure and the Federal Housing Administration (FHA) relied on those maps to provide federal insurance for those mortgages. These discriminatory federal policies in lending practices created significant barriers to mortgage lending for people of color and incentivized white populations to move out of redlined neighborhoods into approved, racially homogenous neighborhoods. Today, these "redlined" areas experience disproportionate amounts of poverty, poor health outcomes, limited

Fig. 3 General view from the water—Fairmount Waterworks, Aquarium Drive, Philadelphia, Philadelphia County, PA. Note dam to the left or north of the waterworks. (Library of Congress Prints and Photographs Division Washington, D.C. 20540 USA http://hdl.loc.gov/loc.pnp/pp.print)

educational attainment, unemployment, and violent crime as compared to other neighborhoods in the city. And, those areas can be overlaid Census-based indicators, which include public assistance usage, poverty rate, the number of female-headed households and people under 18 years old, to discover these are the same areas of the city that have also experienced the highest levels of gun violence and homicide. In fact, Philadelphia has both the highest poverty rate and the highest homicide rate per capita of any of the ten largest cities in the country (Rhynart 2019, 2020).

Philadelphia's Plans for a Sustainable Future

> *The attention of a traveller should be particularly turned, in the first place, to the various works of Nature, to mark the distinctions of the climates he may explore, and to offer such useful observations on the different productions as may occur* (Bartram 1791).

The city has taken great strides over the last decade or so to improve the environmental quality of the city. The city's Office of Sustainability has generated two guidelines to guide the development of building projects within the city. The *Greenworks* plan (Philadelphia's Office of Sustainability 2016) is the city's vision for a sustainable Philadelphia and makes recommendations for accessible food and drinking water, healthy outdoor and indoor air, clean and efficient energy, climate prepared and carbon neutral communities, quality natural resources, accessible, affordable and safe transportation, zero waste, and engaged citizens. *Green City, Clean Waters* (Philadelphia Water Department 2011) is the other guideline and is Philadelphia's plan to reduce the stormwater pollution that is currently entering the City's combined sewer system by encouraging the use of green infrastructure throughout the city. Surprisingly, there are 40,430 lots with no known use in the greater Philadelphia region that could be repurposed as green infrastructure.

The overarching hypothesis of Park in a Truck (PiaT) is that the transformation of Philadelphia's vacant lots to neighborhood parks can be used to revitalize the City's underserved neighborhoods socially, nutritionally, and economically, while simultaneously reducing the burden that toxic runoff originating on impervious surfaces in those same neighborhoods exerts on the City's combined sewer systems that ultimately overflow to the Delaware River Watershed.

Urban parks have been recognized as a new paradigm for community health and wellness. Studies have shown that access to green space promotes healthier eating habits in students and encourages exercise, both of which contribute to healthier youth and less obesity in children and adolescents (Carroll-Scott et al. 2013). Other research results indicate that urban green space may positively affect pregnancy outcomes (Donovan et al. 2011). Studies have also shown that contact with nature and natural daylight improves cognitive functioning. Natural changes in daylight balance the body's circadian rhythm, which determines sleeping and eating patterns, brain wave activity, heart rate, cognitive functioning, and hormone production—virtually all physiological and behavioral parameters (Ellis et al. 2013, 2014).

Research has also demonstrated that contact with the natural environment increases mobility in older adults (Michael et al. 2010; Rosso et al. 2013).

Over two decades of research suggest that patients who can enjoy the view of a garden outside their hospital window, and those who can go outside and experience nature first-hand, heal more quickly and experience marked psychological benefits when compared to those not exposed to a natural environment (Ulrich 1984, 1986; Rubin 1998; Sherman et al. 2005). This symbiotic relationship has also been extended beyond the hospital room to suggest a multitude of cognitive benefits associated with interacting with nature (Berman). Evidence suggests that physiological benefits of exposure to nature can be measured in individuals outside a controlled, healthcare environment. Further, just looking at nature, like walking past a park, can positively impact wellbeing by fostering a sense of nature relatedness (NR), which has been associated with better cognitive, psychological, and physical health and encourages better eating habits (Milliron et al. 2022). For example, dynamic stress responses, using ambulatory heart rate, were monitored among Philadelphia residents as they walked past vacant lots before and after a greening remediation project. These studies showed viewing a green space significantly decreased heart rate when compared to viewing a non-greened vacant lot and when compared to a control view (no vacant lot) (South et al. 2015).

Additionally, the need for urban gardens in areas without access to fresh produce is supported by research that has shown the lack of intake of fresh fruits or vegetables by eating fast-food exclusively or by ingesting off-the-shelf foods results in scurvy in "well-nourished" patients (Al-Dabagh et al. 2013). Although meeting dietary recommendations is challenging for most, especially if the only food source might be a corner market with only processed foods, when compared to non-gardeners, community gardeners report consuming more servings of fruits and vegetables and are more likely to meet national dietary recommendations to consume at least five servings of fruits and vegetables each day. In addition to changes in individual-level behaviors, community gardens may also promote community cohesion, beautification, civic engagement, and social well-being (Blair et al. 1991; Armstrong 2000; Teig et al. 2009; McCormack et al. 2010; Blanck et al. 2011; Litt et al. 2011; Zoellner et al. 2012; Nassauer and Raskin 2014; Alaimo et al. 2016).

Unfortunately, one-hundred million U.S. citizens, over one-quarter of which are children, do not have access to a park within a 10-minute walk of their house. Many of these people are from either marginalized or low-income communities (Shulaker and Stewart 2023). While 30–40% of all food produced never reaches the table, over 49 million individuals report food insecurity and an increasing number of emergency food kitchens and pantries have had to reduce the quantity of food they can make available to marginalized urban areas (Feeding America 2023).

PiaT suggests a paradigm shift in how urban green space is considered. Neighborhood gardens can provide a multi-faceted approach—not only as a safety net but as a solution. Food insecurity in Philadelphia affects 326,390 people with a food insecurity rate of 21.2%. Defined as a person being uncertain of having, or unable to acquire, at times, enough food because of insufficient money or other resources, food insecurity may be reduced among neighborhood residents who have access to an urban farm or community garden (Johnson and Smith 2006; Tsang et al. 2011; Carney et al. 2012; Rabbitt et al. 2023). The production and increased availability of locally grown fresh food promotes self-reliance in communities that have significant food insecurity issues, expands opportunities for gardening and nutrition education among community residents, and increases the availability of locally grown pesticide-free produce.

Biodiversity Is Key to the Responsive Urban Environment

> *Thousands of tired, nerve-shaken over-civilized people are beginning to find out that going to the mountains is going home; that wilderness is a necessity; and that mountain parks and reservations are useful not only as fountains of time and irrigating rivers, but as fountains of life* (Fox 1981).

Biodiversity is rarely considered with respect to the city and the urban condition. However, biodiversity contributes meaningfully to human health and wellbeing globally and its positive effects on the city can be felt when biodiversity is considered as part of the city's urban plan for green public space (World Economic Forum 2022). When we think of cities, probably the last thing we think of is the birds and the bees, but plant life supports insect life which supports animal life. A biodiverse city is a resilient city, in more ways than one.

A city that is biodiverse has lower instances of disease among its population. A biodiverse city adds health to the equation of environmental, social and economic justice. A city with green public spaces can provide a fountainhead of ecosystem services especially clean air and clean water which keeps citizens healthy and prevents cities from flooding. Wetlands can be incorporated into the green public space of city riverbanks to clean hazardous materials from the water and soil in addition to supporting plant, animal and insect life. This is a strategy underway for Philadelphia's two rivers (Kreeger and Padeletti 2011; Partnership for the Delaware Estuary 2012). Biodiversity itself supports the chain of life, starting with healthy soil with micro-organisms that can help support slightly larger creatures, diverse plant life, and so on. Overall, biodiversity leads to resilient cities because of its diversity and its ability to continue functioning due to the redundancy of nature that ensures some organisms will remain functioning even when others fail. Further, biodiversity allows ecosystems to flourish when presented with extreme hazards such as fires and floods.

Biodiversity relies on the connectivity of green spaces, but unfortunately urban development interrupts ecosystems and landscapes, which results in habitat fragmentation. As such, the recent rapid urbanization throughout the world in the last century has caused an extensive loss of biodiversity. This change in land use has significantly damaged ecosystem service functions, including microclimate, recreation, stress reduction and habitat quality for biodiversity, which may further lead to the reduction of pollinators. This is an important factor to recognize because pollinators are essential for the reproduction of many plant species, which in turn provide vital ecosystem services to various ecosystems on the earth (Zhang et al. 2022).

Butterflies and bees play a critical role in ecosystem services. Habitat connectivity is very important for the resilience of insects and pollinators. Pollinators pick up scents and try to find their way to locate the source of the scent. This may be done by following the gradient of a scent emitted from its source, with the assistance of visual images. The linkage of Philadelphia's 40,430 vacant lots into green corridors can greatly improve the connectivity of these floral scents (Fig. 4).

Because pollinators rely on natural vegetation as habitat, the reduction in the connectivity between vegetated areas in cities results in a fragmented ecosystem. However, urban habitats, such as community parks, green roofs, and walls, offer considerable potential for supporting pollinators (Senapathi et al. 2017). Further, urban rivers and their corresponding riparian zones can also become important natural wildlife corridors when considered in urban planning. Since bees and butterflies are closely correlated with flowering vegetation along the riparian zone, a city ecological corridor network could be established along urban rivers (Zhang et al. 2022). Further, restoring biodiversity facilitates other objectives such as carbon sequestration, ecosystem services and socio-economic benefits (Di Sacco et al. 2021).

While green public space and the biodiversity it engenders contribute enormously to the health and wellbeing of a city's citizens, it also helps to prevent crime. Associated with green public space are also community centers and urban gardens that cultivate and maintain these spaces. Surprisingly, farming in the city can engage young people not just with growing food, but also with the entrepreneurship that comes along with harvesting and selling food and food products—which in turn reduces teenage crime and improves eyes on the street, which results in increased feelings of safety and a reduction of gun violence. Further, green public space in the

Fig. 4 A 'network' of parks, Park in a truck toolkit

city encourages healthier eating habits, a healthier lifestyle and exercise, all of which decrease stress. Parks also increase mobility for older adults because green space encourages outdoor activity—even walking slowly through a park will provide health benefits.

There are both environmental and economic benefits when green public space is coupled with green stormwater infrastructure. When green stormwater infrastructure is integrated with green public space it also improves air quality and reduces the heat island effect, which also reduces energy consumption and the carbon footprint of the city resulting in economic benefits. Property near green public space is more valuable, which helps create more jobs because more people want to live nearby, which also makes the city more economically competitive, not to mention beautiful.

Disrupting Who Gets What: Park in a Truck (PiaT) at the Nexus of Social, Economic, Health, and Ecological Justice

> *Nihil de nobis, sine nobis (Nothing about us, without us, is for us)*—Ancient Latin text

> *Go to the people. Live with them. Learn from them. Love them. Start with what they know. Build with what they have. But with the best leaders, when the work is done, the task accomplished, the people will say "We have done this ourselves."*—Lao Tzu, *Tao Te Ching*, ca. 700 BCE

The goal of the Park in a Truck (PiaT) program is to harness the hidden asset found in so many low-income neighborhoods: vacant and unused land. In Philadelphia, there are about 40,000 parcels of open space, primed for improvement. That is 40,000 chances to build safe green places for children to play and run, and everyone else to stroll and exercise. That is 40,000 potential projects to bring communities together to design and build spaces that meet local needs. And that is 40,000 economic catalysts to raise property values, lower health care costs and make Philadelphia a magnet for further investment. No one should ever be far from a safe, high quality green space.

Park in a Truck is an initiative of Thomas Jefferson University's Landscape Architecture Program and Laboratory for Urban and Social Innovation (LUSI). PiaT is a community-operated green network established through low-cost, fast-turnaround renovations of under-utilized lots. Our mission is to improve environmental, social, and physical health in under-resourced neighborhoods, while helping residents lead revitalization efforts (Fig. 4). This open space initiative builds upon many great organizations' ongoing community development work by repurposing underutilized spaces to fill in the gaps. We need safe, high-quality green space because of the social, economic, health, and ecological benefits that public green space can provide urban residents.

PiaT provides social benefits to the neighborhood because parks, in short, get people out of their homes (Fig. 5). The urban park becomes a neighborhood gathering place by providing an outdoor space where people can get together to drink

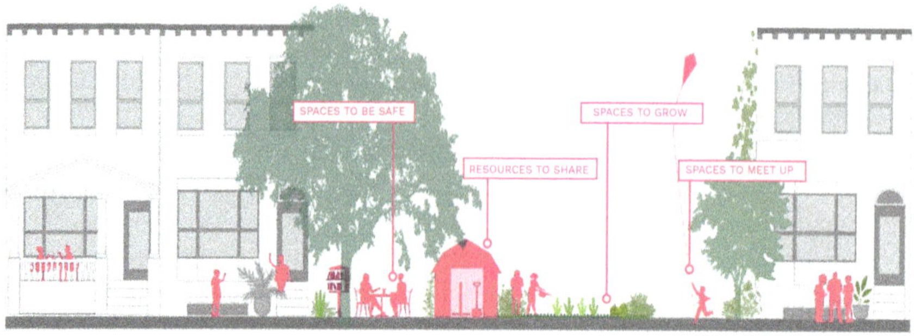

Fig. 5 Social benefits of park in a truck

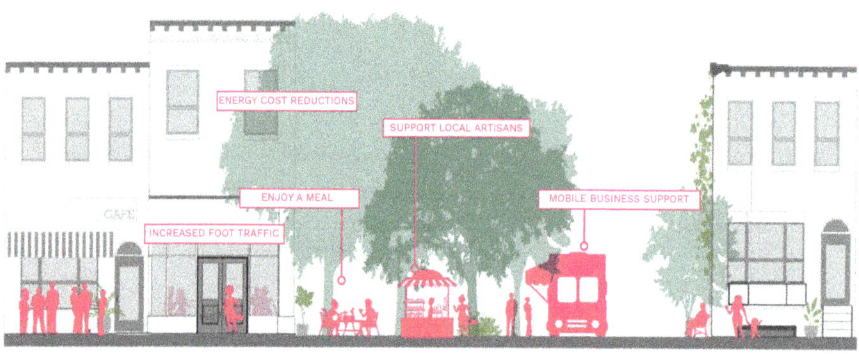

Fig. 6 Economic benefits of park in a truck

coffee, play chess, or for organized neighborhood events. The very act of building their own park helps bring a community together to share, to heal, and to celebrate.

A park can help stimulate the local economy in a number of ways (Fig. 6). Parks can become pop-up markets where residents can sell their crafts. Local businesses thrive when nearby parks attract visitors and customers. Most of all, parks enhance property values and attract families looking for a neighborhood with green space and the other amenities that come with urban parks.

Exposure to green spaces boosts mental health and psychological wellbeing beyond the benefits of physical activity alone (Fig. 7). Parks help reduce blood pressure, increase social wellbeing, and promote positive health behaviors.

Green spaces of every size can provide important ecological services (Fig. 8). Small parks can provide valuable habitat for local and migrating wildlife and serve as important areas for pollinator species. Further, green spaces are essential for absorbing water and decreasing urban flooding. All of which add up to help mitigate climate change.

PiaT is a vision for the city of Philadelphia that can only be fulfilled when the vision itself comes from the community where the urban park will be located. This means including community members every step of the way from the planning

Fig. 7 Health benefits of park in a truck

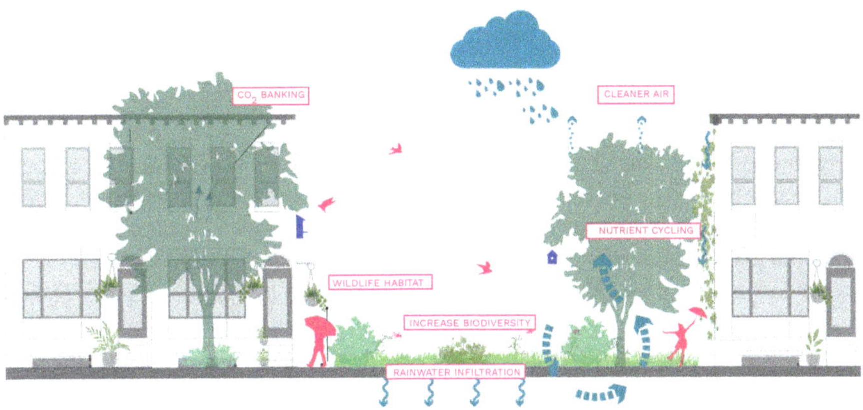

Fig. 8 Ecological benefits of park in a truck

stage through to final completion. Involving the neighborhood in the design of public parks and green space is crucial because it helps foster a sense of community, which is an important indicator of city livability and indirectly contributes to urban economic productivity. When their needs are heard and taken into consideration, the community is more likely to support the project and help to deliver successful outcomes for the future. Increasing engagement increases positive outcomes, including equitable distribution of benefits, knowledge, natural capital, economic sustainability, and community wellbeing, not to mention a more biodiverse and healthful urban environment (Nassauer and Raskin 2014; Wolch et al. 2014; Di Sacco et al. 2021; World Economic Forum 2022). In short, PiaT starts with building trust and a sense of community by engaging directly with the community itself by working closely with individual members of the community—the Toolkit facilitates it all.

The PiaT Toolkit has six chapters, each of which describes a key step to realizing a community park: (1) Acquire, (2) Organize, (3) Assess, (4) Dream, (5) Create, and (6) Sustain. The online Toolkit is hyperlinked to each of these six steps, each of

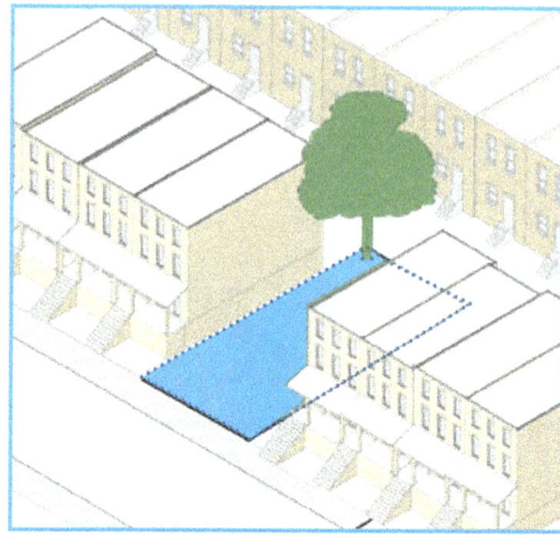

Fig. 9 Mid-block lot—vacant lot

which has a Workbook that has easy to read step by step instructions on how to realize a community park.

The first step is to find a lot. But which lot? While large urban parks can provide many social, ecological, and economic benefits to the residents of an entire city, the neighborhood park can be accessed daily and can become an essential hub of 'social life' that helps build a neighborhood's sense of community. The first step "Acquire" is for park-creators without legal access to land and helps communities identify potential park sites. There can be unused lots in a neighborhood that residents overlook because abandoned but could actually be ideal for a community park. To help residents find their park, the Toolkit identifies three urban conditions in Philadelphia that could be suitable for a neighborhood park (Figs. 9, 10, and 11). Steps are then provided to guide community residents on how to discover who owns the potential urban park site.

The mid-block lot is rectangular in shape and is found in the middle of a residential block. It may have buildings on each side and/or the end. The highest number of available 'vacant' lots in Philadelphia are this category. It may have limitations for some park uses due to its narrow width and/or the lack of sunlight.

The corner lot is rectangular in shape and is found at the end of a residential block. It may have buildings on the side and/or the end. These lots offer high visibility and often have the most amount of sunlight.

The breezeway or alley lot is identifiable by its long and very narrow shape. These locations often connect 'across' a block or connect two blocks. This type of lot was originally set aside for pedestrian connections between streets and/or utility easements.

The second step is "Organize"—a community outreach effort that brings everyone together around a common goal: building their community park. A community working together on a park project is building much more than just a park, they are building stronger community bonds, creating engaging employment opportunities for local youth, and making everyone proud. The community committee likely will have members with multiple skills that balance each other out to

Fig. 10 Corner lot—vacant lot

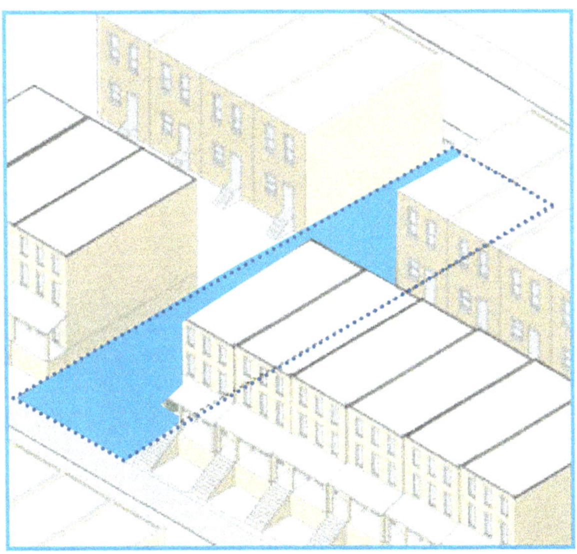

Fig. 11 Breezeway/alley—easement

create a team that can handle the many different organizational tasks in the park-creating project.

The next step is to "Assess" the potential park site and evaluate its potential to ensure that it takes advantage of its existing site conditions. The characteristics that are natural to the site will likely guide its development. For example: What is the path of the sun throughout the day? Where is the site located and is there public transportation nearby? Is there an adjacent existing blank wall suitable to consider for a mural or other decorative treatment? Are there power lines nearby?

The fourth step is when the real fun begins – what is the community's "Dream" for the park? In order to facilitate their dreaming, PiaT provides residents with four themed designs that they can mix and match to help the community define and refine their vision. Each theme supports a different way to use the park from neighborhood gatherings to peace and solitude. Each theme has individual elements that are interchangeable, like benches, tables, or shelters, that residents can include to help customize their park's design. The end result is a custom park best suited to that specific community's needs.

The Edible Theme (Fig. 12) supports community gardens and is a space for sharing food and connecting with neighbors. This vacant lot is most suitable for a corner lot that has sun most of the day with limited trees. It is a great space to support neighborhood meals, recipe swaps and even cooking classes.

The Sanctuary Theme (Fig. 13) is the most versatile of all the park themes as it can accommodate a wide range of park uses and can be located on any lot with any amount of sun and trees. This space emphasizes small spaces to relax and read a book, sit by the calming sound of water, take a class, play board games or talk with a neighbor.

The Nature Theme (Fig. 14) can provide a calming refuge of solitude in the city and a great place for smaller gatherings and outdoor workshops or classrooms. This theme supports nature-based play and can be located on any lot with any amount of sun and trees.

Fig. 12 Edible theme

Fig. 13 Sanctuary theme

Fig. 14 Nature theme

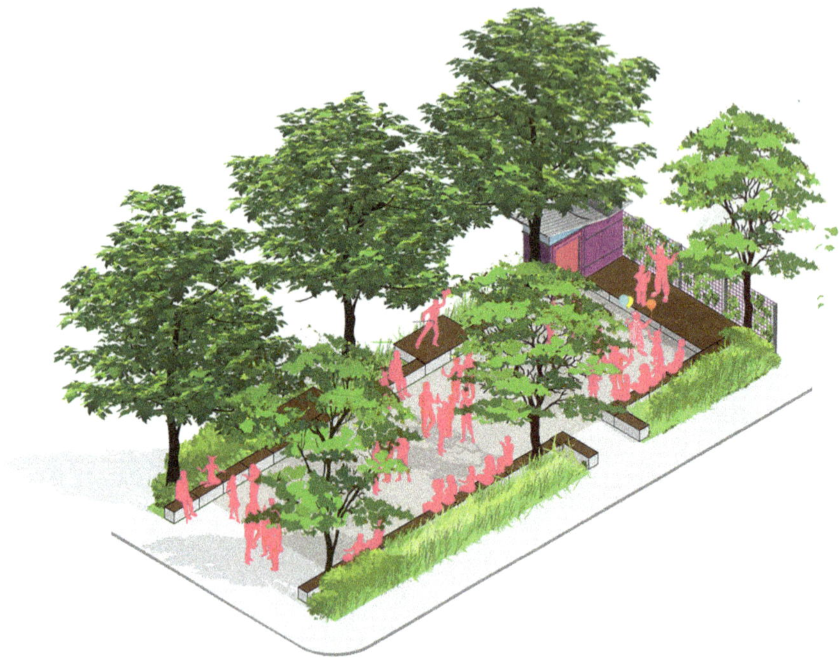

Fig. 15 Event theme

The Event Theme (Fig. 15) is made for having all kinds of neighborhood events throughout the year like music in the park, a spring art fair, a monthly flea market, or a Halloween haunted house. Distinct features include open space that is flexible to support many different events, a small platform for performances, and gallery spaces that can display artwork or school projects. This vacant lot is well suited for a corner lot and could have a few trees and any amount of sun.

Each one of the design templates is comprised of 'park pieces' (Fig. 16). The park pieces include a front, back and frame. Each of these pieces is interchangeable and goes together like pieces in a puzzle. Each community can mix and match the individual pieces to create a unique park for their specific community.

Like the frame around a family photo, the park 'frame' creates the border around the park's interior. This helps distinguish the park from adjacent properties and creates a welcoming location for park entrances, trees, and other features.

The interior of the park is divided into two parts, the 'front' and the 'back'. The front area is the location closest to the park entrance and holds the major fixed elements. Whereas the back is typically open to provide more flexibility of park uses.

The fifth step moves from concept to creation. The "Create" chapter walks community residents through the essential steps of building a park, how to schedule for park construction, and helps the community define tasks for ordering, delivery and assembly of different park elements.

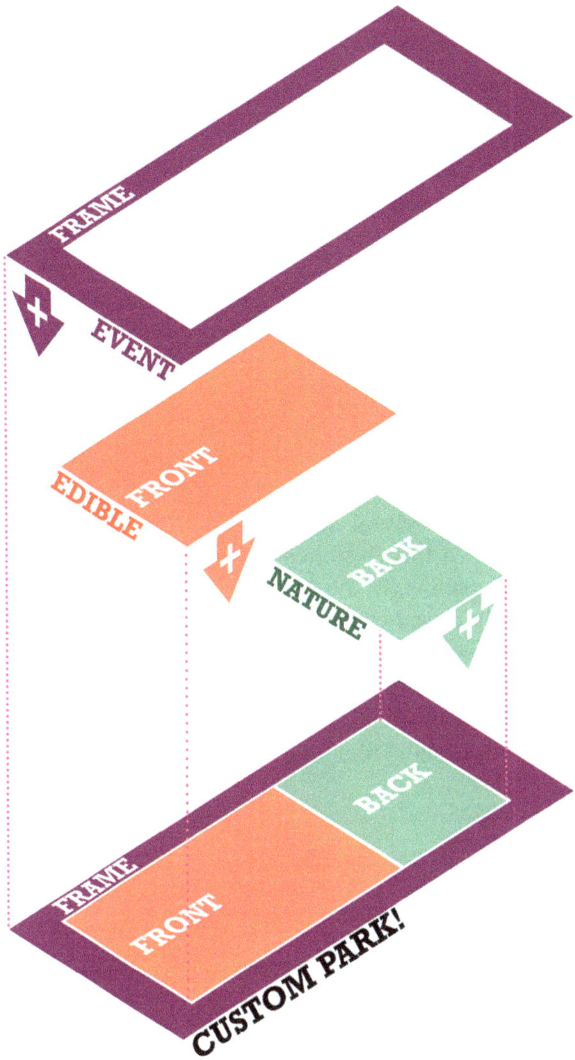

Fig. 16 Park Pieces

This is one of the most important parts of the process because it is the actual construction of the park and is a great way to connect neighbors, friends and others. This process requires dedicated volunteers who can give their time and energy to construct the park so the neighborhood can enjoy the park. But most of all, it is a way to bring the neighborhood together around a common goal that occurs over 6–8 weekends.

The final step guides residents in how to "Sustain" their park by helping them develop an annual workplan to assist them in the ongoing seasonal maintenance of their park. The guidelines list the tasks to be completed each year, month and/or season. In following the workplan for their community park, the residents themselves become stewards of the biodiverse environment they helped create for the city of Philadelphia.

An important component of the "Sustain" step–if not every step of the process–is the Park Ambassador Program. Park Ambassadors are neighborhood youth specifically trained to manage and program park activities. These young people are paid to ensure the park is kept up and becomes an integral part of the community. They perform basic tasks like clean-up and weeding, as well as research into local history, ecological conditions and cultural issues to ensure park programs truly reflect what visitors want. Park Ambassadors provide work and career opportunities to local students critical to the success of the park.

Conclusion

In this chapter, we explored several challenges confronting urban communities like Philadelphia, and detailed their historical, economic, cultural and environmental antecedents. The basic premise of the Park in a Truck program is that engaging the community in a locally driven effort to design, build and maintain neighborhood parks utilizing unused open space will increase economic opportunities, promote social cohesion, improve health outcomes and provide high-quality outdoor space for humans, plants and pollinators. The Park in a Truck Toolkit provides a time-tested process that any community group can follow to build their own neighborhood parks.

References

Alaimo K, Beavers AW, Crawford C et al (2016) Amplifying health through community gardens: a framework for advancing multicomponent, behaviorally based neighborhood interventions. Curr Envir Health Rpt 3:302–312. https://doi.org/10.1007/s40572-016-0105-0

Al-Dabagh A, Milliron B-J, Strowd L, Feldman SR (2013) A disease of the present: Scurvy in "well-nourished" patients. J Am Acad Dermatol 69:e246–e247. https://doi.org/10.1016/j.jaad.2013.04.051

Armstrong D (2000) A survey of community gardens in upstate New York: implications for health promotion and community development. Health Place 6:319–327. https://doi.org/10.1016/S1353-8292(00)00013-7

Bartram W (1791) Travels through North and South Carolina, Georgia, east and west Florida, the Cherokee country, the extensive territories of the Muscogulges or Creek Confederacy, and the country of the Chactaws. James & Johnson, Philadelphia

Blair D, Giesecke CC, Sherman S (1991) A dietary, social and economic evaluation of the Philadelphia urban gardening project. J Nutr Educ 23:161–167. https://doi.org/10.1016/S0022-3182(12)81191-5

Blanck HM, Thompson OM, Nebeling L, Yaroch AL (2011) Improving fruit and vegetable consumption: use of farm-to-consumer venues among US adults. Prev Chronic Dis 8:A49

Carney PA, Hamada JL, Rdesinski R et al (2012) Impact of a community gardening project on vegetable intake, food security and family relationships: a community-based participatory research study. J Community Health 37:874–881. https://doi.org/10.1007/s10900-011-9522-z

Carroll-Scott A, Gilstad-Hayden K, Rosenthal L et al (2013) Disentangling neighborhood contextual associations with child body mass index, diet, and physical activity: the role of built, socioeconomic, and social environments. Soc Sci Med 95:106–114. https://doi.org/10.1016/j.socscimed.2013.04.003

Di Sacco A, Hardwick KA, Blakesley D et al (2021) Ten golden rules for reforestation to optimize carbon sequestration, biodiversity recovery and livelihood benefits. Glob Change Biol 27:1328–1348. https://doi.org/10.1111/gcb.15498

Donovan GH, Michael YL, Butry DT et al (2011) Urban trees and the risk of poor birth outcomes. Health Place 17:390–393. https://doi.org/10.1016/j.healthplace.2010.11.004

Ellis EV, Gonzalez EW, Kratzer DA et al (2013) Auto-tuning daylight with LEDs: sustainable lightung for health and wellbeing. In: Proceedings of the 2013 ARCC spring research conference, pp 465–473

Ellis EV, Gonzalez EW, Kratzer DA, McEachron DL (2014) EBD using daylight-mimicking LEDs for improved health outcomes in older adults at St Francis. In: Proceedings of the 2014 ARCC/EAAE international conference on architectural research, pp 275–285

Feeding America (2023) Charitable food assistance participation in 2022. In: feedingamerica.org. Accessed 2 May 2024

Fox SR (1981) John Muir and his legacy: the American conservation movement, 1st edn. Little Brown, Boston

Frumkin H (2013) The evidence of nature and the nature of evidence. Am J Prev Med 44:196–197. https://doi.org/10.1016/j.amepre.2012.10.016

Johnson DB, Smith LT (2006) Testing the recommendations of the Washington state nutrition and physical activity plan: the Moses Lake case study. Prev Chronic Dis 3:A59

Kitchel E, Clark M, Tevlin K, et al (2002) A report on the state of the Schuylkill River Watershed. The Conservation Fund for the Schuylkill River Watershed Initiative

Kreeger D, Padeletti AT (2011) The condition of tidal freshwater wetlands in Pennsylvania. Partnership for the Delaware Estuary

Litt JS, Soobader M-J, Turbin MS et al (2011) The influence of social involvement, neighborhood aesthetics, and community garden participation on fruit and vegetable consumption. Am J Public Health 101:1466–1473. https://doi.org/10.2105/AJPH.2010.300111

McCormack LA, Laska MN, Larson NI, Story M (2010) Review of the nutritional implications of farmers' markets and community gardens: a call for evaluation and research efforts. J Am Diet Assoc 110:399–408. https://doi.org/10.1016/j.jada.2009.11.023

Michael YL, Perdue LA, Orwoll ES et al (2010) Physical activity resources and changes in walking in a cohort of older men. Am J Public Health 100:654–660. https://doi.org/10.2105/AJPH.2009.172031

Milliron B-J, Ward D, Granche J et al (2022) Nature relatedness is positively associated with dietary diversity and fruit and vegetable intake in an urban population. Am J Health Promot 36:1019–1024. https://doi.org/10.1177/08901171221086941

Nassauer JI, Raskin J (2014) Urban vacancy and land use legacies: a frontier for urban ecological research, design, and planning. Landsc Urban Plan 125:245–253. https://doi.org/10.1016/j.landurbplan.2013.10.008

Partnership for the Delaware Estuary (2012) Freshwater Mussel Recovery Program in the Delaware Estuary

Philadelphia Water Department (2011) Green city clean waters. In: water.phila.gov/green-city/. https://water.phila.gov/green-city/

Philadelphia's Office of Sustainability (2016) Greenworks. A vision for sustainable Philadelphia, City of Philadelphia, Office of Sustainability

PWD (2023) Drinking water quality report. Philadelphia Water Department, Philadelphia

Rabbitt MP, Hales LJ, Burke MP et al (2023) Household food security in the United States in 2022. Economic Research Service, U.S. Department of Agriculture, Washington, DC

Rhynart R (2019) Report on the economic impact of homicides. City of Philadelphia, Office of the City Controller, Philadelphia

Rhynart R (2020) Mapping the impact of structural racism in Philadelphia. City of Philadelphia, Office of the City Controller, Philadelphia

Rosso AL, Grubesic TH, Auchincloss AH et al (2013) Neighborhood amenities and mobility in older adults. Am J Epidemiol 178:761–769. https://doi.org/10.1093/aje/kwt032

Rubin HR (1998) Status report—an investigation to determine whether the built environment affects patients' medical outcomes. J Healthc Des 10:11–13

Senapathi D, Goddard MA, Kunin WE, Baldock KCR (2017) Landscape impacts on pollinator communities in temperate systems: evidence and knowledge gaps. Funct Ecol 31:26–37. https://doi.org/10.1111/1365-2435.12809

Sherman SA, Varni JW, Ulrich RS, Malcarne VL (2005) Post-occupancy evaluation of healing gardens in a pediatric cancer center. Landsc Urban Plan 73:167–183. https://doi.org/10.1016/j.landurbplan.2004.11.013

Shulaker B, Stewart K (2023) A policy roadmap for building equitable park systems. In: nextcity.org. Accessed 2 May 2024

South EC, Kondo MC, Cheney RA, Branas CC (2015) Neighborhood blight, stress, and health: a walking trial of urban greening and ambulatory heart rate. Am J Public Health 105:909–913. https://doi.org/10.2105/AJPH.2014.302526

Teig E, Amulya J, Bardwell L et al (2009) Collective efficacy in Denver, Colorado: strengthening neighborhoods and health through community gardens. Health Place 15:1115–1122. https://doi.org/10.1016/j.healthplace.2009.06.003

Tsang S, Holt AM, Azevedo E (2011) An assessment of the barriers to accessing food among food-insecure people in Cobourg, Ontario. Chronic Dis Inj Can 31:121–128

Ulrich RS (1984) View through a window may influence recovery from surgery. Science 224:420–421. https://doi.org/10.1126/science.6143402

Ulrich RS (1986) Human responses to vegetation and landscapes. Landsc Urban Plan 13:29–44. https://doi.org/10.1016/0169-2046(86)90005-8

Weigley RF (ed) (1982) Philadelphia: a 300-year history. Norton, New York

Wolch JR, Byrne J, Newell JP (2014) Urban green space, public health, and environmental justice: The challenge of making cities 'just green enough'. Landsc Urban Plan 125:234–244. https://doi.org/10.1016/j.landurbplan.2014.01.017

World Economic Forum (2022) BiodiverCities by 2030: transforming cities' relationship with nature. Cologny/Geneva

Zhang X, Zhang L, Wang Y et al (2022) Pollinators and urban riparian vegetation: important contributors to urban diversity conservation. Environ Sci Eur 34:78. https://doi.org/10.1186/s12302-022-00661-9

Zoellner J, Zanko A, Price B et al (2012) Exploring community gardens in a health disparate population: findings from a mixed methods pilot study. CPR 6:153–165. https://doi.org/10.1353/cpr.2012.0014

Technology, Governance and Community Engagement

The Future of Urban Greening Strategies Between Planning, Shared Governance Models and Technologies

Eugenio Morello

Abstract Urban greening strategies offer significant potential to enhance natural capital in cities, benefiting both the environment and inhabitants. However, transitioning from concrete-dominated landscapes to greener urban spaces poses multifaceted challenges, requiring a paradigm shift to avoid "green gentrification" and achieve equitable outcomes. This chapter explores three key lenses to address primary greening challenges in cities.

Firstly, spatial challenges demand innovative solutions for planning and designing greening projects within limited urban space. Brownfield recovery, densification of green spaces, including nano-urban solutions, require special attention. In addition, the increasing focus on urban biodiversity demands the implementation of greening solutions at various scales.

Secondly, governing green spaces requires robust decision-making processes involving diverse stakeholders and competing interests. Building consensus and raising awareness about the value of urban greening is crucial. Scaling up Nature-based Solutions (NbS) while avoiding superficial "greenwashing" necessitates vigilant monitoring and support from local governments. Recognizing the agency of non-human entities and including them as stakeholders and right holders in decision-making processes safeguards nature's rights and interests.

Thirdly, technology plays a critical role in supporting nature in urban areas. Integrating engineering advancements into NbS and Green and Blue Infrastructure enhances their vigor and performance. Technology-driven approaches like the "Internet of Nature" and citizen science engage communities in preserving and monitoring green spaces.

In conclusion, successfully addressing the challenges of urban greening demands comprehensive spatial planning, inclusive governance models, and deep technology integration for a transformative journey toward sustainable development, nurturing natural capital within a holistic urban health framework.

E. Morello (✉)
Department of Architecture and Urban Studies, Laboratorio di Simulazione Urbana Fausto Curti, Politecnico di Milano, Milan, Italy
e-mail: eugenio.morello@polimi.it

G. Iannaccone, E. V. Ellis (eds.), *Towards a Nature-Positive Built Environment*, Research for Development, https://doi.org/10.1007/978-3-031-95292-0_7

Keywords Urban greening · Nature-based solutions · Internet-of-nature · Co-creation methods · City governance

Introduction

Urban greening regeneration strategies hold significant potential for enhancing natural capital in cities, offering numerous benefits for both people and the environment. These strategies have been vastly boosted by global policies to react to climate change emergency and biodiversity loss, acknowledging the important role that cities play in driving transition. For instance, European policies are at the forefront for the green transition through the European Green Deal (European Commission 2019). The Deal promotes a societal transformational change and focuses on the role of nature as an ally against the impact of climate change. Restoring nature and preserving biodiversity are specifically addressed by the EU Biodiversity Strategy for 2030 (European Commission 2020). As part of this strategy, Europe committed to put forward a proposal for legally binding nature restoration targets to repair degraded ecosystems. The Nature Restoration Law, approved by the European Parliament in 2023 and finally adopted by the European Council in 2024, envisions an increasingly pressing commitment to the recognition of nature's role in our lives. Contact with nature provides evident and multiple co-benefits to people and in the urban setting, from social and psychological benefits (Chiesura 2004), to health impact (Hartig et al. 2014), to physiological reliefs under heat stress (Shashua-Bar et al. 2011), and more.

In particular, the recent promotion of the concept of Nature-based Solutions (NbS) is a tangible way for moving "from gray to green" (European Commission, Directorate-General for Research and Innovation 2015; Kabisch et al. 2016, 2022), enhancing ecosystem services in cities and developing the evidence-base for the benefits of nature (Lafortezza and Sanesi 2019). Through the European Commission's Framework Program Horizon 2020, the European Council of Research has traced a roadmap for implementing NbS in cities, opening several opportunities for research, business and governance (Davies et al. 2021; European Commission, Directorate-General for Research and Innovation et al. 2020; Maes and Jacobs 2017). In particular, NbS have been an occasion to tackle urban regeneration through shared governance and co-creation approaches, relying on the urban living lab as a method to ensure multi-stakeholder participation (Lupp et al. 2020). The experience and institutionalization of NbS is now mature, compared to 2015 (Mendes et al. 2020), when the concept was first introduced into European policies.

Despite the multi-functionalities of NbS creating potential for trade-offs in cities (Bush and Doyon 2019; Dorst et al. 2019), and the rich experience gained in NbS implementations over the last decade, transitioning "from green to gray" poses multifaceted challenges, including standards, regulation, socio-economic factors, financeability, and innovation (Staddon et al. 2018). Nor can this green transition be understood solely as an elitist process of "green gentrification", which exclusively concerns the richest cities in the northern hemisphere (Kim and Wu 2022). Instead, to avoid exclusion, an overall and radical paradigm shift in society is needed in approaching the theme of nature in urban environments, profoundly changing the way of thinking, planning, designing, making decisions and governing, within a framework of profound cultural revolution, which sees the collaboration of people, technologies and nature in all the practices. This chapter investigates three key lenses to address the primary challenges in greening cities, to provide some perspectives for the urban design and planning practices: space, governance and technology.

Spatial Challenges of Greening Cities: Finding Space for Nature

The first challenge reminds us that space matters. The same is true for urban planning work when it comes to the arrangement of non-human natural elements in space, whether biotic or abiotic components.

Today, we are asked to plan and design green solutions within the constraints of limited urban space as an infill strategy. Cities are established environments that have been strongly influenced by the automobile revolution of the 20th century. For instance, thanks to the spread and freedom of movement by automobiles, city boundaries finally expanded isotropically in all directions, and urban sprawl could flourish. In addition, streets have been transformed into functional spaces for motorized transportation, reducing the vocation of public space as a social sphere. The car literally reconfigured urban life, transforming social life and citizenship into a "civil society of automobility" (Sheller and Urry 2000). Asphalt and impermeable surfaces provided the most efficient way to alleviate traffic.

Thus, if green has not been planned from scratch in cities, it becomes very difficult to implement it at a later stage. Nevertheless, we have some opportunities ahead of us. Firstly, decreasing society's dependence on the car goes hand in hand with reclaiming public space in a new conceptualization of the city of the future. This recaptured public space is an opportunity to incorporate nature, e.g., permeable soils and vegetation, as key elements for a new experience in the social sphere.

Secondly, there are numerous opportunities for urban regeneration, such as the rehabilitation of brownfield sites and abandoned buildings, both manufacturing and housing. More recently—and even more so in the future—disuse will involve tertiary buildings, offices and commercial spaces, in a changing society that increasingly relies on e-commerce and remote working. The "on-life" society will profoundly affect people's relationship with space (Floridi 2015), and this is a huge opportunity for nature to recapture decommissioned space from people moving and working in cities. Indeed, there is a tendency to reduce human entropy in urban space in terms of movement and consumption, thus reinforcing the circular metabolism model. Perhaps we are moving toward a society that is slower in the physical world and faster in the online world.

However, reclamation of brownfields for regeneration of public and social spaces is often expensive. Reclaiming and clearing land to accommodate new green areas requires various resources: financial resources, time resources and governance models (see the next section) (Atkinson et al. 2014). Regeneration of brownfields also calls for experimentation with innovative approaches and technologies (e.g., bioremediation and phytoremediation) for green redevelopment toward healthy soil, healthy nature, and healthy cities in general (see the third section). In terms of planning and design, however, the regeneration of brownfields requires a specific spatio-temporal conceptualization, based on both the work of nature in reclaiming polluted soils and the work of humans in resettling on the regenerated land.

In addition, new design solutions are needed to densify green spaces at all scales, from the macroscale of regional corridors to nano-urban solutions at the building component level. "From gray to green" approaches can invade all scales of action in urban transformation processes. NbS have proven to be effective and can pervasively conquer urban spaces. Nature must become a parasite on the urban landscape at all possible levels. For instance, the growing focus on urban biodiversity increasingly justifies the implementation of green solutions at different scales. In fact, connecting with nature can occur in a variety of spatial contexts and modes, from direct and continuous ecological corridors (e.g., regional and urban parks), to stepping stones (e.g., Miyawaki's compact and dense urban forests), to widespread and pervasive point measures (e.g., bee hotels, bee bricks) (Miyawaki 1998). All the occasions to embed nature-supportive actions are relevant in urban planning. Hence, planning and designing taking into account urban biodiversity requires new spatial models and approaches, beyond simple urban greening models.

From Governing Nature to Governing with Nature in Cities

Governance aspects in NbS implementation pose a major challenge. Indeed, governing green spaces in urban settings requires robust decision-making processes because of the intricate web of competing stakeholders, interests and land uses. Non-human elements in cities represent "the other". Until we overcome this detachment from the ecosystem and feel part of nature (we are nature), it will be impossible to move to a green city paradigm. We must rethink society and design cities to accommodate human, plant and animal communities, and more generally non-human elements, such as air, soil and water, which are actually complex systems incorporating biotic and abiotic components. This conceptualization of the recipients of urban planning and design is deeply linked to the concept of "One Health", which exalts the nexus between humans, the environment and living systems (Cassidy et al. 2021). The concept was later transferred to the urban planning discipline (Murray et al. 2022), building on a solid foundation of nature-sensitive urban design and planning theories. In her definition of an integral city, Marilyn Hamilton argues that a city exists as a complete living system in the context of a specific natural environment, climate and ecology (Hamilton 2008). Marina Alberti envisions a new era of co-evolution between humans and nature for planetary resilience based on hybrid ecosystems (Alberti 2018).

Building consensus and raising awareness about the value of urban greening is vital to amplify its social and environmental impact. Unlike climate mitigation solutions (e.g., the generation of renewable energy from photovoltaic panels and the straightforward business model based on the payback time of the financial investment through the production of clean energy), the benefits and impacts of greening initiatives are often intangible and long-term, hard to monetize with traditional business models (Toxopeus and Polzin 2021). Hence, this makes it challenging to garner political support, since the return on the investment does not allow for immediate visibility of the cause-effect relation. One possibility is to focus on the insurance value of NbS investments as a key to both increase awareness and facilitate evaluation of interventions (Denjean et al. 2017).

The upscaling and mainstreaming of NbS must be vigilantly supported and monitored by local governments to avoid superficial greenwashing with minimal societal impact (Dushkova and Haase 2020). "Eyes on green" is the first step to enhance nature in cities in the long term. Therefore, monitoring and evaluation practices should assess and validate the design and selection of species in terms of durability, social acceptance, and impact on biodiversity over time, among other things. To be established, co-monitoring communities require awareness raising on the relevance of nature in cities, not only in terms of recognizing the multiple co-benefits generated by nature for human beings, but also in terms of approach to nature, towards a new peer alliance with non-human communities for multispecies coexistence.

More specifically, people should get involved in all the phases of urban greening, from conception and design of solutions to the implementation and maintenance of nature in place. Co-creation approaches in urban living lab settings with public-private-people partnerships demonstrated to be a powerful way to establish long term take caring of nature and consolidate trust in local government at the local scale of communities (Ferreira et al. 2020; van der Jagt et al. 2019) and achieve transformational change in society (von Wirth et al. 2019). The Horizon 2020 research programme promoted by the European Commission initiated several projects in cities to diffuse NbS through co-creation approaches (O'Sullivan et al. 2020), and the living lab model was further confirmed in the following Horizon Europe research programme. Through these pilot projects across Europe, cities could test co-creation procedures with multiple stakeholders, generate capabilities in the shared governance of NbS, sustain new competences in businesses and professions around NbS implementation, raise awareness in civil society in general. The challenge now is to transform pilot tests into new routines that can permeate all procedures and decision-making moments to intercept all urban transformations to always evaluate green alternatives to gray solutions.

Regarding how to participate in urban greening processes, stakeholder involvement is the first step to be drafted. Multi-stakeholder partnerships and clear positioning of the landowner with respect to the common interest are key aspects of place-based action. The next radical paradigm shift will be the direct involvement of non-human entities in decision-making processes. The fact that the environment should be given legal rights is now a well-established concept, and a growing number of jurisdictions around the world now recognize the rights of nature (Chapron et al. 2019). One further step forward is the recognition of agency in terms of thinking and acting of non-human entities, which has implications for their moral status (Sebo 2017). Technologies that allow us to give nature a voice are becoming increasingly popular (see next section). From a procedural point of view, governments are implementing several solutions to give agency to non-human entities. This can occur at different levels. In fact, cities are introducing "green guarantors", i.e., figures appointed with standing in for nature at decision-making tables. For instance, the Zoöp model calls for the inclusion of a representative of non-human life voices and interests within an organization's board of directors (Nieuwe Instituut 2022). It is an effective strategy to give the voiceless a chance to vote, and also to force decision makers to put themselves in the shoes of the non-human and thus make their unheard needs and rights explicit. In so doing, we are overcoming the traditional quadruple helix approach to co-creation, embracing a "quintuple helix"—or simply a multi-stakeholder—approach, whereby the non-human entities are equally represented in decision-making (Carayannis et al. 2012).

At the closer level of interaction with nature in participatory processes, co-creation methods are also attempting to engage non-human entities in interspecies collaborations (Herrmann-Pillath et al. 2023). Role-playing and immersion methods to take on the point of view of the non-human enable deep reflection with respect to the inclusion of different needs and perspectives (Hernandez-Santin et al. 2023). It means, in fact, defining design requirements to make places accessible and equipped for all. All this finally reminds us that the city is designed not only for humans, but for extended inhabitants and users, recognizing the importance of interdependencies and their ability to define the design of place.

Technology Augmenting Nature and Socio-Technical Approaches to Nature

To involve and preserve non-human entities in decision-making processes on urban transformations, in addition to shared governance and participatory tools, technology also plays an increasingly important role. Indeed, sensing supports us in giving plants a voice. Increasingly sophisticated sensors reveal not only the water content of plants, but also its health or the chemical reaction to exposure to music or noise. Making plant "consciousness" finally visible makes us more involved in advocating for their causes and rights.

Following the technological advancements, new concepts are emerging, like the "The Internet of Bio-Nano Things" (IoBNT) and "Internet of Nature" (IoN). IoBNT is a paradigm based on biological embedded computing devices, enabling applications like intra-body sensing and actuation networks, and environmental control need (Akyildiz et al. 2015). Bio-Nano Things communicate with each other and interact through exchanged information. The IoN aims to bridge the gap between greener and smarter cities by promoting self-sufficiency and resilience in urban ecosystem management through ecosystem intelligence (Galle et al. 2019). This technology-enabled sensing contributes to a comprehensive reconceptualization of nature's agency, revealing the complexity of interspecies dependencies. Hence, thanks to the digital transition, digital twins of non-human entities will facilitate a deeper interaction with human beings in the phygital world and in the multiverse.

Plant life co-evolves with human beings. A new symbiosis with human beings might promote specific vegetative species at the expense of others, as in fact has always been the case in agriculture. Plants are now "augmented" through technology "in" and "for" green (Mahmoud et al. 2024). On one hand, technology directly embedded "in" green, like sensors and Bio-Nano Things, act as a prosthesis for plants. In fact, engineering advances support and prolong the life of plants. On the other end, technology "for" green helps to remotely map and monitor the condition and performance of nature. For instance, technology engages citizens and local governments in nature monitoring and conservation through citizen science tools, digital mobile applications, location-based services, GIS platforms, and remote sensing.

Conclusion: Towards an Urban Green Horizon

In conclusion, addressing the challenges of urban greening implementation demands comprehensive spatial planning, inclusive governance models, and deep technology integration. Embracing these three critical lenses together, cities can enhance capacities in society and embark on a transformative journey toward sustainable development. Hence, a new alliance with non-human communities requires a socio-technical and a socio-ecological approach to urban transformation decisions. The deployment of new digital technologies integrated with nature requires expanded societal involvement, the development of new ontologies and protocols to recognize and enhance the agency of non-human communities. With a forward-looking perspective, the enhancement of natural capital should take place within a holistic urban health framework. Only a new understanding of societal values inspired by the integrated health of the environment, people, and organisms can enable a paradigm shift in approaches to city planning and management for and with the natural capital.

Acknowledgment Funder: Project funded under the National Recovery and Resilience Plan (NRRP), Mission 4 Component 2 Investment 1.4—Call for tender No. 3138 of 16 December 2021, rectified by Decree n.3175 of 18 December 2021 of Italian Ministry of University and Research funded by the European Union—NextGenerationEU; Award Number: Project code CN_00000033, Concession Decree No. 1034 of 17 June 2022 adopted by the Italian Ministry of University and Research, CUP D43C22001250001, Project title "National Biodiversity Future Center - NBFC".

References

Akyildiz IF, Pierobon M, Balasubramaniam S, Koucheryavy Y (2015) The internet of bio-nano things. IEEE Commun Mag 53(3):32–40. https://doi.org/10.1109/MCOM.2015.7060516

Alberti M (2018) Cities that think like planets. University of Washington Press

Atkinson G, Doick KJ, Burningham K, France C (2014) Brownfield regeneration to greenspace: delivery of project objectives for social and environmental gain. Urban Forest Urban Green 13(3):586–594. https://doi.org/10.1016/j.ufug.2013.04.002

Bush J, Doyon A (2019) Building urban resilience with nature-based solutions: how can urban planning contribute? Cities 95:102483. https://doi.org/10.1016/j.cities.2019.102483

Carayannis EG, Barth TD, Campbell DF (2012) The quintuple helix innovation model: global warming as a challenge and driver for innovation. J Innov Entrepreneur 1(2). https://doi.org/10.1186/2192-5372-1-2

Cassidy A, Bresalier M, Woods A (2021) One health in history. In: Zinsstag J, Schelling E, Crump L, Whittaker M, Tanner M, Stephen C (eds) One health: the theory and practice of integrated health approaches, 2nd ed., pp. 1–14. CABI. https://doi.org/10.1079/9781789242577.0001

Chapron G, Epstein Y, López-Bao JV (2019) A rights revolution for nature: introduction of legal rights for nature could protect natural systems from destruction. Science 363(6434):1392–1393. https://doi.org/10.1126/science.aav5601

Chiesura A (2004) The role of urban parks for the sustainable city. Landsc Urban Plan 68(1):129–138. https://doi.org/10.1016/j.landurbplan.2003.08.003

Davies C, Chen WY, Sanesi G, Lafortezza R (2021) The European Union roadmap for implementing nature-based solutions: a review. Environ Sci Policy 121:49–67. https://doi.org/10.1016/j.envsci.2021.03.018

Denjean B, Altamirano MA, Graveline N, Giordano R, van der Keur P, Moncoulon D, Weinberg J, Máñez Costa M, Kozinc Z, Mulligan M, Pengal P, Matthews J, van Cauwenbergh N, López Gunn E, Bresch DN (2017) Natural assurance scheme: a level playing field framework for green-grey infrastructure development. Environ Res 159:24–38. https://doi.org/10.1016/j.envres.2017.07.006

Dorst H, van der Jagt A, Raven R, Runhaar H (2019) Urban greening through nature-based solutions—key characteristics of an emerging concept. Sustain Cities Soc 49:101620. https://doi.org/10.1016/j.scs.2019.101620

Dushkova D, Haase D (2020) Not simply green: nature-based solutions as a concept and practical approach for sustainability studies and planning agendas in cities. Land 9(19):1–37. https://doi.org/10.3390/land9010019

European Commission (2019) The European Green Deal (COM(2019) 640). https://www.eumonitor.eu/9353000/1/j9vvik7m1c3gyxp/vl4cnhyp1ort

European Commission (2020) EU biodiversity strategy for 2030 bringing nature back into our lives (COM(2020) 380 final). https://eur-lex.europa.eu/legal-content/EN/TXT/?uri=celex%3A52020DC0380

European Commission, Directorate-General for Research and Innovation (2015) Towards an EU research and innovation policy agenda for nature-based solutions & re-naturing cities: final report of the Horizon 2020 expert group on 'Nature based solutions and re naturing cities'. Publications Office of the European Union. https://doi.org/10.2777/765301

European Commission, Directorate-General for Research and Innovation, Bulkeley H, Naumann S, Vojinovic Z, Calfapietra C, Whiteoak K, Freitas T, Vandewoestijne S, Wild T (2020) Nature-based solutions: state of the art in EU funded projects. Publications Office of the European Union. https://doi.org/10.2777/236007

Ferreira V, Barreira AP, Loures L, Antunes D, Panagopoulos T (2020) Stakeholders' engagement on nature-based solutions: a systematic literature review. Sustainability 12(2):2. https://doi.org/10.3390/su12020640

Floridi L (ed) (2015) The online manifesto. Springer International Publishing. https://doi.org/10.1007/978-3-319-04093-6

Galle NJ, Nitoslawski SA, Pilla F (2019) The internet of nature: how taking nature online can shape urban ecosystems. Anthropocene Rev 6(3):279–287. https://doi.org/10.1177/2053019619877103

Hamilton M (2008) Integral city: evolutionary intelligences for the human hive. New Society Publishers

Hartig T, Mitchell R, de Vries S, Frumkin H (2014) Nature and health. Ann Rev Public Health 35:207–228. https://doi.org/10.1146/annurev-publhealth-032013-182443

Hernandez-Santin C, Amati M, Bekessy S, Desha C (2023) Integrating biodiversity as a non-human stakeholder within urban development. Landsc Urban Plan 232:104678. https://doi.org/10.1016/j.landurbplan.2022.104678

Herrmann-Pillath C, Sarkki S, Maran T, Soini K, Hiedanpää J (2023) Nature-based solutions as more-than-human art: co-evolutionary and co-creative design approaches. Nat Based Solut 4:100081. https://doi.org/10.1016/j.nbsj.2023.100081

Kabisch N, Frantzeskaki N, Pauleit S, Naumann S, Davis M, Artmann M, Haase D, Knapp S, Korn H, Stadler J, Zaunberger K, Bonn A (2016) Nature-based solutions to climate change mitigation and adaptation in urban areas: perspectives on indicators, knowledge gaps, barriers, and opportunities for action. Ecol Soc 21(2) https://www.jstor.org/stable/26270403

Kabisch N, Frantzeskaki N, Hansen R (2022) Principles for urban nature-based solutions. Ambio 51(6):1388–1401. https://doi.org/10.1007/s13280-021-01685-w

Kim SK, Wu L (2022) Do the characteristics of new green space contribute to gentrification? Urban Studies 59(2):360–380. https://doi.org/10.1177/0042098021989951

Lafortezza R, Sanesi G (2019) Nature-based solutions: settling the issue of sustainable urbanization. Environ Res 172:394–398. https://doi.org/10.1016/j.envres.2018.12.063

Lupp G, Huang JJ, Oen A, Pauleit S (2020) Living labs—a concept for co-designing nature-based solutions. Sustainability 13(1):188

Maes J, Jacobs S (2017) Nature-based solutions for Europe's sustainable development. Conserv Lett 10(1):121–124. https://doi.org/10.1111/conl.12216

Mahmoud I, Morello E, Bisello A, Kolokotsa D (2024) Augmented nature-based solutions: a possible taxonomy of technologies "in" and "for" urban greening strategies. In: Bisello A, Vettorato D, Bottero M, Kolokotsa D (eds) Smart and sustainable planning for cities and regions. Springer, pp 125–139. https://doi.org/10.1007/978-3-031-39206-1_10

Mendes R, Fidélis T, Roebeling P, Teles F (2020) The institutionalization of nature-based solutions—a discourse analysis of emergent literature. Resources 9(1):1. https://doi.org/10.3390/resources9010006

Miyawaki A (1998) Restoration of urban green environments based on the theories of vegetation ecology1. Ecol Eng 11(1):157–165. https://doi.org/10.1016/S0925-8574(98)00033-0

Murray MH, Buckley J, Byers KA, Fake K, Lehrer E, Magle SB, Stone C, Tuten H, Schell C (2022) One health for all: advancing human and ecosystem health in cities by integrating an environmental justice lens. Ann Rev Ecol Evol Syst 53(1):403–426

Nieuwe Instituut (2022) Zoöp model. Nieuwe Instituut. https://nieuweinstituut.nl/en/projects/zoop/zoop-model. Accessed 14 February, 2024

O'Sullivan F, Mell I, Clement S (2020) Novel solutions or rebranded approaches: evaluating the use of nature-based solutions (NBS) in Europe. Front Sustain Cities 2. https://doi.org/10.3389/frsc.2020.572527

Sebo J (2017) Agency and moral status. J Moral Philos 14(1):1–22. https://doi.org/10.1163/17455243-46810046

Shashua-Bar L, Pearlmutter D, Erell E (2011) The influence of trees and grass on outdoor thermal comfort in a hot-arid environment. Int J Climatol 31(10):1498–1506. https://doi.org/10.1002/joc.2177

Sheller M, Urry J (2000) The City and the Car. Int J Urban Regional Res 24(4):737–757. https://doi.org/10.1111/1468-2427.00276

Staddon C, Ward S, De Vito L, Zuniga-Teran A, Gerlak AK, Schoeman Y, Hart A, Booth G (2018) Contributions of green infrastructure to enhancing urban resilience. Environ Syst Decis 38(3):330–338. https://doi.org/10.1007/s10669-018-9702-9

Toxopeus H, Polzin F (2021) Reviewing financing barriers and strategies for urban nature-based solutions. J Environ Manage 289:112371. https://doi.org/10.1016/j.jenvman.2021.112371
van der Jagt APN, Smith M, Ambrose-Oji B, Konijnendijk CC, Giannico V, Haase D, Lafortezza R, Nastran M, Pintar M, Železnikar Š, Cvejić R (2019) Co-creating urban green infrastructure connecting people and nature: a guiding framework and approach. J Environ Manage 233:757–767. https://doi.org/10.1016/j.jenvman.2018.09.083
von Wirth T, Fuenfschilling L, Frantzeskaki N, Coenen L (2019) Impacts of urban living labs on sustainability transitions: mechanisms and strategies for systemic change through experimentation. Eur Plan Studies 27(2):229–257. https://doi.org/10.1080/09654313.2018.1504895

Climate Resilient Coastal Protection: Guidelines for Development of Multifunctional Infrastructure

Fatemeh Nasrollahi and Franco Montalto

Abstract Climate Resilient Development (CRD) refers to decisions, policies, and strategies that build urban resilience by simultaneously reducing exposure and vulnerability to climate hazards (adaptation), cutting back greenhouse gas (GHG) emissions (mitigation), and advancing related Sustainable Development Goals (SDGs). Crucial in the effort to promote a nature-positive built environment, CRD is facilitated through cooperation and by governments at all levels working with communities, civil society, educational bodies, scientific and other institutions, media, investors, businesses, and marginalized groups to identify synergistic strategies. Implementation of CRD is often challenging due to conflicting needs and objectives. However, there is now a growing body of scientific evidence suggesting the potential for synergies between efforts to advance adaptation, mitigation, and sustainability goals in multifunctional resilience projects, especially in cities. This chapter explores CRD in the context of flood risk management (FRM). After developing a conceptual framework within which to consider the ability of FRM to promote CRD, we critically review the leading FRM strategies under consideration in Eastwick, an environmental justice community located in the southwest corner of Philadelphia (USA). The framework and case study are used in a broader discussion of how FRM can advance CRD in nature-positive cities.

Keywords Climate resilient development · Sustainable development goals · Flood risk management · Adaptation

F. Nasrollahi · F. Montalto (✉)
The Sustainable Water Resource Engineering Laboratory, Drexel University, Philadelphia, Pennsylvania, USA
e-mail: fn72@drexel.edu; fam26@drexel.edu

G. Iannaccone, E. V. Ellis (eds.), *Towards a Nature-Positive Built Environment*, Research for Development, https://doi.org/10.1007/978-3-031-95292-0_8

Introduction to Climate Resilient Development

Climate Resilient Development (CRD) refers to decisions, policies, and strategies that build urban resilience by *simultaneously* reducing exposure and vulnerability to climate hazards (adaptation), cutting back greenhouse gas (GHG) emissions (mitigation), and advancing other related Sustainable Development Goals (SDGs). CRD is facilitated through cooperation and by governments at all levels working with communities, civil society, educational bodies, scientific and other institutions, media, investors, businesses, and marginalized groups (Intergovernmental Panel On Climate Change (IPCC) 2023) to identify synergistic strategies to advance nature-positive outcomes in the built environment.

Implementation of CRD is challenging due to a wide range of global trends. Over the last few years, health, climate, biodiversity, geopolitical, and military crises have halted progress towards the SDGs (Sachs et al. 2022), and climate adaptation plans currently fall short of what is needed to respond to even present-day climate risks (Intergovernmental Panel On Climate Change (IPCC) 2023, p. 202). In the future, if global warming exceeds 2 °C, a window that is rapidly closing, CRD will become impossible in some regions of the world (Intergovernmental Panel On Climate Change (IPCC) 2023). Average annual GHG emissions during 2010–2019 were higher than in any previous decade, and under all future emissions scenarios, choices and opportunities for sustainable development are now limited (Birkmann et al. 2022).

Despite these challenges, there is a growing body of scientific evidence suggesting the potential for synergies between efforts to advance adaptation, mitigation, and sustainability goals in multifunctional resilience projects (Grafakos et al. 2019), including those focused on managing flood risks. For example, nature-based flood mitigation solutions, like coastal dunes and wetlands, can reduce the impacts of climate hazards (e.g., waves and storm surges) while also sequestering carbon; net-zero buildings can reduce vulnerability to infrastructure system failures (e.g., power lines downed during a wind storm), while also cutting energy demand; dense, mixed-use developments can reduce commuter trips (and associated air pollution), while conserving natural open spaces and the ecosystem services they provide. Applied to a particular place, CRD projects integrate a variety of mitigation actions (involving energy, building, transportation, and water resources) and adaptation actions (involving infrastructure, natural resources, health, and consumption among others) in ways that produce synergistic social, ecological, and economic outcomes (Grafakos et al. 2019)

Research on CRD is relatively modest but has been growing, particularly after its conceptual introduction in scholarly literature and development program documents over a decade ago (IPCC 2014). Earlier IPCC Assessment Reports explored the links between climate change, adaptation, and sustainable development, with CRD becoming more prevalent in recent years. The concept has evolved to not only address how climate risk undermines CRD but also how current development patterns contribute to climate risk. A key interest is in characterizing systemic transitions and societal transformations needed to advance CRD, with a focus on how different actors engage to pursue policy and practice consistent with its goals (Martin et al. 2022).

Developed urban areas are the optimal scale at which to innovate, implement, and assess CRD strategies, and specifically within the context of flood risk management (FRM). Climate change has caused observable negative impacts on human health, livelihoods, and key infrastructure in cities and due to their proliferation, urban infrastructure systems can no longer be viewed as separate from the natural systems they modify (Chester et al. 2021). Urban infrastructure decisions carry implications for regionally proximate environments, but also remote locations linked to them by flows of resources, services, and people (Donaghy 2012). Davis and Caldeira (2010), for example, found that 23% of global CO_2 emissions were traded internationally. Because they house most of the world's population and function as key nodes of resource consumption and innovation, cities are also rapidly becoming laboratories for resilience, both in theory and in practice (Meerow et al. 2016). Their position on the front lines of climate change provides them with the opportune level of governance to incubate, implement, and evaluate integrated mitigation and adaptation solutions (Rosenzweig et al. 2024).

Within cities, management of flood risks has become a key global challenge, exacerbated by climate change, urbanization, and aging infrastructure (O'Donnell and Thorne 2020). In the US, disadvantaged communities and racial and ethnic minorities are disproportionately affected by flooding (Committee on Urban Flooding in the United States et al. 2019), and reducing flood risks is difficult due to the US regulatory context, in which decisions regarding the management of water quality and water quantity are subject to different sets of regulations and actions at the federal, state, local, and individual scale.

FRM includes a variety of structural and non-structural approaches (Rosenzweig and Solecki 2018) to reducing flood risks, each with its own focus, strategies, objectives, and relevance to CRD. A high-level distinction can be made between the Resistance and Resilience approaches (Jongman 2018). The Resistance approach typically involves creating barriers such as levees, floodwalls, or other physical infrastructures to prevent water from entering certain areas (Karamouz and Mohammadi 2020). The emphasis is on protecting the status quo. The Resilience approach, by contrast, encompasses a suite of incremental adaptation and long-term transformation strategies designed specifically to respond to changing hazard profiles. Within the Resilience approach, Adaptation measures include strategic modifications to the built environment to prepare it for a specific future flood hazard condition. Future risks are understood to be different from those experienced in the past and the adaptation approach emphasizes incremental adjustments to the existing built environment to meet those needs. These could include strategies like flood-proofing buildings, creating spaces that can withstand flooding without significant damage, and ensuring that critical infrastructure can continue to operate during and after a flood event (Webber et al. 2021). Going much further in planning for long term resilience, the Transformation approach is motivated by the belief that the future is dynamic and uncertain, and it emphasizes new ways of structuring the built environment in this evolving context. The Transformation approach could involve completely rethinking and redesigning communities and systems with the anticipation of future flood risks in mind, potentially including relocation or significant changes in land use (Meng et al. 2020).

Assessing which flood risk management approach is more consistent with the goals of CRD is only possible through an evaluation of site-specific conditions that ideally take place before a specific FRM approach is selected. For example, consider a project in which an earthen berm embankment is used to reduce fluvial flood risk. The berm is constructed by grading back, and thus restoring, the river's floodplain. It diverts riverine floodwaters away from a vulnerable riparian community and into a restored wetland. The CRD score of this approach to FRM is likely to be high because it uses local, and low-carbon materials, to protect a vulnerable community, while advanced ecological restoration goals, and generally making the community more sustainable. However, if instead of an earthen berm, an alternative design provided the same degree of protection of the riparian community but utilized a concrete flood wall and armored creek bank with the unintentional effect of induced flooding in a downstream community, the CRD score of the same resistance approach would be much lower. The project's positive adaptation value might be offset by the negatively valued use of carbon-intensive concrete, and its displacement of flood hazards to new communities. Through the lens of CRD, the first strategy results in the greatest number of impacts and is thus preferable. The second approach is maladaptive. This paper introduces a new framework that can be used as a guide to evaluate the ability of different FRM strategies to locally advance CRD and thus promote a more nature-positive built environment. After introducing the framework, we demonstrate its applicability by comparing three flood risk management strategies currently under consideration in Eastwick, a flood-prone environmental justice community in Philadelphia, Pennsylvania (USA). Areas for additional research into specific ways that FRM can advance CRD are identified at the end of the chapter.

Methodology. Framework Development

Climate Action is one of the 17 SDGs (Fig. 1) that collectively make up the core of the 2030 Agenda for Sustainable Development, adopted by all United Nations Member States in 2015 . Floods are specifically mentioned in SDG 11 (sustainable cities and communities) and SDG 13 (climate action) (Nurlambang 2022), but the occurrence of floods is undermining progress on many of the SDGs (Vaughan and Corps 2019).

The explicit goal of CRD is to simultaneously advance climate adaptation, mitigation, and sustainability. Because climate adaptation and climate mitigation are included in SDG13, the complete list of SDGs represents a blueprint for CRD. However, all climate adaptation efforts result in trade-offs with and among the SDGs (Calvin et al. 2023), and FRM is no exception. As in the example

Fig. 1 The UN Sustainable Development Goals (SDGs). Source: UN (2024)

provided above, FRM strategies that advance one SDG may create negative consequences for others. FRM projects that significantly increase the risk of negative outcomes, increase greenhouse gas emissions, increase vulnerability, lead to inequitable outcomes, or diminish welfare are referred to as maladaptive (Intergovernmental Panel On Climate Change (Ipcc) 2023).

Table 1 provides a list of questions that can be used as a guide to explore linkages between alternative FRM strategies and CRD. FRM strategies with greater positive impacts on the SDGs are considered more exemplary of the goals of CRD. FRM strategies that obstruct progress on a significant number of SDGs may be considered maladaptive. Because all SDGs may not be relevant in the context of a specific project, the full list of SDGs has been aggregated into five indices that can be used to evaluate the ability of a particular FRM strategy to advance CRD.

- **Index 1** represents the social dimensions of sustainability and includes SDG2, SDG3, SDG4, SDG5, SDG10, SDG16, and SDG17.
- **Index 2** represents the ecological dimensions of sustainability and includes SDG6, SDG7, SDG14, and SDG15
- **Index 3** represents the economic dimensions of sustainability and includes SDG1, SDG8, and SDG9.
- **Index 4** represents the impact of the project on the emission of greenhouse gases (SDG 13a: mitigation)
- **Index 5** represents the extent to which the project helps to prepare the community for climate change (SDG13b: adaptation)

Table 1 FRM-CRD framework including questions that can be used to evaluate the potential of different FRMs to promote CRD. Relationships shown in the table are informed in part by (Vaughan and Corps 2019) and Committee on Urban Flooding in the United States et al. (2019)

SDG	Does the FRM strategy…
1. No poverty (Econ)	…increase the community's ability to access resources needed to rebuild and recover after a flood? Does it enable local residents to get to work during a flood?
2. Zero hunger (Soc)	… divert resources away from other community needs such as associated with food provision? Does it negatively impact local residents' ability to grow or obtain food?
3. Good health and well-being (Soc)	… reduce the direct (e.g. drowning, heart attacks, injuries, and hypothermia) and indirect (e.g. waterborne infectious disease, mental health problems, respiratory problems, allergies) impacts of flooding?
4. Quality education (Soc)	… protect school buildings from damage? Does it ensure student access to schools during a flood? Does it promote awareness of flood risks in the subject community?
5. Gender equality (Soc)	…ensure that women are not more exposed to flood risks than men? Does it reduce the risk of livelihood loss and post-flood gender-based violence?
6. Clean water and sanitation (Env)	… reduce the possibility that floods contaminate drinking water sources, pollute surface waters, and/or damage infrastructure assets?
7. Affordable and clean energy (Env)	… require significant amounts of energy for pumping? Does it minimize disruption to the energy supply?
8. Decent work and economic growth (Econ)	… reduce economic losses and advance economic development in the subject community? Does it minimize disruption or elimination of jobs?
9. Industry, innovation, and infrastructure (Econ)	…reduce the potential for flood damage to vital infrastructure systems? Does it allow communities to continue to develop along trajectories they define for themselves?
10. Reduced inequality (Soc)	… reduce flood risks for the most vulnerable? Does it avoid geographically displacing flood risks from one location to another? Does it enable the most vulnerable to recover?
11. Sustainable Cities and Communities (Env)	… make the subject community more safe, resilient, and sustainable? Does it comport multiple co-benefits? Does it help the community to build back better?
12. Responsible consumption and production (Env)	… utilize natural and built resources efficiently?
13a. Climate action: mitigation (Mit)	…require excessive amounts of embodied carbon and energy during its full life cycle? Does it require pumping?
13b. Climate action: adaptation (Adap)	…prepare the subject community for future climate changed conditions? Does conceptualization of the project consider a range of future conditions?
14. Life below water (Env)	… enhance, create, or restore aquatic ecological systems? Does it elevate ecosystem services provided by aquatic ecosystems?
15. Life on land (Env)	… enhance, create, or restore terrestrial ecological systems? Does it elevate ecosystem services provided by land-based ecosystems?

(continued)

Table 1 (continued)

SDG	Does the FRM strategy…
16. Peace, justice, and strong institutions (Soc)	… promote collaboration that is vertical (e.g. across federal, state, and local governance systems) and horizontal (between government, and local communities, private sector, and others). Does it help to build understanding and consensus between different flooding stakeholders and decision-makers?
17. Partnerships for the goals (Soc)	… promote collaboration and coordination regarding SDGs? Is it likely to produce multiple co-benefits?

Evaluation Methodologies

The techniques used to evaluate the impact that a particular FRM strategy will have on a particular SDG vary based on the project and the resources available. The techniques can be either qualitative or quantitative and may involve course or fine-grained levels of analysis. In most cases, their evaluation requires deep local knowledge, requiring that authentic stakeholder engagement and a collaborative approach that considers the diverse perspectives of community members and other governmental and non-governmental stakeholders. Even if the answers to these questions cannot be substantiated with quantitative analyses, consideration of all of them is a helpful way of embedding CRD considerations into the decision-making process. A brief overview of some of the methods that can be applied in evaluating each of the five indices is provided below:

Index 1 focuses on the social factors and includes SDGs that can be quantitatively evaluated using asset and vulnerability mapping, or qualitatively evaluated through semi-structured interviews, surveys, or other community engagement processes. Evaluation of this metric would also assess how the FRM impacts distributive, contextual, and procedural measures of equity (McDermott et al. 2013), and the extent to which it fosters collaboration and coordination among stakeholders, who were also engaged in its conception co-production process.

Index 2 focuses on the environmental impacts of the FRM strategy. Evaluation of Index 2 could simply involve a qualitative comparison of the positive and negative impacts of the project on natural resources, and various forms of biota. It could also include a more in-depth valuation of the monetary and non-monetary impacts of the project on supporting, regulating, provisioning, and supporting ecosystem services. Computer modeling of the physical system may also assist in evaluating Index 2. Hydrologic and hydraulic (H&H) models can be used to simulate flowrates, sediment transport, and surface inundation patterns, all of which can be particularly useful in evaluating environmental impacts.

Index 3 focuses on the economic dimensions of sustainability and could include attempts to compare the economic benefits and costs associated with the project. Where appropriate data sets are available, this analysis could include more formal life cycle and benefit cost computations, utilizing the outputs of H&H models which can be input into models like HAZUS to estimate damages to buildings and their contents, number of displaced individuals, and the volume of debris generated from particular events under different FRM regimes.

Index 4 focuses specifically on the impact of the FRM project on greenhouse gas emissions. At the most basic level, this evaluation could involve creating a bill of materials associated with each FRM alternative and then creating an eco-audit to quantify the embodied energy and GHG footprint associated with the construction materials (Wegst et al. 2010). A more detailed analysis would involve a full-blown consequential Life Cycle Assessment (Ekvall and Weidema 2004) that tracks energy and GHG associated with the project's full life cycle. Either a cradle-to-gate analysis, which includes the extraction of the requisite materials, their transportation to the construction sites, and the construction itself, or a cradle-to-grave analysis, which also includes energy and emissions associated with operation and decommissioning of the project are possible. We note that published research on the use of LCA to evaluate FRM strategies is currently limited.

Index 5 focuses specifically on whether the FRM project helps the subject community to adapt to climate change. Quantitatively, Index 5 can be assessed using dynamic H&H models forced by future climate time series (e.g. multiplicative change factors applied to historical hyetographs, and additive change factors applied to historical tidal boundary conditions). Qualitative evaluations would simply investigate whether climate change was factored in some meaningful way into the project conceptualization and design, and whether the FRM strategy results in the need to displace the resident population.

Once Indices 1–5 have been evaluated, the results can be visualized using spider diagrams (Fig. 2). The FRM that results in the largest web would be the option that best exemplifies the goals of CRD.

Case Study: Eastwick Community in Philadelphia

To demonstrate the framework presented in the previous section, we focus on FRM strategies currently under consideration in Eastwick, a flood prone community of southwest Philadelphia, Pennsylvania.

Eastwick, as depicted in Fig. 3, is a mostly Black environmental justice community located near the confluence of the Darby and Cobbs creeks, about 5 km from their confluence with the Delaware River in southeast Pennsylvania (USA). The community has a fraught history that includes redlining, a failed urban renewal project, illegal dumping, and frequent flooding due to its low-lying position within the FEMA 100 and 500-year floodplain as shown in Fig. 3. Eastwick lies at the downstream end of the 205 km^2 watershed of the Darby and Cobbs Creeks, and adjacent to the 1.2 km^2 John Heinz National Wildlife Refuge. Residents have had to navigate recovery from flooding and other environmental hazards without sufficient federal assistance, highlighting disparities in disaster recovery aid. It is difficult to separate the need for FRM from Eastwick's other immediate and long-term environmental justice challenges. The complex and interconnected implications of flooding in Eastwick make it an ideal case study for CRD.

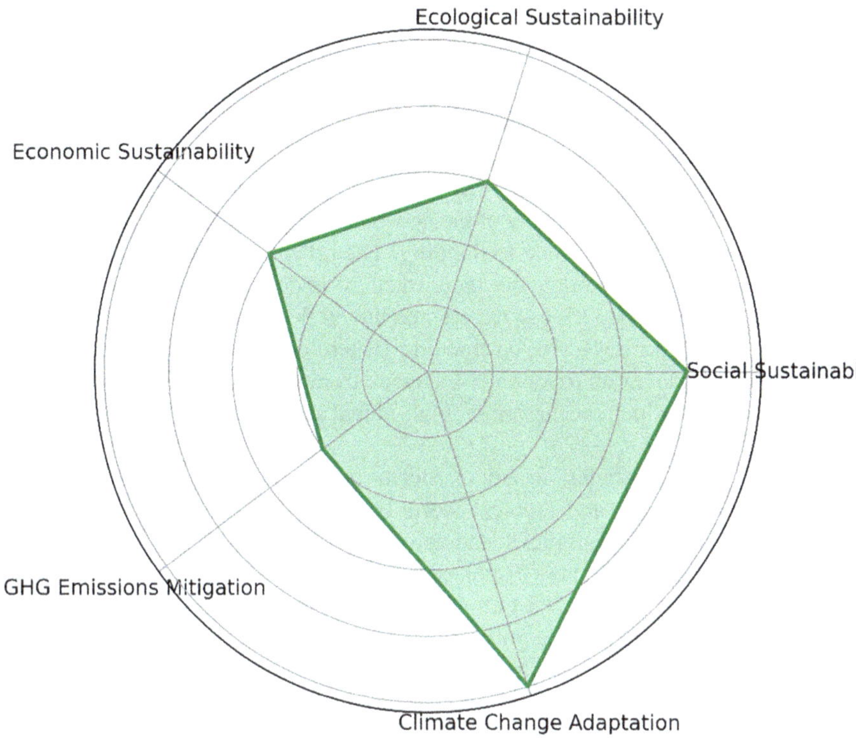

Fig. 2 Sample spider diagram used to evaluate the FRM option that brings the greatest CRD

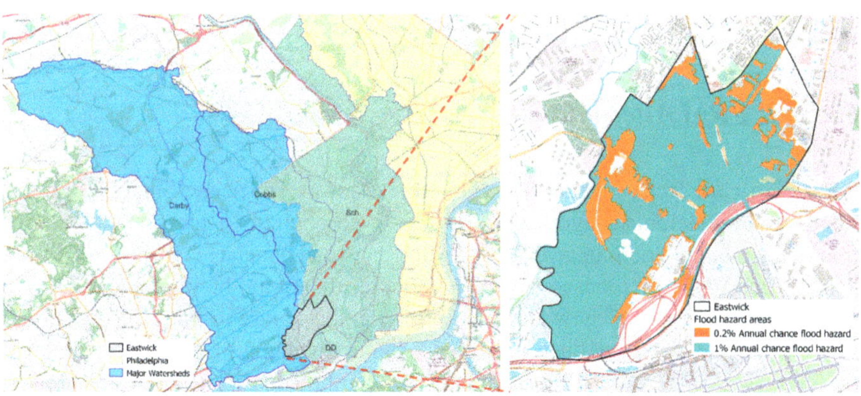

Fig. 3 Study area, Eastwick

Eastwick's Flooding Context

Flooding in Eastwick is intertwined with its history of development, and its geographical position. Originally marshland that flooded benignly, the land was eventually dyked for agriculture, before being extensively filled for low density residential development. When the community was redeveloped in the mid-20th century as part of an urban renewal project, many of the new buildings were built on unstable soils, which became even more prone to flooding. The urban renewal project was never completed, leaving three relatively large tracts of undeveloped in the heart of the community. Today, Eastwick can experience fluvial flooding (when the Darby and/or Cobbs Creek overflow), coastal flooding (when high tides and/or storm surges travelling up the Delaware River enter Eastwick from its southern border), and compound flooding (which occur when high tides and precipitation occur at the same time).

In the last few decades, infamous storms such as Hurricane Floyd (1999), Hurricane Irene (2011), and Tropical Storm Isaias (2020) left permanent marks on the community, proding significant damage to homes and infrastructure, disruptions to the local economy and many displaced residents. The most recent extreme flood, TS Isaias, deposited nearly 203 mm of rain in the upper suburban portions of the Darby Cobbs watershed, while only 63.5 mm fell in Eastwick where flooding was worst (Fig. 4a). Water levels in the Darby and Cobbs creeks were elevated through two high tide cycles (Fig. 4b), inundating large parts of Eastwick under about a meter of water. Residents were evacuated and when the flood waters receded, several blocks were rendered uninhabitable. Such dynamics underscore the importance of upstream-downstream collaboration in addressing local flooding problems. Climate change is likely to exacerbate these conditions in the years to come.

Current FRM Alternatives

After Hurricane Irene's (fluvial) flood and just before Hurricane Sandy's (coastal) flood, the Philadelphia Mayor called for the development of a comprehensive solution to Eastwick's flooding problems. To date, no such plan has materialized. Public deliberation typically focuses on FRM strategies that could be implemented in the downstream portions of the watershed (e.g. in and around Eastwick). Downstream FRM strategies that have been proposed include diversion of more stormwater to the local wastewater treatment plants, increasing the capacity of the local groundwater pump stations, and the construction of a levee along Cobbs Creek, a concept that the Army Corps of Engineers has been exploring for nearly a decade (Gates 2012). The City of Philadelphia has also been considering buyouts of properties that flood regularly as part of a long-term strategy for reducing flood impacts throughout the city (Winberg 2020).

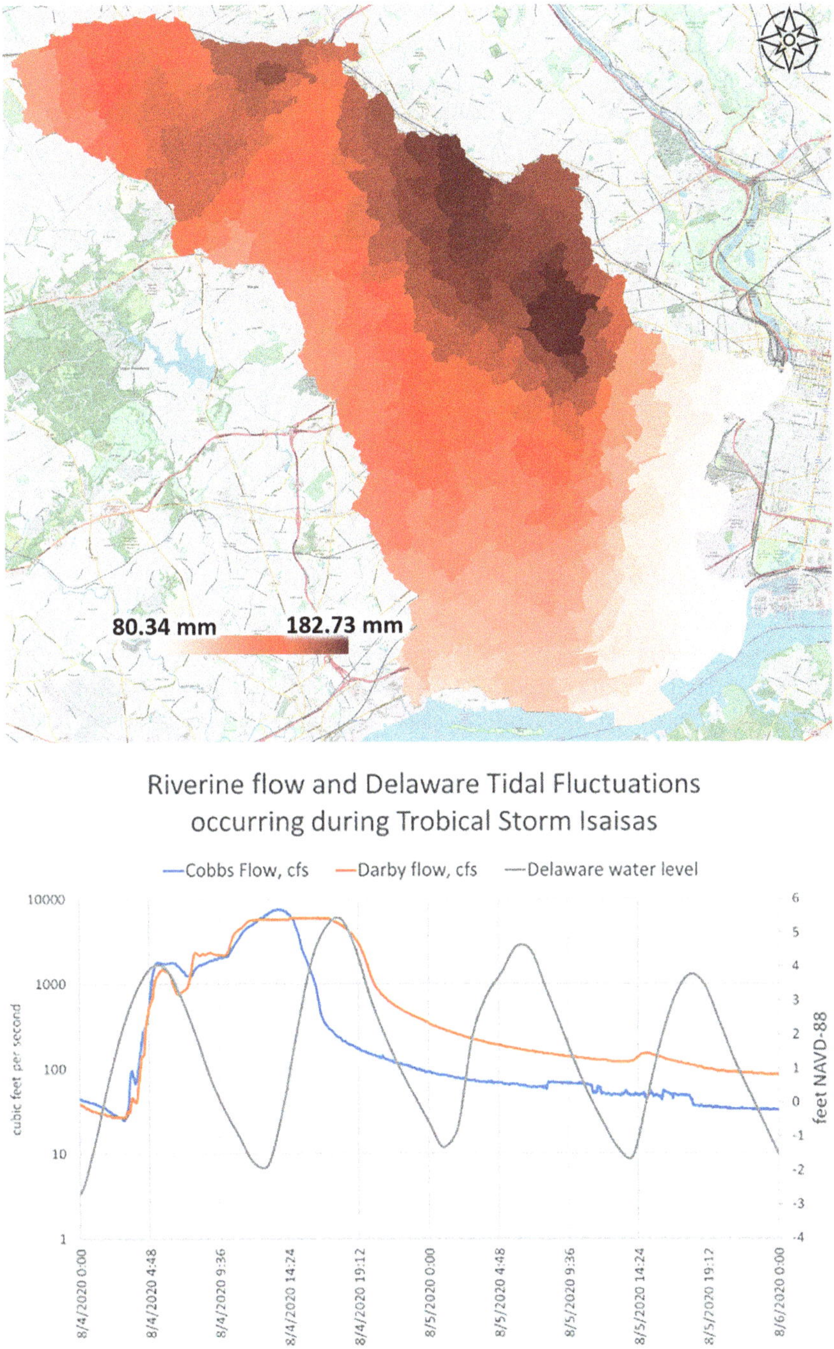

Fig. 4 (**a**) Precipitation in the Darby-Cobbs Watershed and (**b**) Riverine and Tidal fluctuations during Tropical Storm Isaias

Consensus around a comprehensive FRM solution is complicated by regional interests that advocate, on the one hand, for new economic development to support the existing community and create jobs, and on the other, for the restoration, enhancement, and creation of wetlands on the undeveloped tracts of land left in the community (Interface Studio 2019). The Heinz Wildlife Refuge has expressed interest in expanding into the vacant land parcels left over from the incomplete urban renewal project and has advocated for a "nature-based approach" to FRM.

The recently launched "Eastwick: From Recovery to Resilience" (ERR) initiative by the City of Philadelphia aims to build consensus by centering the Eastwick community in deliberation and decision-making around FRM. Managed by the City's Office of Sustainability (OOS), the plan seeks to elevated lived experience of flood survivors in the development of a vision for Eastwick's flood resilient future. ERR seeks to reduce the impact of flooding on residents' lives, property, and well-being through a comprehensive approach that involves local, state, and federal efforts (McGraw et al. 2022).

Adaptation Options

ERR seeks to develop both immediate (e.g. short-term interim) and long term FRM solutions. Among the long-term FRM solutions, three discrete alternatives are central to the current public deliberation. These include the construction of a levee on lower Cobbs Creek, a managed retreat strategy featuring a landswap scheme, and a nature-based watershed-scale approach focusing on upstream green stormwater infrastructure. Each of these options is described physically in the sections that follow.

The Levee Approach

In partnership with the Philadelphia Water Department, the US Army Corps of engineering has selected a levee on the lower Cobbs Creek as its Tentatively Selected Plan (TSP) for reducing flooding in Eastwick (Fig. 5). The levee would be constructed along the east bank of Cobbs Creek, within the city-owned Eastwick Regional Park and just north of the recently capped Clearview Landfill. The top of the levee is proposed at an elevation of 7.5 m (24.7 feet NAVD88), with a section designed for managed overtopping at a slightly lower elevation of 7.2 m (23.7 feet NAVD-88). This TSP aims to mitigate flood risks for events up to the 1% Annual Exceedance Probability (AEP), with overtopping expected during more significant flood events, and additional optimization of the design expected in subsequent project phases (USACE Philadelphia District 2023).

The levee would primarily reduce fluvial flooding in Eastwick due to high water levels in Cobbs Creek (as occurred during TS Isaias, for example). However, as acknowledged by the (USACE Philadelphia District 2023), by keeping Cobbs Creek

Fig. 5 Location and alignment of the levee

from overtopping into Eastwick, the levee would induce flooding at points upstream and downstream and on the west side of Cobbs Creek. The levee would be less effective in reducing flooding during pluvial, coastal, and compound events.

Managed Retreat Through a Landswap

A "landswap" is defined by Georgetown Climate Center (Georgetown Climate Center 2020) as "the exchange or "swap" of title to land in perpetuity between two or more property owners. It typically centers on an in-kind exchange of property between parties instead of the purchase of land. In the context of FRM, landswaps are used to move people from regions of high flood risk to regions of lower flood risk, often within the same community. They are often accompanied by ecological restoration or re-naturalization of the previously developed land. As such, landswaps are one approach to operationalizing a FRM concept often referred to as managed retreat" (Haasnoot et al. 2021).

The Eastwick landswap concept was conceived by Eastwick United Community Development Corporation, translated into a physical design by Drexel University design students, and eventually refined into a concept plan by a Philadelphia-based urban design firm (Olin). The plan is only feasible because the city owns the three large tracts of vacant land, left over from the incomplete urban renewal project. Through the proposed landswap, flood-prone Eastwick homeowners who live close

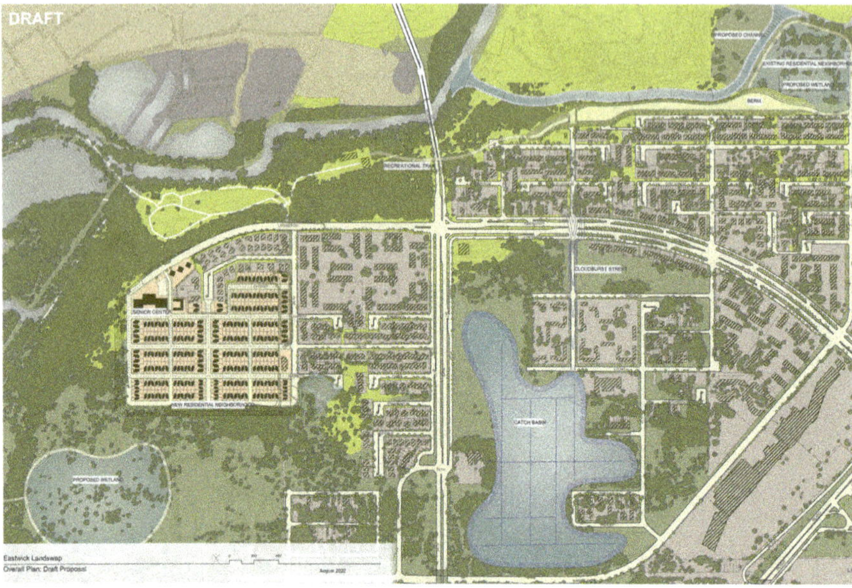

Fig. 6 Landswap draft concept

to Cobbs Creek would be allowed to swap title with the city and move to new flood-resilient housing constructed along Lindbergh Boulevard and South 86th street (Fig. 6) on the higher elevation land, most of which is above the 500 year floodplain (Fig. 7). The design is attractive to some Eastwick residents because it would enable them to reduce their flood risk without having to permanently leave their community.

Managing Stormwater Throughout the Watershed with Green Stormwater Infrastructure

The last long-term adaptation option seeks to reduce downstream flooding through the decentralized management of stormwater throughout the 31 municipalities that constitute the upstream portions of the Darby-Cobbs watershed. These towns are required to manage stormwater to comply with federal (e.g. Clean Water Act), state (e.g. PA Stormwater Management Act of 1978—Act 167, Fig. 8), and county clean water regulations. Because water quality impairment of the Darby and Cobbs Creeks is primarily due to sediment inputs, stormwater management goals—articulated through Pollutant Reduction Plans (PRPs)—strive to reduce sediment loading. As of this writing, 26 of the 31 municipalities had publicly published these plans, making up roughly 4/5 of the land area within the watershed. Nine of the 26 municipalities (Upper Darby, East Lansdowne, Yeadon, Darby Township, Collingdale, Morton, Sharon Hill, Darby Borough, Glenoldon) collaborated to form the Eastern Delaware County Stormwater Collective (EDCSC) and submitted one PRP. The

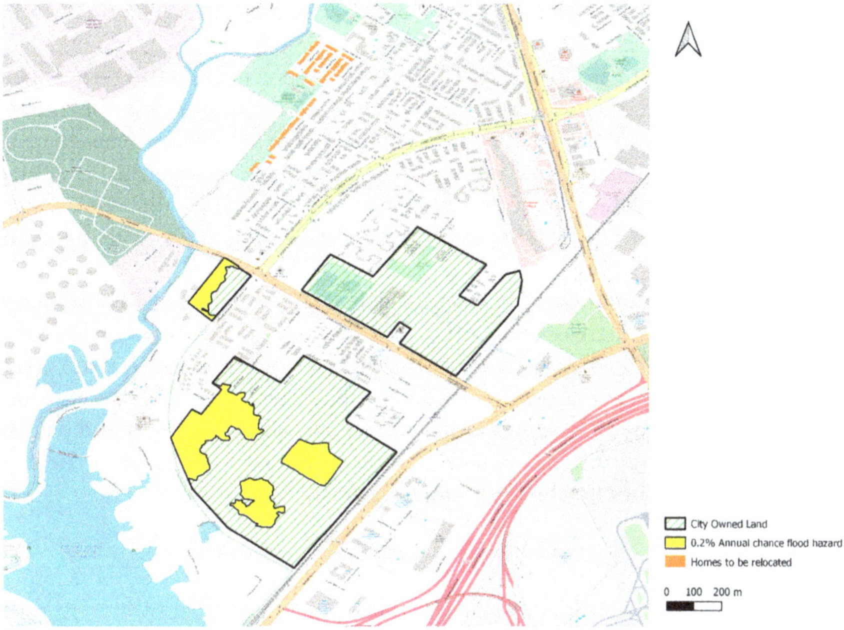

Fig. 7 City owned lands and flood prone areas within Eastwick

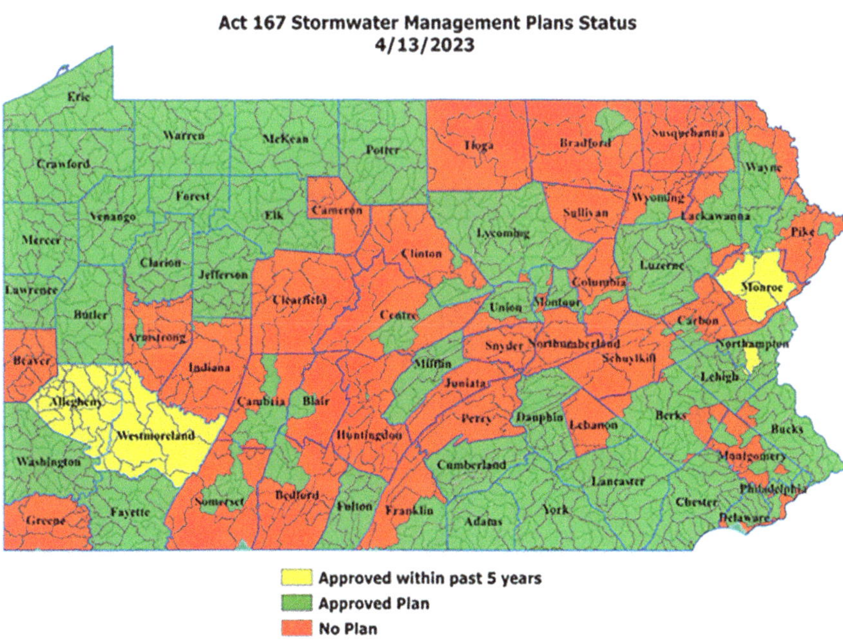

Fig. 8 Act 167 Plan status

municipalities are required to update their PRPs every 5 years, A total of 18 PRPs were reviewed by the authors.

Based on the type of stormwater best management practices (BMPs)—and their areal extent—that are included in the current round of PRPs, at best, the current round of PRPs suggests that nature-based approaches to stormwater management, including GSI could treat 12% of impervious areas and 5% of watershed as shown in Table 2. H&H modeling conducted by the authors (Nasrollahi et al. 2025), suggests that this level of GSI is inadequate to reduce flooding in Eastwick under an event like TS Isaias.

As described in greater detail in Nasrollahi et al. (2025), the same H&H model was used to gradually increase the level of GSI introduced into the watershed until an adequate level of FRM was achieved. To assess adequacy, the peak flood depth simulated with the model at the point where Cobbs Creek typically overtops into Eastwick was compared to the local land surface elevation (Fig. 9). Per the cross sections shown in Fig. 9, Cobbs Creek starts to overflow when stage is equal to 4.5 m and water reaches first building when stage is 4.75 m.

Table 2 Final % treatment of the watershed and impervious surfaces based on the current round of the PRPs

Summary	
WS area	170 km^2
WS imp area	65 km^2
GSI area	8 km^2
% WS area treated	5%
% imp area treated	12%

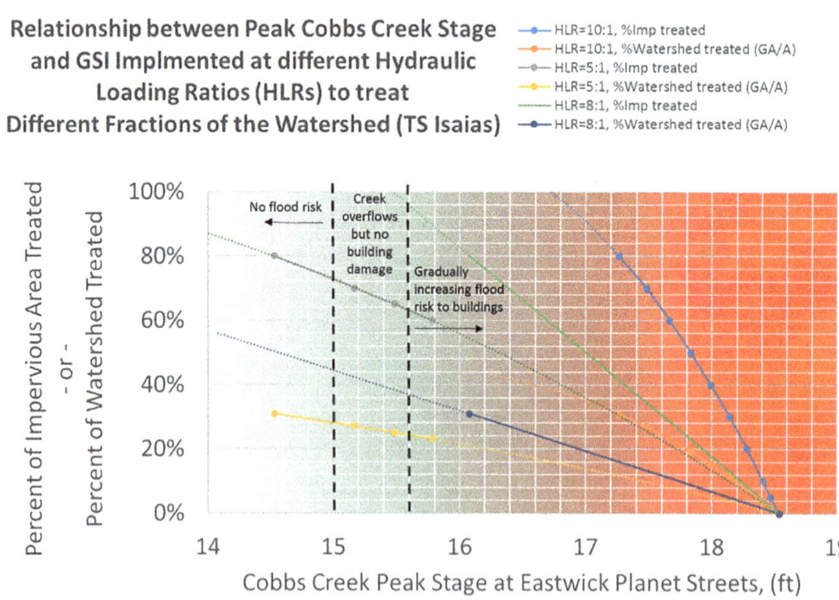

Fig. 9 GSI scenarios

Table 3 GSI modeling assumptions

Prototypical GSI type	Prototypical GSI area	HLR range	% Watershed treated	% Impervious area treated
Bioretention	300 sf	5–10	0–40%	0–80%

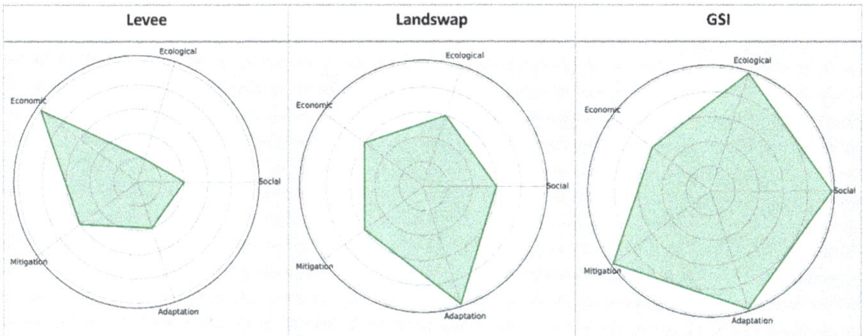

Fig. 10 Spider diagram used to evaluate the FRM option that brings the greatest CRD

To simplify the H&H modeling exercise, all of the additional upstream GSI was assumed to be in the form of replicate bioretention facilities, each 300 sf in area, that receive runoff from catchment areas that are between 5 and 10 times their size (e.g. their hydraulic loading ratios—or HLRs were varied to between 5–10. Multiple model runs were performed treating 0–40% of the total watershed (or 0–80% of the impervious areas)—see Table 3. For each scenario, the model was used to simulate peak stream flow in Cobbs Creek near Eastwick as would have resulted during precipitation and tidal conditions as they actually occurred during TS Isaias. These simulations were used to create Fig. 10. The figure depicts the peak flood depth associated with different GSI scenarios, and its associated relationship to flood risk. Based on this analysis, the GSI that would be required to reduce flooding during TS Isaias to levels similar to those achieved by the USACE levee would need to treat 65% of the impervious area—or 25% of watershed treated, at HLRs of 5 or below.

Results: Classification of Adaptation Strategies in the Field of Flood Risk Reduction Strategies and Their Philosophical Approaches

In Table 4, the three different FRM strategies are evaluated using the FRM-CRD framework developed earlier in the paper. Each of the adaptation scenarios is scored on a scale from -2 to 2, where -2 represents the lowest contribution to the relevant SDG and 2 represents the highest. Figure 10 presents the results graphically using spider diagrams. The narrative that follows justifies the scores provided.

Table 4 Synthesis of the FRM capability in promoting SDGs

SDG	Levee	Landswap	Upstream GSI
1. No poverty (Econ)	0	1	1
2. Zero hunger (Soc)	0	0	1
3. Good health and well-being (Soc)	0	0	0
4. Quality education (Soc)	0	1	0
5. Gender equality (Soc)	0	1	0
6. Clean water and sanitation (Env)	−1	1	2
7. Affordable and clean energy (Env)	0	0	0
8. Decent work and economic growth (Econ)	1	1	1
9. Industry, innovation, and infrastructure (Econ)	3	1	1
10. Reduced inequality (Soc)	0	1	2
11. Sustainable Cities and Communities (Env)	1	2	2
12. Responsible consumption and production (Env)	0	1	2
13a. Climate action: mitigation (Mit)	−1	−1	1
13b. Climate action: adaptation (Adap)	−1	2	2
14. Life below water (Env)	−1	0	1
15. Life on land (Env)	−1	0	2
16. Peace, justice, and strong institutions (Soc)	1	1	2
17. Partnerships for the goals (Soc)	1	0	2
Index 1	2	4	7
Index 2	−2	4	9
Index 3	4	3	3
Index 4	−1	−1	1
Index 5	−1	2	2
Overall score	2	12	22

The levee strategy is primarily focused on immediate flood protection through structural means, which aligns well with SDGs focused on industry, innovation, and infrastructure (SDG 9) and sustainable cities and communities (SDG 11). However, it falls short in promoting environmental sustainability goals, such as life below water (SDG 14) and life on land (SDG 15), due to potential negative impacts on natural water flows and ecosystems. Its focus on a singular, physical barrier limits broader contributions to social and ecological sustainability, emphasizing resistance over resilience and adaptability. The construction of a levee is generally focused on protecting a specific area or community; thus the stakeholder engagement is more localized (SDG 16). Residents of the affected area, Eastwick, local businesses, and municipal governments are the primary stakeholders involved. While there may be consultations with environmental experts and regulatory agencies, the scale and scope of engagement are usually narrower compared to watershed-wide initiatives (SDG 17). The focus of levee is often on engineering solutions, risk assessment, and property protection, with less emphasis on broader ecological and social impacts (SDG 1–7 & 10). Additionally, levee can alter the natural flow of rivers, potentially harming aquatic ecosystems downstream (SDG 14). It can also create barriers that disrupt the movement of wildlife, leading to fragmented habitats. Preventing floods by a levee also means preventing the natural delivery of sediments to floodplains, which can affect the fertility of these areas and the ecosystems they support (SDG 1,2 & 6).

The landswap strategy addresses multiple SDGs by combining social, economic, and environmental objectives. It promotes sustainable cities and communities (SDG 11) by relocating residents to safer areas, potentially reducing inequalities (SDG 10) and enhancing community resilience. By allowing for the restoration of natural floodplains and habitats, it contributes positively to life on land (SDG 15) and clean water and sanitation (SDG 6). This strategy embodies a holistic approach to CRD, integrating adaptation and mitigation efforts with sustainable development goals. However, its complexity and the need for extensive stakeholder engagement could present challenges. Landswap program involves the residents directly affected by the relocation, local government entities, and most likely housing and urban development agencies. This approach requires intensive consultation with the community members who are being asked to move, as well as with those in the receiving areas. Landswap can be complex, involving negotiations over property values, relocation assistance, and the development of new infrastructure. While the engagement is intensive among the impacted parties, it might not be as broad in terms of the diversity of stakeholders compared to upstream restoration efforts (SDG 16 & 17).

Upstream GSI aligns closely with a broad spectrum of SDGs by promoting ecological sustainability (e.g., SDG 14 and 15), enhancing community resilience (SDG 11), and supporting climate action through both mitigation and adaptation (SDG 13). This strategy emphasizes working with natural systems to manage flood risks, offering benefits like improved water quality, increased biodiversity, and reduced urban heat island effects. It embodies the principles of CRD by fostering synergies between urban resilience, environmental sustainability, and social well-being. GSI's scalable and multifunctional nature allows for extensive community engagement and collaboration (SDG 17), making it the most comprehensive in achieving the SDGs. This strategy typically requires a wide range of stakeholders to be involved due to its impact on the watershed level. It encompasses not only the immediate community but also upstream and downstream communities, local governments, environmental organizations, businesses, and potentially federal agencies. The collaborative nature of watershed restoration initiatives encourages participation from a diverse group of stakeholders to address water quality, habitat restoration, and flood mitigation. Engaging a broad array of stakeholders allows for a more comprehensive understanding of the ecological, social, and economic implications of the projects (SDG 8, 9, 11, 16 & 17). It also helps in identifying and leveraging synergies between water management, biodiversity conservation, and community resilience. By filtering pollutants through natural processes, green infrastructure improves the quality of water entering streams and rivers (SDG 6). GSI also increases biodiversity. Restoring natural habitats through upstream restoration can increase the diversity of plant and animal species in the area (SDG 15). These measures can also reduce the intensity and frequency of floods, protecting both natural and human communities by absorbing and slowing down stormwater runoff.

Moreover, vegetated areas contribute to carbon dioxide absorption, playing a role in climate change mitigation (SDG 13a). Green spaces can help moderate local temperatures, providing cooling effects which are beneficial in urban areas.

Discussion

The application of the FRM-CRD framework in Eastwick, as detailed above, revisits the critical interplay between FRM and CRD within the context of urban resilience, sustainability, and the broader spectrum of the SDGs. By applying the framework to Eastwick—a community marked by its environmental justice challenges and vulnerability to flooding—the analysis delves into the nuances of selecting adaptation strategies that are not only effective in mitigating immediate flood risks but also align with long-term sustainability and resilience goals.

The FRM-CRD framework illuminates the multifaceted impacts of different FRM strategies on achieving CRD, with a particular focus on social, ecological, economic dimensions, greenhouse gas emissions, and climate adaptation. This comprehensive approach underscores the necessity of a balanced assessment that takes into account the potential for synergies among adaptation, mitigation, and sustainability efforts. In Eastwick, the comparison of levee construction, landswap, and upstream GSI reveals varied implications for advancing CRD, each with distinct strengths and trade-offs.

The levee option, while offering a structural solution to riverine flooding, demonstrates limitations in contributing to broader environmental and social sustainability goals. It underscores a traditional resistance approach that prioritizes immediate protection but may inadvertently exacerbate ecological disruptions and restrict the community's adaptive capacity to future climate scenarios. This strategy's narrow focus on physical barriers highlights the critical need for integrating FRM decisions with wider environmental and social considerations to ensure they do not inadvertently hinder long-term resilience and sustainability objectives.

Decisions based on cost alone might prioritize the most economically efficient options without fully considering the long-term social, ecological, and economic impacts. Constructing a levee might appear as a cost-effective solution to protect against flooding in the short term. However, this approach could have significant negative impacts on natural ecosystems, reduce biodiversity, and even displace water problems to other areas, leading to increased vulnerability elsewhere.

Similarly, decisions driven solely by ease of implementation might favor solutions that can be quickly deployed or require the least change to existing infrastructure and social systems. While this might offer immediate relief in some cases, it often fails to address the root causes of vulnerability to climate hazards and can lead to solutions that are unsustainable in the long term.

In contrast to levee, the landswap and GSI strategies offer more holistic solutions that align closely with the principles of CRD. The landswap initiative, by facilitating the relocation of vulnerable communities to safer areas and enabling ecological restoration of floodplains, embodies a transformative approach that addresses social equity, ecosystem health, and community resilience. Similarly, upstream GSI represents a paradigm shift towards embracing nature-based solutions that work with, rather than against, natural hydrological processes. This approach not only mitigates flood risks but also enhances biodiversity, water quality, and social well-being, showcasing the power of multifunctional resilience projects in advancing a wide array of SDGs.

The application of the FRM-CRD framework in Eastwick serves as a compelling case study for other urban areas grappling with similar challenges. It emphasizes the importance of adopting a systems perspective that considers the interconnectedness of urban infrastructure, natural ecosystems, and social systems. By fostering synergistic strategies that bridge the gap between adaptation, mitigation, and sustainability, cities can navigate the complexities of climate change and urbanization more effectively. Moreover, the Eastwick case study highlights the critical role of community engagement and equity in the development and implementation of FRM strategies, ensuring that resilience-building efforts are inclusive and just.

In moving forward, it is imperative to further explore the dynamic interplay between different FRM strategies and their long-term implications for CRD in Eastwick and beyond. This entails a deeper dive into quantitative and qualitative analyses to evaluate the performance of adaptation options under varying future climate scenarios, their scalability, and their potential to address compound flood risks. Such efforts should be grounded in a participatory approach that leverages local knowledge and prioritizes the needs and aspirations of the community.

Conclusion

In concluding this analysis of CRD within the context of FRM, particularly as it applies to the urban landscape of Eastwick, Philadelphia, a crucial question arises: Do we talk about CRD more than we work to implement it? This question prompts us to assess not only our goals but also our willingness to confront and overcome the myriad barriers that stand between current practices and the transformative outcomes CRD promises.

Our discourse on CRD, despite being ambitious and comprehensive, often belies the complexity and resistance encountered in the process of translating these ideas into tangible actions. The case study of Eastwick serves as a microcosm for the broader challenges faced globally in urban resilience and sustainability efforts. It highlights the conflict between the desire for innovative, multi-purpose resilience strategies and the real-world constraints of time, logistics, and aversion to change that often limit their adoption.

The evidence presented suggests a path toward transformational outcomes through the adoption of strategies that not only mitigate immediate flood risks but also advance broader sustainability and resilience goals. These outcomes, however, are contingent upon our collective willingness to invest in the long-term and often intangible benefits of CRD. This entails a readiness to rethink traditional approaches to urban planning, to embrace more complex, nature-based solutions, and to foster a more inclusive and participatory decision-making process.

However, the resistance of existing systems, the appeal of cost-effective but shortsighted solutions, and the difficulty of uniting various stakeholders often weaken the motivation for change. The question, then, is not just whether we value the ideals of CRD, but whether we possess the collective will and foresight to implement them amidst these challenges.

As we navigate the interim, striving to bridge the gap between our current state and the aspirational goals of CRD, a pragmatic approach is warranted. This involves leveraging every available opportunity to pilot and scale up small, incremental adaptations that align with CRD principles, even as we work to build the broader consensus and systemic changes necessary for more ambitious transformations. Engaging communities, fostering interdisciplinary collaborations, and securing the commitment of political and economic stakeholders are critical steps in this journey.

References

Birkmann J, Jamshed A, McMillan JM et al (2022) Understanding human vulnerability to climate change: a global perspective on index validation for adaptation planning. Sci Total Environ 803:150065. https://doi.org/10.1016/j.scitotenv.2021.150065

Calvin K, Dasgupta D, Krinner G et al (2023) IPCC, 2023: Climate change 2023: synthesis report. In: Core Writing Team, Lee H, Romero J (eds) Contribution of working groups I, II and III to the sixth assessment report of the intergovernmental panel on climate change. IPCC, Geneva, Switzerland. Intergovernmental Panel on Climate Change (IPCC)

Chester, M., Underwood, B.S., Allenby, B. et al (2021) Infrastructure resilience to navigate increasingly uncertain and complex conditions in the Anthropocene. npj Urban Sustain 1, 4. https://doi.org/10.1038/s42949-021-00016-y

Committee on Urban Flooding in the United States, Program on Risk, Resilience, and Extreme Events, Policy and Global Affairs et al (2019) Framing the challenge of urban flooding in the United States. National Academies Press, Washington, DC

Davis SJ, Caldeira K (2010) Consumption-based accounting of CO_2 emissions. Proc Natl Acad Sci U S A 107:5687–5692. https://doi.org/10.1073/pnas.0906974107

Donaghy K (2012) Managing change in urban infrastructure systems. In: Brooks N, Donaghy K, Knaap G (eds) The Oxford handbook of urban economics and planning, 1st edn. Oxford University Press, pp 80–97

Ekvall T, Weidema BP (2004) System boundaries and input data in consequential life cycle inventory analysis. Int J LCA 9:161–171. https://doi.org/10.1007/BF02994190

Gates KP (2012) What it would take to build an earthen berm for flood control on a section of Cobbs Creek in Eastwick. In: whyy.org. https://whyy.org/articles/what-it-would-take-build-earthen-berm-flood-control-section-cobbs-creek-eastwick/. Accessed 5 May 2024

Georgetown Climate Center (2020) Managed retreat toolkit. Land swaps. In: georgetownclimate.org. https://www.georgetownclimate.org/adaptation/toolkits/managed-retreat-toolkit/land-swaps.html

Grafakos S, Trigg K, Landauer M et al (2019) Analytical framework to evaluate the level of integration of climate adaptation and mitigation in cities. Clim Change 154:87–106. https://doi.org/10.1007/s10584-019-02394-w

Haasnoot M, Lawrence J, Magnan AK (2021) Pathways to coastal retreat. Science 372:1287–1290. https://doi.org/10.1126/science.abi6594

Interface Studio. 2019. Lower Eastwick Public Land Strategy. March 25. https://ytirohtua.xyz/wpcontent/uploads/2019/03/Lower-Eastwick-Public-Land-Strategy-March-25-2019.pdf

Intergovernmental Panel On Climate Change (IPCC) (2023) Climate change 2022—impacts, adaptation and vulnerability: working group II contribution to the sixth assessment report of the intergovernmental panel on climate change, 1st edn. Cambridge University Press

IPCC (ed) (2014) Climate change 2013: the physical science basis. Cambridge University Press, New York

Jongman B (2018) Effective adaptation to rising flood risk. Nat Commun 9:1986. https://doi.org/10.1038/s41467-018-04396-1

Karamouz M, Mohammadi K (2020) Nonstationary based framework for performance enhancement of coastal flood mitigation strategies. J Hydrol Eng 25:04020020. https://doi.org/10.1061/(ASCE)HE.1943-5584.0001918

Martin MA, Boakye EA, Boyd E et al (2022) Ten new insights in climate science 2022. Glob Sustain 5:e20. https://doi.org/10.1017/sus.2022.17

McDermott M, Mahanty S, Schreckenberg K (2013) Examining equity: a multidimensional framework for assessing equity in payments for ecosystem services. Environ Sci Policy 33:416–427. https://doi.org/10.1016/j.envsci.2012.10.006

McGraw D, Tangtrakul K, Quinland T (2022) "Eastwick: From Recovery to Resilience is a Multi-Agency Initiative Coordinated by the City." City of Philadelphia. Last modified November 9, 2022; accessed May 12, 2023. Available at: https://www.phila.gov/2022-11-09-eastwick-fromrecovery-to-resilience/

Meerow S, Newell JP, Stults M (2016) Defining urban resilience: a review. Landsc Urban Plan 147:38–49. https://doi.org/10.1016/j.landurbplan.2015.11.011

Meng M, Dabrowski M, Stead D (2020) Enhancing flood resilience and climate adaptation: the state of the art and new directions for spatial planning. Sustainability 12:7864. https://doi.org/10.3390/su12197864

Nasrollahi F, Orton P, Montalto F (2025) Modeling the Effectiveness of Alternative Flood Adaptation Strategies Subject to Future Compound Climate Risks. Land, 14(9), 1832. https://doi.org/10.3390/land14091832

Nurlambang T (2022) Sustainable development in the face of floods and rising sea levels: Jakarta case study. In: sdg-action.org. https://sdg-action.org/sustainable-development-in-the-face-of-floods-and-rising-sea-levels-jakarta-case-study%EF%BF%BC/. Accessed 5 May 2024

O'Donnell EC, Thorne CR (2020) Drivers of future urban flood riskPhil. Trans. R. Soc. A.378: 20190216. http://doi.org/10.1098/rsta.2019.0216

Rosenzweig C, Solecki W (2018) Action pathways for transforming cities. Nat Clim Change 8:756–759. https://doi.org/10.1038/s41558-018-0267-x

Rosenzweig B, Montalto FA, Orton P et al (2024) NPCC4: climate change and New York city's flood risk. Annals of New York Academy of Science

Sachs J, Kroll C, Lafortune G et al (2022) Sustainable development report 2022, 1st edn. Cambridge University Press

USACE Philadelphia District (2023) Eastwick Flood Risk Management Feasibility Study. Hydrology & hydraulics documentation. Sub-appendix I. US Army Corps of Engineering

Vaughan A, Corps M (2019) Reaching the sustainable development goals. The need for building resilience, Flood Resilience Alliance

Webber JL, Chen AS, Stevens J et al (2021) Targeting property flood resilience in flood risk management. J Flood Risk Manage 14:e12723. https://doi.org/10.1111/jfr3.12723

Wegst U, Barr C, Montalto F (2010) Eco-audit of seven green infrastructure practices. In: Brdiget maintenance, safety, management and life-cycle optimization. Frangopol, Suase & Kusko

Winberg M (2020) Philly's long-term flooding plan could involve relocating parts of neighborhoods. In: billypenn.org. https://billypenn.com/2020/02/13/phillys-long-term-flooding-plan-could-involve-relocating-parts-of-neighborhoods/

Urban (and) Nature: The Activation of Nature-Based Solutions as a Silver Bullet for Urban Resilience. Specific Insights from the Industrial Zone of the City of Bolzano

Francisca Tapia, Silvia Croce, and Sonja Gantioler

Abstract Activating Nature-based Solutions (NbS) in urban areas can be a promising strategy for enhancing urban resilience to climate change and other environmental challenges. NbS involves protecting, conserving, restoring, sustainably using, and managing natural or modified ecosystems, to provide ecosystem services that benefit human well-being and support biodiversity and sustainable development. However, the activation is often piece-meal-like, fast-tracked, highly selective, and partially inclusive, evidencing a large gap between theory and practice. This chapter first offers insights into the various concepts adopted and ways forward described in activating NbS in cities, from the strategic planning of green infrastructure to the co-creation of solutions. It then looks specifically at the activation of NbS through the implementation of green roofs in the industrial district of Bolzano South, Italy. Green roofs are an example of an NbS that can play a critical role in enhancing urban resilience, by reducing the urban heat island effect, improving air quality, managing stormwater runoff, and enhancing biodiversity. At the same time, depending on how it is activated, it can pose significant barriers and limitations to the benefits it offers. This study identifies environmental benefits, such as urban heat island mitigation, through urban and microclimate simulations in an industrial setting. It also proposes a co-participatory service design process that engages local stakeholders to ensure the long-term sustainability of nature-based solutions.

Keywords Nature-based solutions · Green roof · Co-creation · Urban resilience

F. Tapia (✉)
Marcel Breuer Doctoral School, Faculty of Engineering and Information Technology, University of Pécs, Advanced Building and Urban Design, Budapest, Hungary
e-mail: francisca.tapia@abud.hu

S. Croce · S. Gantioler
Institute for Renewable Energy, Eurac Research, Bolzano, Autonome Provinz Bozen—Südtirol, Italy
e-mail: silvia.croce@mrenergy.it; sonia.gantioler@eurac.edu

G. Iannaccone, E. V. Ellis (eds.), *Towards a Nature-Positive Built Environment*, Research for Development, https://doi.org/10.1007/978-3-031-95292-0_9

Nature-Based Solutions and the Resilience of the Built Environment

Nature-based Solutions or NbS were initially coined as a policy concept. This particularly referred to the climate policy arena, where they were termed a conduit to integrate considerations on biodiversity conservation into climate change mitigation and adaptation measures. '*Working with nature*' as regards natural carbon capture and storage or building on the Ecosystem-based Adaptation (EbA) approach (e.g. wetlands and flooding) was deemed an effective as well as efficient means in tackling the twin grand challenges of climate change and biodiversity loss simultaneously—at EU level (e.g. EU climate adaptation strategy) as well as in the international arena (e.g. nature-based carbon offsets). At the same time, NbS are expected to result into additional environmental, social and economic benefits. In this regard, the concept seemingly relies on the EcoSystem Services (ESS) approach. It entered international biodiversity policy in the mid-2000s to classify, quantify and partly monetize the benefits expected to result from conserving or restoring ecosystems in particular to business and the economy but also to society at large (e.g. provisioning services such as food or raw materials to regulating services such as air purification and carbon sequestration) (Reid et al. 2005).

Launched by the International Union for the Conservation of Nature (IUCN), the most prominent definition of NbS does not mention ESS explicitly, describing them rather widely as '*actions to protect, sustainably manage, and restore natural or modified ecosystems, that address societal challenges effectively and adaptively, simultaneously providing human well-being and biodiversity benefits* (Cohen-Shacham et al. 2016).

Since then, diverse definitions have evolved. Most express the common goal of working with nature to address several societal challenges simultaneously. It has been expanded to the 'big three', beyond climate change and biodiversity loss to also include human well-being (e.g. health impacts), or even to sustainable development contributions (e.g. countervailing socio-economic inequality or achievement of the Sustainable Development Goals, SDGs). However often the finally adopted concept puts an emphasis on a defined aspect, depending on the problem 'nature' is expected to 'solve', e.g. natural climate solutions, NbS for water or 'digitalized' Nb solutions. The definition, for example, provided by of the EU Horizon 2020 expert group to be included in the relevant European Research and Innovation Programme, magnifies it to all solutions '*inspired by nature*' (e.g. encompassing biomimicry). It also highlights the importance of cost-effectiveness, and re-introduces ecosystem services and the support of their delivery as part of the definition (European Commission 2015). This suggests a specific interest into technological or engineered innovation for defined (business) purposes, albeit also more than others emphasizing the role of NbS for cities. More recently, building on the IUCN definition, the UN Environment Assembly followed suit with a resolution on NbS, though further emphasizing the sustainable use of ecosystems and, interestingly, separating between anticipating human well-being, ecosystem services, resilience and lastly biodiversity benefits (UNEP 2022).

As such, not surprisingly, the scope of the use of the NbS concept has widened, to the extent that some of its initial adopters have coined it an umbrella term for many of the previously evolving notions such as EbA and ESS (Cohen-Shacham et al. 2016; Dumitru et al. 2020). It is argued that this wide scope might have hampered its uptake and putting the concept effectively into practice (Pauleit et al. 2017). Others state that for example in urban environments its novelty can be questioned, given the long-standing history of urban green space design and planning, but also due to the micro-interventions pursued as part of European NbS projects so far (O'Sullivan et al. 2020).

Nevertheless, the wide range of research and innovation projects financed by the EU's Horizon 2020 program resulted into the NbS concept gaining momentum, especially as regards their consideration in urban planning and the built environment (European Commission 2021). According to Eggermont et al. (2015) and followed by Somarakis et al. (2019), NbS can be distinguished according to their degree of management, resulting into 3 broad types: Type 1 refer to those with minimum or no management required, focused on conservation and protection (e.g. protected city biotope); type 2 include the development of multi-functional ecosystems, including restoration and management (e.g. urban forests); and type 3 are mainly those creating new ecosystems (e.g. green roofs, green walls). This classification helps making evident that many of the EU financed NbS projects, in particular in the built environment, contained interventions strongly aimed at the engineering of 'new ecosystems', especially to target sometimes very specific urban problems (e.g. water retention), and in close interaction with technological innovations (e.g. linked to smart cities) and the identification of new business opportunities (e.g. bio-circular economy). Biodiversity conservation unfortunately, mostly remained a by-product of the activities, if at all considered.

This is likely the result of the focus of the definition and related criteria applied to the topic calls, as well as of the proposal evaluations by a defined research community. In a more recent effort in mapping the NbS scope, a study reveals how NbS research communities have predominantly evolved around the European Commission's framework, in addition to the ones by the Intergovernmental Panel on Climate Change (IPCC) and the Millennium Ecosystem Assessment (MEA), now represented by the assessment reports released by the Intergovernmental Science-Policy Platform on Biodiversity and Ecosystem Services (IPBES) (Cousins 2021). Not surprisingly, the most prominent network has thus been fashioned in relation to urban applications, and to ecosystem services and the measurement and valuation of co-benefits (Cousins 2021).

In addition, the previously suggested piece-meal like implementation of NbS potentially links to the missed opportunity of the NbS concept in taking duly into account key principles of another policy concept adopted to integrate biodiversity conservation especially into land use policy and territorial planning: Green Infrastructure (GI). Different to GI, NbS definitions fail to explicitly refer to network and connectivity aspects and do not call for a strategic planning at various scales, which is inherently part of the GI definition (Gantioler 2019). This might have seriously hampered the up-scaling of NbS, as thus mostly focused on (demo-site) interventions at building or open space scale and some neighborhood approaches in cities.

However, different to other approaches, and this might be its largest potential, the NbS concept puts emphasis on a '*problem to be solved*', rather than a society and economy '*to be served by nature*'. Although this might have seemed especially appealing to engineering disciplines and those operating on the built environment, it offers the chance of a shift of perspective to 'needs', rather than 'mere' (consumer) preferences, still mostly of human beings but also with the opportunity of widening the scope to consider flora and fauna and ecosystems more widely (e.g. regarding urban thermal comfort).

In order to achieve a strategic, and less piece-meal-like, fast-tracked or highly selective activation of NbS, a hierarchical four-tier system is suggested (Gantioler et al. 2023). It is recommended to refer to the 'activation' of NbS rather than to their 'development', to take into account the multitude and diversity of possible interventions as well as its various stages (e.g. design, implementation and maintenance). At the first level, the challenges NbS are intended to address require scrutiny (see Fig. 1), especially considering their systemic nature, i.e. their various interdependencies both in relation to synergies as well as trade-offs. Systemic does not imply the same as holistic, as it does not suggest looking at the whole system but at an entry point in a complex system of interactions between environmental, social and for example the urban built environment, and following the various connections.

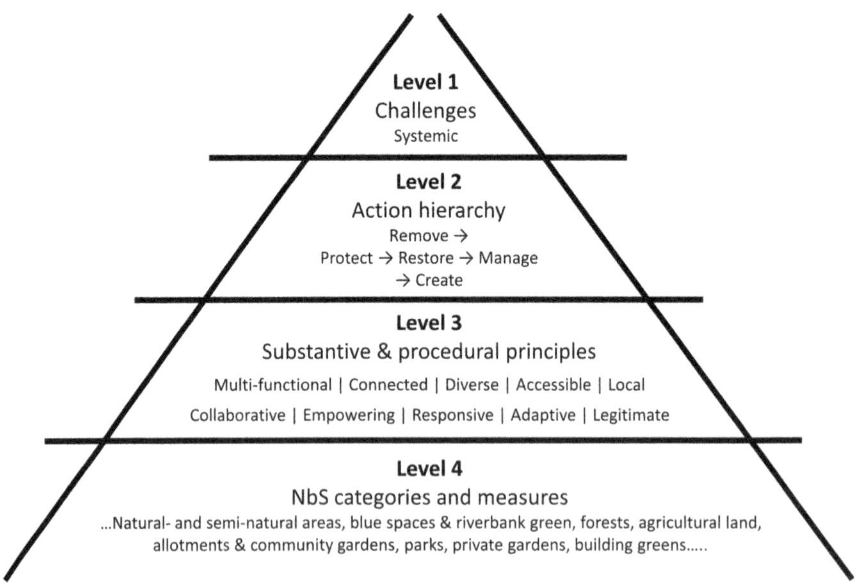

Fig. 1 A 4-tier approach to activating NbS. Source: Building on Gantioler et al. (2023) and Pico et al. (2023)

At Level 2, the 4-tier system introduces a clear action hierarchy, which so far NbS definitions have shied away to adopt. It requires to first identify actions that remove the problem at its root, as NbS might not always be the best option to achieve major changes (e.g. fossil fuel outsourced for climate change mitigation or reducing traffic loads to combat air pollution in cities). This is followed by a clear priority given to the protection of elements (e.g. protected areas, heritage trees), restoration (e.g. wetlands), sustainable management (e.g. mowing) and lastly to creation (e.g. green walls), to avoid 'end-of-pipe' solutions potentially detrimental to biodiversity conservation. The line cannot necessarily be drawn neatly, as different actions can be applied at one site and at different scales. However, it should help avoid calling for example the creation of a green roof on a newly constructed building that replaces an existing urban green area, even if an abandoned one, a NbS. Level 3 introduces various principles to be respected, from the multi-functionality of solutions to their connectiveness, and level 4 invites to study and activate various NbS categories and measures.

To synthesize, activating the NbS potential implies taking into due account several challenges related to climate change, biodiversity loss and well-being, and their systemic nature; prioritizing effort according to the action hierarchy of first removing the problem and lastly the creation of new solutions, considering a wider package of NbS categories and measures; and respecting key substantive as well as procedural principles to ensure their strategic activation. Especially procedural principles have an important role to play. Considering the complexity of the challenges impacting the built environment, the involvement of a diversity of actors and knowledge becomes key. It requires the NbS activation to be collaborative (e.g. co-design), empowering (e.g. opportunities for less powerful to be heard), responsive (e.g. to local context), adaptive (e.g. continuous learning) and legitimate (e.g. transparent) (Pico et al. 2023).

The following sections introduce and discuss the activation of NbS for the city of Bolzano and in particular its industrial zone 'Bolzano South'. Cities are perceived as human-made strongholds of built environment, not affected by nature, and this even more applies to their industrial production zones. Activating NbS also implies making visible ecological conditions and how they impact the built environment as well as the people living and working within. In the case of the city of Bolzano (Italy) this especially refers to aspects of thermal comfort, considering how a changing climate is affecting the city with increased heatwaves, and the built environment of the industrial zone being especially responsible for exacerbating the urban heat island effect in on-going urban transformation processes. Of particular interest is the effective role and impact of green roofs as newly created NbS and 'green surface use' solutions, given how the city has been promoting their development for newly constructed or re-developed buildings, e.g. based on the use of the certification instrument of Reduced Building Impact ('Riduzione dell'impatto edilizio' or R.I.E). In order to better understand how such engineered solutions also follow defined procedural principles the service design approach has tested, discussing with experts and defined stakeholders, to gain a better understanding of how the planning of green roofs as green infrastructure can contribute to the development of an eco-industrial community.

The City of Bolzano, Its Landscapes and Industrial Neighborhood 'Bolzano South'

The city of Bolzano (UTM 46°29′53.8″ N, 11°21′17.1″ E) is located in the north-east of Italy, in the centre of the south-eastern Alps, in the administrative Province of Bolzano or South Tyrol. It has a population of about 105,700 inhabitants, with a high population density of 2000 inhabitants per km^2. Bolzano lies 265 m above sea level in a basin surrounded by four mountain ranges. The city is crossed by the Talvera river, which flows into the Isarco river within the city limits. The water-courses are a characteristic element of the city; the main urban parks and the main cycle paths develop along their banks. The considerable height of the mountains surrounding the city makes it difficult to balance the currents and humidity. As a result, Bolzano is characterised by a humid continental climate, Dfb according to the Köppen-Geiger classification (Kottek et al. 2006) with strong seasonal variations. Due to its location and climatic characteristics, the city is often affected by the presence of high temperatures and heat waves in summer (Papathoma-Köhle et al. 2014), and it is one of the hottest Italian cities, with (dry-bulb) air temperatures (T_{air}) often exceeding 35 °C, with maximum peaks reaching 40 °C. Thus, not surprisingly, the effects of climate change are visible in the city. The analysis of the historical data series from 1950 to 2019 (Autonomous Province of Bolzano 2020) shows an increase in the mean annual T_{air} of more than 3 °C (from 10.7 °C in 1977 to 13.9 °C in 2014 and 2015) (Lobaccaro et al. 2018). Furthermore, in recent decades the rainfall occurred mostly in the form of intense precipitation. Especially during summer, these intense precipitation phenomena are accompanied by severe storms, often causing flooding, landslides and other damages in the urban area. Climatologists assume that, with the rise of temperatures, the number and intensity of thunderstorms will also increase, with a consequent intensification of the risk of flooding (Eurac Research 2018). On the contrary, the analysis of the data of the local aquifer in the period 2008–2016 shows no significant reductions; hence, risks of freshwater scarcity due to drought are not foreseen within the next years (Municipality of Bolzano 2020). The urban area of Bolzano is also affected by the urban heat island (UHI) effect (i.e. a marked rise in air temperature in urban areas compared to the surrounding environment), which exacerbates summer temperatures. A study analysing the impact of urbanisation on the intensity of UHI under summer conditions, comparing temperatures in Bolzano with those in surrounding rural areas, shows peaks of 5.5 °C difference at night on days with low cloud cover in August 2015, while the daytime average was around 1.5 °C difference (Pappaccogli 2019).

Bolzano's environmental ranking in the Ecosistema Urbano 2023 report has dropped from first to ninth place. This comprehensive analysis by Legambiente,

Ambiente Italia, and Il Sole 24 Ore evaluates 105 provincial capitals in five macro areas: air quality, water management, waste, mobility, and overall environmental conditions. Despite this decline, Bolzano still maintains a position within the top ten overall. The city excels in efficient land use (4th place, index of 8.5), total greenery (5th place, 225.8 m^2 per inhabitant), and land use considering per capita consumption over 5 years (5th place).

However, challenges remain in terms of average NO_2 concentration (82nd place, 26.8 µg/mc), number of trees (69th place, 12 per 100 inhabitants), and domestic water consumption (65th place, 151 L per inhabitant per day). Bolzano's overall standing highlights the need to address specific environmental indicators to improve sustainability and performance.

Notable decisions and initiatives in the field of urban planning include the adoption of the Masterplan of the City of Bolzano in 2009. The Masterplan (Fig. 2) focuses on landscape conservation, expansion guidelines, and the protection of open spaces, emphasizing soil protection, the integration of the landscape theme in planning, the improvement of infrastructure and mobility, and the importance of urban quality and public spaces. It also included a strategy for the protection and enhancement of the ecological network, which prioritizes public green areas and provides recommendations for private green spaces, contributing to the city's resilience and biodiversity. In 2020, the City Council approved the Municipal Green Regulation, which introduces innovative guidelines for the development and management of green areas. It aims to protect the territory, increase the functionality and accessibility, and promote biodiversity through environmentally friendly practices, tree protection, and the establishment of a Technical Committee for Green Areas (Municipality of Bolzano 2020). Finally, the New Municipal Green Plan has been approved in 2022 (LAND 2021). It is an essential urban planning tool, as it defines

Fig. 2 Structural and strategic design of the city, Masterplan 2009. Source: Extract from Bozen Greenery Plan by LAND Italia srl

an organic intervention programme for the quantitative and qualitative development of urban, peri-urban and natural green areas in the medium and long term, in line with the socio-economic development and urban transformation of the municipal territory. As shown in Fig. 3, the analysis accompanying the plan indicated that the extent of privately owned green spaces is greater than that of public spaces, which are often fragmented, suggesting the need for a balance. Public gardens and green areas play a crucial role in shaping the city's green landscape and promoting overall wellbeing, with proposed actions including seamless connections with public streets, preservation within existing buildings, support for private garden owners, promotion of native plant species, ecological connectivity, and management of invasive species.

The main industrial area of Bolzano is located in the southern part of the city, and covers an area of almost 4 km^2, representing 5.7% of the total municipal area, and 31% of the total built-up area of the city (Fig. 4). Due to its characteristics, Bolzano South represents an opportunity for the integration of NbS in terms of sustainable urban development. The Municipality of Bolzano is strongly committed to improving the environmental conditions of the district through various actions. Indeed, the area hosts many companies and workers (i.e. more than 2500 companies and over 20,000 commuters), playing a crucial role in the city's growth. The municipality's objectives prioritize infrastructure, additional services, and the revitalization of abandoned areas. Furthermore, the Green Plan identifies the district as the largest

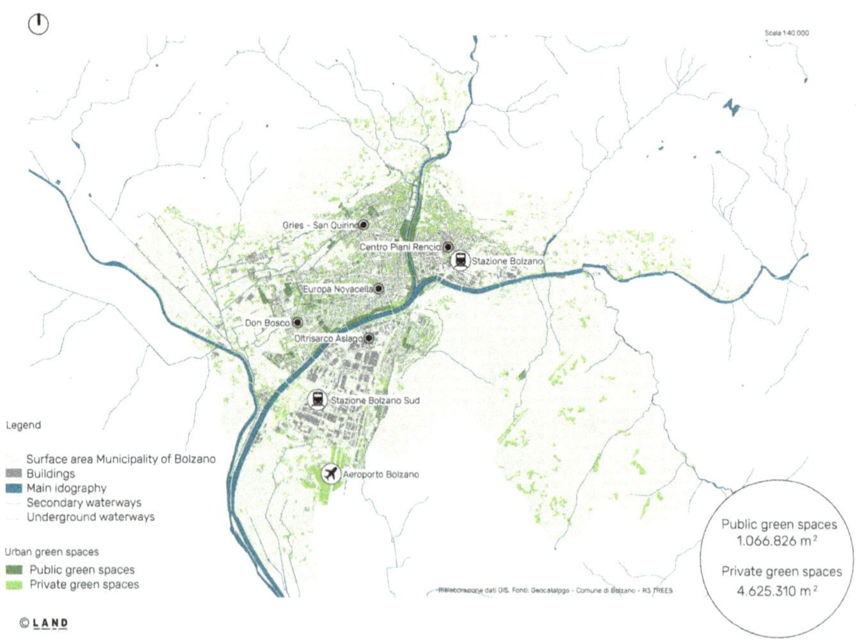

Fig. 3 Location of public and private green spaces. Source: Extract from Bozen Greenery Plan by LAND Italia srl

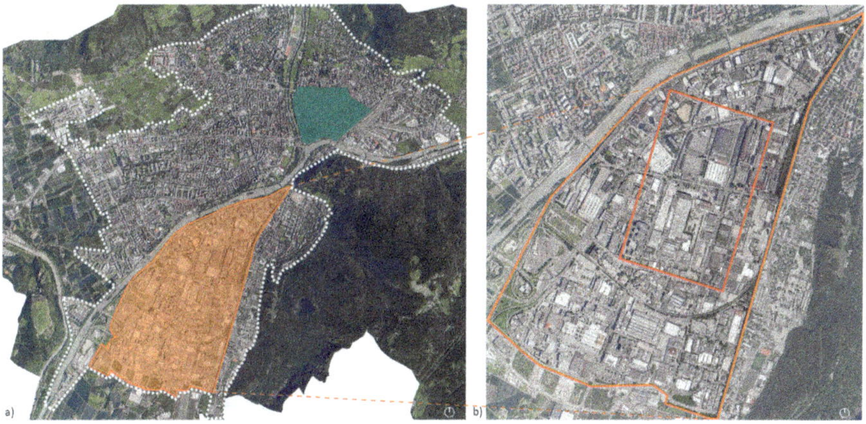

Fig. 4 (**a**) Bolzano South industrial area (orange) compared to the territorial extension of Bolzano (white dashed line) and its city centre (blue). (**b**) Aerial view of the industrial area; the case study area is highlighted in red. Elaborated by the authors from Google Earth

urban area in Bolzano without public green spaces, and a relevant area for intervention.

The Municipality has been working for several years to promote the diffusion of green infrastructure in Bolzano South to address the problems related to the UHI and summer overheating, the high level of traffic in the area and the consequent air pollution, and the storm-water management. To this aim, one of the tools used by the Municipality of Bolzano is the R.I.E., an environmental quality index that assesses the impact of construction works on the permeability of the soil and green areas; it is mandatory for new buildings and the renovation of existing buildings, as well as for any type of intervention on external surfaces exposed to precipitation (e.g. terraces, courtyards, roofs, etc.). The calculation of the index is mainly based on the permeability of materials, which are classified according to the type of urban surface, its characteristics, and its runoff coefficient. The municipal building code defines minimum R.I.E. reference values depending on the land use. In industrial or productive areas, the R.I.E. must be at least 1.5 (in residential areas it is 4) and the ongoing discussion on green infrastructure in Bolzano South includes the possibility of increasing the minimum value to reduce the degree of soil sealing in the district. In this context, horizontal greening systems—also known as green roofs—are considered the most appropriate solution to achieve both the municipal objectives and respect the required minimum R.I.E. value for several reasons: (i) roofs represent the most unused surfaces in Bolzano South, (ii) there are few open spaces at ground level suitable for the creation of green or permeable areas, and (iii) the road network in the area can hardly be modified.

In 2018, the Municipality of Bolzano, in collaboration with the local innovation center IDM, launched the Green Roof Initiative. It aimed to analyse the potential of transforming city roofs into green spaces, providing opportunities for relaxation and business, and improving the city's microclimate and economic potential. As shown

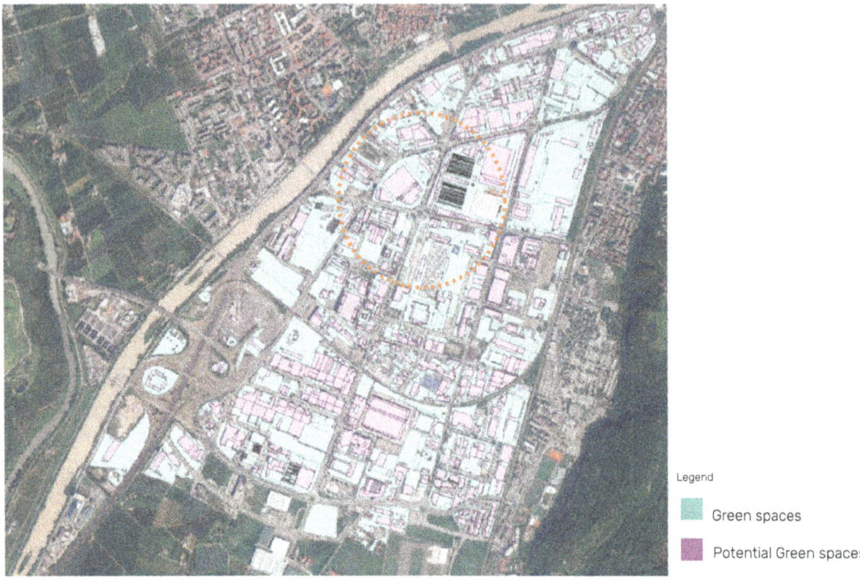

Fig. 5 Analysis of surfaces potentially suitable for green infrastructure in Bolzano South. Source: Municipality of Bolzano

in Fig. 5, the areas—including roofs—potentially suitable for hosting green infrastructures in Bolzano South have been analysed. The Sustainable Energy and Climate Action Plan 2020 also proposes greening initiatives (Municipality of Bolzano 2020). Furthermore, in 2020 the Municipality commissioned Eurac Research to carry out a study on the possibilities and drawbacks of the diffusion of horizontal greening systems in the district. The main objective of the collaboration was to encourage private individuals, in particular the owners of warehouses and factories in the area, to integrate green roofs on existing buildings. An entrepreneurial discovery process was developed to promote the use of green roofs as ecological/green infrastructure, through interviews and workshops.

Surface Uses and Microclimate in Bolzano South: The Potential of Green Roofs

One of the first approaches adopted in the analysis of the Bolzano South case study area was to focus on urban surfaces, the definition of their uses, and the impact on microclimate and human thermal comfort conditions. Urban surfaces include all the physical and morphological features of the built environment (i.e. horizontal and vertical surfaces of the ground and of the building envelope) that affect radiative, thermal, and hydrological processes, can be characterized by different materials, the presence of vegetation, and can host multiple functions (Croce and Vettorato 2021).

As a fundamental element of the built environment, the design of urban surfaces could have a significant impact on improving environmental quality and increasing resilience to climate change in urban areas (Leistner et al. 2019). This is particularly true when urban surfaces are not only considered in terms of man-made materials, but their scope is extended in the context of NbS applications. Indeed, urban surfaces and their characteristics have a significant impact on environmental conditions and quality of life in urban areas. The replacement of natural, porous surfaces with mineral materials is responsible for the UHI; while the low albedo (reflectivity to solar radiation) of the materials increases the absorption of solar radiation and contributes to the elevated surface temperatures observed in cities. Elevated air temperatures not only exacerbate outdoor thermal discomfort, causing adverse effects on human health, but also increase energy consumption for cooling, leading to peak electricity demand in hot seasons (Santamouris et al. 2017). Furthermore, soil sealing intensifies flood risk by reducing infiltration, consequently amplifying the run-off of rainwater (Tsoka et al. 2018). In this context, a move towards considering the incorporation of NbS on urban surfaces as a key element could contribute to reducing environmental impacts and provide a range of associated benefits.

In recent years, urban surfaces have shifted from being seen as a cost burden due to maintenance to a crucial opportunity. Urban areas are particularly vulnerable in three sectors—food, energy, and water—and the exploitation of under-utilized urban surfaces could potentially play a key role in promoting resource efficiency and minimizing a range of associated environmental impacts (Kellett 2011; Corcelli et al. 2019). To this end, it is important to transition from designing urban surfaces solely for single long-term purposes to considering them holistically in terms of functionality and flexibility, quality and productivity (Kellett 2011; Henzler et al. 2019). In this framework, the term 'urban surface use' describes the comprehensive way in which urban surfaces are designed and used; it encompasses all materials and solutions that can be used to build or applied to urban surfaces. Surface uses can be grouped into five main clusters: three related to NbS, namely (i) green solutions, (ii) water solutions, (iii) urban agriculture, and two more ‘artificial’ ones (iv) cool materials and innovative solutions, and (v) renewable energy systems (Croce and Vettorato 2021). In urban areas, NbS can be applied to ground surfaces (e.g. parks and urban forests), street furniture, and building elements (i.e. green roofs and facades), and improve urban environmental conditions through various processes, such as direct sun shading, ambient cooling from evapotranspiration, mitigation of air movement, improvement of local air quality, storm-water management, and regulation of heat exchange through the building envelope leading to reduced cooling energy consumption and corresponding greenhouse gas emissions (Vuckovic et al. 2017).

An analysis of the urban surfaces in the district of Bolzano-South (Fig. 6) shows that the area is characterized by a high proportion of impervious surfaces (permeable materials represent only 7% of the total horizontal surface) and by the presence of buildings with a large surface area. In fact, 69% of the surfaces are horizontal and roofs account for 44% of the total.

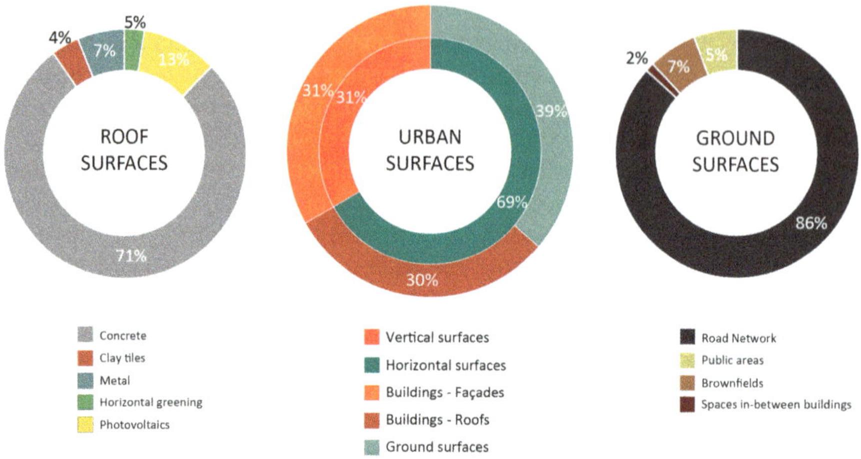

Fig. 6 Share of urban surfaces in *Bolzano South* case study area. Percentages over total surface area in the district (center), total roof area (left), and total ground area (right). Source: Silvia Croce

The district is delimited to the west by the Talvera river and the highway linking Italy and Austria, and to the east by the railway and the mountainside. These important boundaries have had a strong influence on the development of the whole area. In fact, the industrial structure, as it has expanded, has incorporated the main roads, resulting in a high level of traffic and very limited possibilities for redevelopment and improvement of the road network due to the presence of buildings on all sides. Due to the high number of workers, there are cars parked along all the roads, but the lack of parking spaces is one of the problems of the district. The area is characterized by a lack of public green spaces, and a very scattered presence of roadside planting, but its characteristics make it difficult to increase the amount of vegetation on the ground.

Microclimate Analysis in Bolzano South: Evaluating the Influence of Green Roofs

The main objective of the microclimate analysis was to develop a detailed model to analyse the urban microclimate of Bolzano South, and to estimate the impact of the proposed green interventions on human outdoor thermal comfort. This was achieved by taking into account a number of morphological (e.g., building layout, surface materials, vegetation and soil types, and project site parameters) and climatic (e.g., initial meteorological conditions) factors that influence local climatic conditions.

The study analysed the current conditions of the area and proposed a strategy for implementation, which includes the incorporation of green roofs in line with the Municipal Green Plan. The main objective was to evaluate the impact of vegetation on both air temperature and overall human outdoor thermal comfort in urban environments, as assessed by the Universal Thermal Climate Index (UTCI). An area in the centre of the district was selected for analysis (shown in Fig. 4). This exemplifies the common urban characteristics of Bolzano South—streets with high traffic levels, limited presence of permeable surfaces, sizable industrial structures (with heights between 8 and 12 m), and a lack of greenery. Two scenarios were considered: (i) a current scenario, and (ii) a green scenario. The study focused firstly on analysing of the current summer climate conditions in the district, and secondly on the potential reduction in air temperatures and improvement in outdoor thermal comfort due to green roofs.

The analysis was conducted using ENVI-met V. 4.4, a numerical model able to simulate the interactions between different urban surfaces, vegetation, and the atmosphere. ENVI-met analyzes the effects of small-scale changes in urban design (e.g. trees, greening, new buildings) on the microclimate under different mesoscale conditions and evaluates the impacts on human thermal comfort (Huttner and Bruse 2009). The simulation input data and scenario are described in Table 1; for a comprehensive understanding of the analysis and methodology, additional details can be found in (Tapia 2019).

The model in the current scenario predicts the temperatures of a typical hot summer day (Fig. 7). At 6:00 a.m., T_{air} is between 24.27 °C and 26.89 °C. By 10:00 a.m., the temperature rises to a range of 30.16 °C to 32.79 °C. The most severe conditions occur in the afternoon; at 2:00 p.m. T_{air} peaks at 34.26 °C and reaches a maximum of 37.10 °C. This is the highest temperature recorded. Figure 7 illustrates the

Table 1 Description of simulation inputs and configuration

Study area:	Block in the centre of Bolzano South
Data sources:	Aerial images and Geographic Information System (GIS) data
Tool	ENVIMET V4.4
Size of the model:	around 2 km × 0.5 km
Grid resolution:	0.5 m in space and 1–5 s in time
Date of simulation:	24 July 2019 (hottest day in summer)
Duration:	44 h, with 20 h initialization time
Initial meteorological conditions:	Wind speed at 10 m: 1.00 m/s; Wind direction: 135.00° (0 = from North, 180 = from South); Roughness length at measurement site: 0.020; air temperature: 22.15 °C (minimum at 04:00 a.m.), 36.75 °C (maximum at 03:00 p.m.); Relative humidity: 31.65 % (minimum at 03:00 p.m.), 82.53% (maximum at 04:00 a.m.)
Microclimate parameters analysed	Air temperature, relative humidity, wind direction, wind speed
Assessment of human thermal comfort	UTCI (Universal Thermal Climate Index)
Comparison	Current scenario vs. Green scenario
NbS intervention:	Extensive Green roof (20 cm thick)

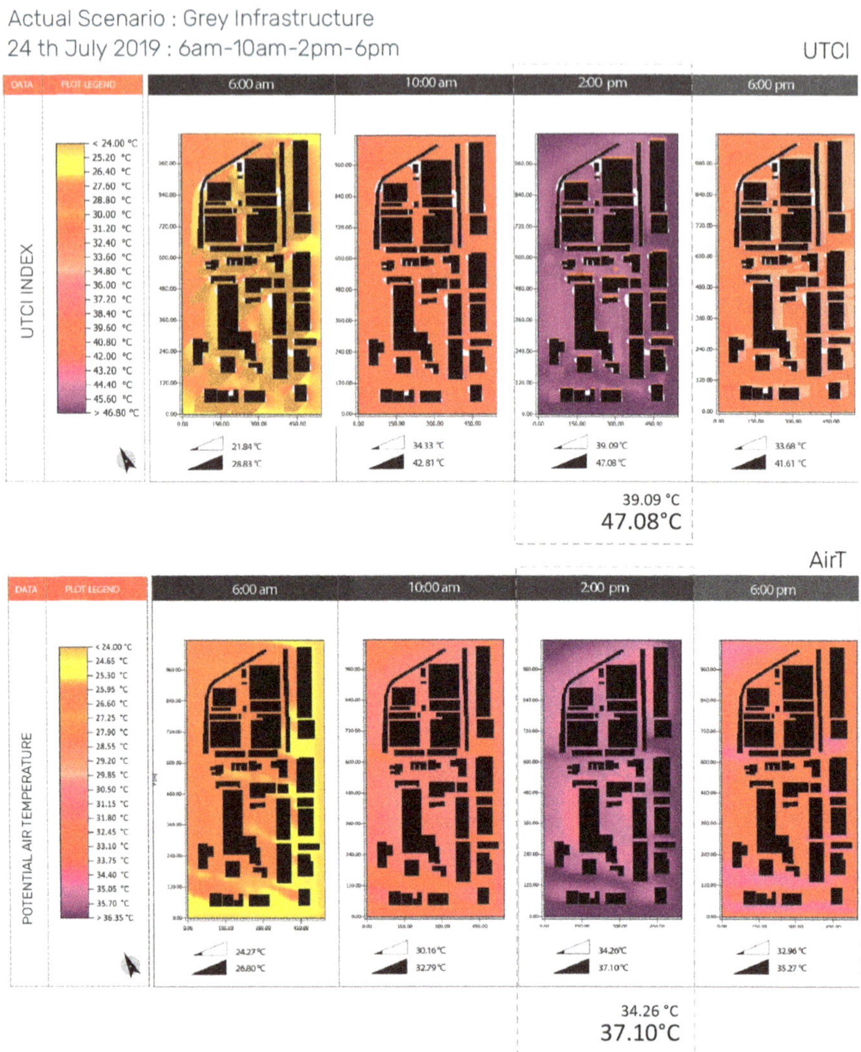

Fig. 7 Results for the current scenario on the summer day of the 24th of July, 2019, at 6 a.m., 10 a.m., 2 p.m. and 6 p.m. Air temperature (above) and UTCI index (below) at pedestrian level. Source: Francisca Tapia

challenges posed by the high summer temperatures, which can be uncomfortable for pedestrians, especially during afternoon activities in and around buildings. The thermal behaviour of urban surfaces influences local microclimates. The highest daytime temperatures are associated with the presence of impervious, artificial materials such as concrete and asphalt, as the properties of these surfaces lead to increased heat storage and surface temperatures. Similarly, impermeable, and compacted soils retain heat and contribute to warming, albeit to a lesser extent than artificial materials. Conversely, vegetated, and permeable surfaces provide cooler conditions, but can be still hotter than natural landscapes (Rakha et al. 2017). Understanding these surface characteristics is crucial for addressing overheating and the UHI effect, especially in industrial areas like Bolzano South. UTCI value confirm extreme heat stress conditions for pedestrians, with maximum values ranging from 39 °C to 47 °C during the afternoon at 2:00 p.m. High UTCI persists until 6:00 p.m., with a maximum of 41.61 °C. Prolonged periods of high temperatures can induce discomfort for pedestrians and significantly affect our overall well-being (Franck et al. 2013).

Improving Outdoor Comfort: Advantages of Green Roof Implementation

The analysis of the green scenario provides valuable information for developing effective strategies to improve thermal comfort and environmental quality in urban areas. Two parameters have been considered to compare the results between the two scenarios and to analyse the benefits of green practices in the case study area. In Fig. 8, the absolute difference in air temperature highlights the variations, especially during the critical afternoon period. The green scenario shows that the integration of green roofs on buildings results in an average air temperature difference at pedestrian level of—0.40 °C at 2:00 p.m. The cooling effect of green roofs may not be readily apparent as they are located at a height of approximately 15 m above pedestrian level. Studies in the literature have shown that extensive green roofs, with a substrate depth of approximately 20 cm, have the potential to reduce pedestrian-level temperatures by up to 0.7 °C (Rakha et al. 2017). The analysis of the UTCI index indicates that green roofs have a limited cooling effect at pedestrian level. At 2:00 p.m., the hottest time of the day, there is an improvement of −0.8 °C in human thermal perception (Fig. 8). UTCI is an approved method for determining the human thermal index, and the results can be used to create human bioclimatic maps. These maps are a valuable tool for analysing the outdoor thermal effects of an area.

In the green scenario, UTCI values at 2:00 p.m. range from 38.24 °C to 46.40 °C. At these temperatures, pedestrians experience a moderate level of heat stress, which is mitigated by ground surfaces, green roofs, and spaces such as surrounding trees. In the morning hours at 9 a.m., the analysis shows a slight change, with temperatures ranging from 20 °C to 28.83 °C in areas near buildings providing shading, as well as trees located in courtyards. Figure 8 illustrates the absolute difference during the morning, with a difference of −1.89 °C and −0.58 °C. According to the UTCI thermal stress categories (Bröde et al. 2011), these temperatures fall within the no thermal stress category. However, in the late afternoon at 6 p.m., in the green scenario, the analysis indicates a minimum temperature of 33.01 °C and a maximum temperature of 41.09 °C. This represents a decrease of −0.62 °C at the

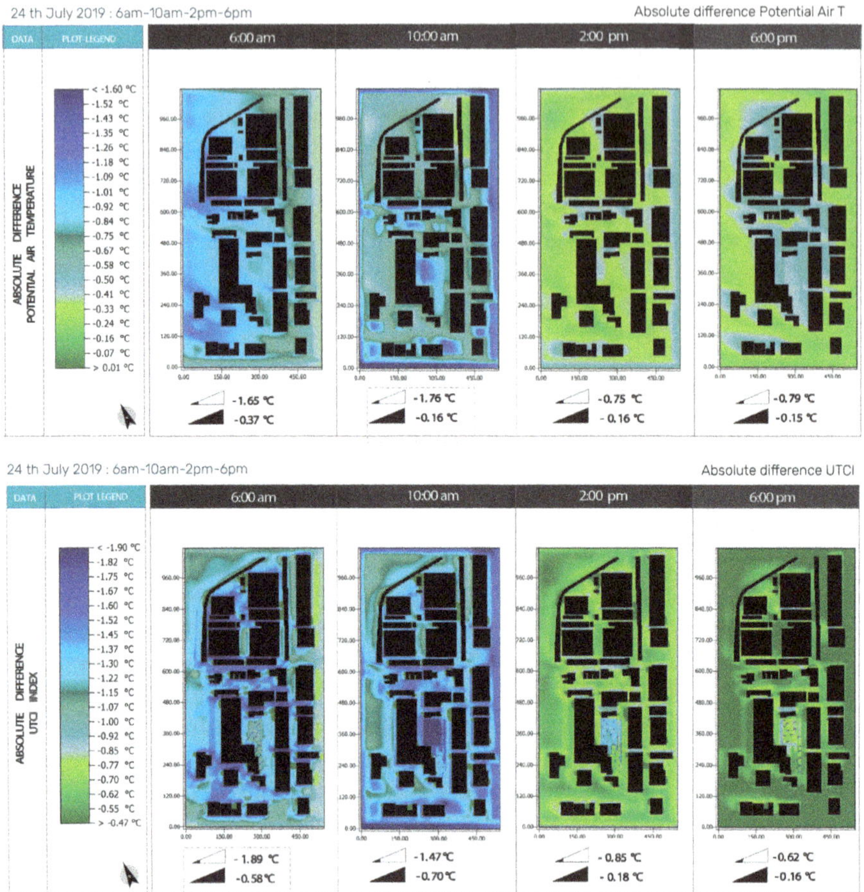

Fig. 8 Results for the green scenario on the summer day of the 24th of July, 2019. Absolute difference of air temperature (above) and UTCI (below) at 6 a.m., 10 a.m., 2 p.m. and 6 p.m. Source: Francisca Tapia

minimum and −0.16 °C at the maximum compared to the current scenario (Fig. 8). These values suggest a significant heat stress for pedestrians.

The results show that while green roofs can enhance air temperature and outdoor comfort, their integration with other NbS such as tree planting, water bodies, and urban parks is recommended to provide further benefits at the pedestrian level. In particular, trees provide cooling through evapotranspiration and shade during the day, improving outdoor comfort and contributing to the overall microclimate amelioration. Green roofs also contribute to strategies combating urban heat islands by increasing permeable surfaces and evapotranspiration.

The potential for creating cool areas of different sizes, shapes, orientations, and connectivity on rooftop levels positions green roofs not only to improve the comfort of building occupants but also to provide recreational opportunities. In future, envisioning the roofs integrated with an urban park linking various green roofs could further contribute to a holistic and sustainable urban environment.

Although less appreciated than other technologies competing for roof space, such as solar active systems, green roofs and energy generation can effectively complement each other. In the industrial area of Bolzano, several buildings already have solar panels on their roofs, so it is proposed to combine the efforts and benefits of these technologies. In this context, implementing bio-solar roofs can be a suitable alternative for existing buildings that already use solar panels, allowing for sustainable practices and maximizing available resources (Fleck et al. 2022) A comprehensive assessment should also include factors such as biodiversity enhancement, aesthetic improvement, and ecological balance maintenance. This highlights the multiple benefits of green roofs in urban environments.

Co-design, Service Design & Stakeholder Engagement

Green infrastructure (GI) as NbS is considered an integral part of urban planning and a positive approach to the application of green spaces. The concept of NbS suggests that the implementation of GI is a shared challenge for academia, public sectors, and practical organizations involved in green space matters. It serves to foster interdisciplinary and political collaboration in promoting environmental consciousness (Ferreira et al. 2020).

In further exploring the effectiveness of green infrastructure in the industrial neighbourhood, a service design approach is considered useful as a catalyst for planning NbS such as green roofs. Service design, as defined by Delft University of Technology and Wageningen University & Research, (Van der Meulen 2019) serves as a backbone for policymaking that facilitates the initiation and sustainable implementation of nature-based solutions. Using service design methodology for GI improves environmental quality in neighbourhoods and aids decision-making in urban and regional planning. Treating green infrastructure as a service involves stakeholders in environmental and greenspace planning at different levels (Elderbrock et al. 2020). This approach fosters diverse partnerships and synergies within and between the public and private sectors, ensuring that NbS are not superimposed but collaboratively developed. Given that NbS aim to benefit the environment, society, and the economy, service design becomes instrumental in establishing connections between these sectors. The establishment of various partnership models represents a new synergy between the public and private sectors, moving away from a model imposed solely by the government alone.

To apply service design as an integrated methodology, a framework is proposed that recognizes the public and private sectors as key partners in addressing the aforementioned challenges. This framework establishes connections between technical assessments and stakeholders, utilizing data from previous city masterplans as valuable guides for proposal implementation. Green roof initiatives need to be tailored to meet the specific needs of communities, going beyond a simple investment policy imposed by local governments.

Documentation, policies, and plans, such as the Masterplan of Bolzano and municipal regulations, play a crucial role in comprehending the tools and resources required for implementing new initiatives. These sources serve as valuable references for crafting proposals, including the co-planning proposal derived from the Municipality's 2019 green roof plan. The service design approach adopts a forward-looking perspective for sustaining these initiatives, outlining objectives and strategies collaboratively with relevant stakeholders and in consideration of available time and resources. It delineates the scope of actions, establishes targets, and underscores key steps for implementing the green roof plan within a specific timeframe. In the case of Bolzano South, service design plays a crucial role in establishing a collaborative platform between the Municipality, academia and private institutions. This platform is closely linked to an infrastructural framework where expert assessments, stakeholder participation, and political involvement are essential for developing effective actions and maintenance plans (see Fig. 9). The infrastructure must

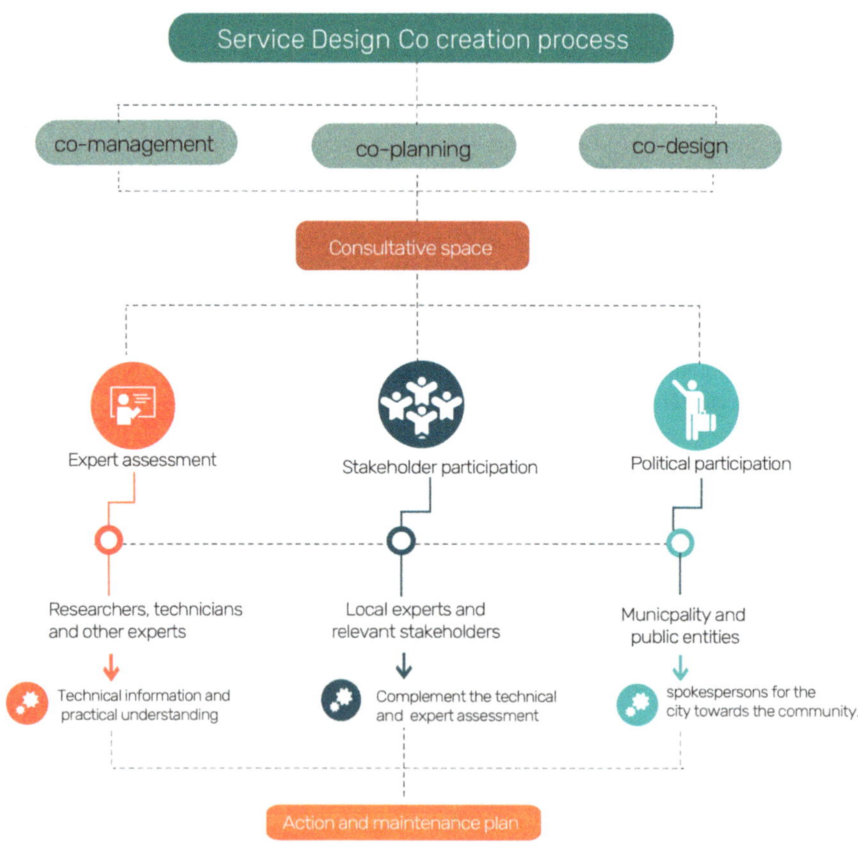

Fig. 9 Service design co creation process. Source: Francisca Tapia

be customized to meet the actual needs of communities, going beyond a mere investment policy imposed by local governments. By incorporating service design methodology into the implementation of green infrastructure, not only do neighbourhood green initiatives receive a boost, but it also cultivates more informed decision-making in urban and regional environmental planning (Willems et al. 2020).

To formulate an initial proposal for the service design approach in Bolzano South, three interviews were conducted with key institutions in the public and private sectors; detailed information can be found in (Tapia 2019).The purpose of these interviews was to gather insights on the status of green initiatives in the city and the potential benefits of green roofs in the industrial area, including their social, economic, and environmental contributions. Additionally, the interviews identified challenges and barriers that the green roof proposal can address, as well as the requirements for implementing this urban green initiative in the future.

The engagement with local stakeholders in Bolzano South included conducting interviews and workshops with municipal representatives, private entities, and professionals involved in green roof design. The focus of these engagements was on the development of horizontal greening systems on buildings, the role of the municipality, and the significance of vegetation as ecological infrastructure.

Stakeholders' insights revealed a collective understanding of the detrimental effects of UHI and overheating and highlighted the importance of creating an ecological infrastructure to improve the overall quality of life. Common findings from the three interviews were identified. For instance, it is crucial to establish strategies and adopt policies that are in line with green infrastructure and biodiversity, while also encouraging investments in NbS, to ensure a successful implementation and actively involve the community in the process. Past participatory initiatives in the city have highlighted certain challenges to implementation, including limited technical expertise, insufficient collaboration among various municipal departments (e.g., urban greening and water), and the necessity for enhanced communication strategies with residents.

Regarding the proposal for the green roof, stakeholders expressed the opinion that small individual green spaces alone would not be adequate for this purpose. They suggested that implementing a larger system or redesigning larger roof areas, in conjunction with utilizing public land, would offer greater benefits. Green roofs were recognized as a key measure to reduce soil sealing and increase water retention, improve building insulation, and add aesthetic and branding value to structures. Reduction of UHI was not perceived as a major benefit. This likely resulted into a preference for intensive green roofs due to their better thermal performance and accessibility compared to extensive ones, but this was linked to concerns about maintenance requirements and costs. While acknowledging conflicts with solar systems, stakeholders recognized the positive effects of vegetation in improving the efficiency of PV panels. However, the high costs associated with the development of an integrated configuration and the increased maintenance requirements were highlighted. Regarding the objective of reducing overheating and UHI in Bolzano South, the stakeholders agreed that horizontal greening offers less possibilities than vegetation on the ground. Therefore, the importance of integrated urban planning measures was highlighted, requiring coordinated efforts on both roofs and ground spaces.

Considering the viewpoints obtained from the interviewees and the conclusions drawn, it is important to highlight the concept of Service Design as a proposed approach that involves networking and assigning specific tasks to each participant within a suggested timeframe. The service design proposal provided serves as a schematic model and does not cover management and maintenance plans, as these aspects go beyond the study's scope. However, it is advisable that when employing this approach in the NbS co-creation process, it can be advantageous for formulating initial operational models (Tapia 2019). The findings from the microclimate analysis were supported by the results of interviews, highlighting the significance of collaboration between public and private sectors in promoting green practices within communities. The service design methodology has proven to be successful by offering a structure to identify initial maintenance tasks and ensure the long-term viability of new green initiatives in the industrial neighbourhood. To optimize the effectiveness of NbS initiatives, it is advisable to institute a coordination board team employing service design methodology. The active engagement of research institutes and the ongoing enhancement of staff skills through tailored training programs are imperative components. Motivating personnel is equally pivotal for realizing successful outcomes. In conclusion, embracing a co-design approach such as service design not only amplifies the efficacy of green infrastructure initiatives but also contributes to their enduring sustainability, thereby mitigating the need for expensive retrofits in the future (See Fig. 10).

Lessons Learned and Conclusions

The landscape and environmental analysis for Bolzano revealed a city facing the environmental challenge of decreasing thermal comfort due to the changing climate as well as urban transformation processes, despite its commitment to sustainability. The scarcity of public green spaces, ranging from 11% to 1% across neighbourhoods, is a critical concern, especially when compared to the more abundant private green spaces, varying from 17% to 6%. The presence of significant river corridors, such as the Talvera, Isarco, and Adige rivers, underscores the crucial role of green spaces as NbS in Bolzano's ecological and urban development. However, challenges persist, particularly in the city center as well as in the industrial area, where high construction density has led to the fragmentation of public green spaces or creation of 'concrete deserts' exacerbating the urban heat island effect. For a sustainable and resilient future, addressing these deficiencies and enhancing the provision of public green spaces as NbS should be a priority in Bolzano's urban planning and development strategies. This not only aligns with legal standards but also ensures a healthier and more vibrant (working) environment for the city's (temporary) residents.

The definition of urban surface uses, in particular to host NbS, is a multidisciplinary and complex process, which requires both quantitative (e.g.

Fig. 10 View of green-roof implementation at building and pedestrian level. Source: Francisca Tapia

microclimate analyses) and qualitative data to make informed decisions. It also involves different disciplines: urban planning, architecture, urban climatology, and building physics. Depending on the site-specific objectives, several other disciplines may be required, ranging from biology to economics, and sociology. In fact, in the Bolzano South case study, limiting the analysis of the potential of green roofs to pedestrian comfort would have led to the idea that they would not bring enough benefit to the area. However, the involvement of stakeholders showed that they were fully aware of the limited impact of green roofs on the microclimate, and that their decision to implement them was mainly linked to other benefits, such as improving air quality, reducing runoff and improving runoff quality, reducing energy consumption of buildings, and recreation for users. Rather than describing green roofs as a solution to address one specific challenge, the results highlight the importance of considering their multi-functionality. Continuous collaboration between private and public entities is essential to activate this multi-functionality, aligning strategies for defined function in particular with biodiversity goals. NbS have the potential to benefit the environment, society, and the economy, connecting the public and private sectors. Service design plays a crucial role in facilitating these connections.

Involving stakeholders from the beginning allows for shared responsibilities, ownership, and long-term sustainability. Early collaboration between public and private sectors improves success rates and emphasizes the importance of Service Design in development. Recognizing the importance of previous urban planning tools, service design is emerging as a participatory approach, but it requires further development and a detailed co-planning framework for effective implementation.

The results of the microclimatic analysis are limited to the presence of green roofs on neighboring buildings, elevated at least 10 m above street level. The implementation of green roofs reduces the air temperature during the hottest hour of the day by an average of −0.40 °C, with a maximum reduction of almost −0.80 °C compared to the current scenario without greenery. While this reduction may not seem significant, incorporating vegetation at ground level could have an even greater positive impact on temperature. In the context of warmer industrial zones in urban landscapes, it is important to recognize the value of increasing vegetation and creating urban parks to enhance the overall quality of outdoor spaces. It is therefore clear that there is a need to integrate greenery into these areas.

As regards the activation of green roofs as a nature-based solution, it remains a valuable option in particularly densely built environments, although, as the microclimate analysis has already shown, they develop their full potential if combined with additional solutions such as green walls or even urban green spaces on the ground. They should not be considered a silver bullet for all different challenges, but if planned carefully, strategically and multifunctionally they offer a light of hope in sometimes ‘desperate’ situations.

References

Balvedi N, Giglio T (2023) Influence of green roof systems on the energy performance of buildings and their surroundings. J Build Eng 70:106430. https://doi.org/10.1016/j.jobe.2023.106430

Bröde, P., Krüger, E., & Rossi, F. (2011) Assessment of urban outdoor thermal comfort by the Universal Thermal Climate Index (UTCI).

Chokhachian A, Santucci D, Auer T (2017) A human-centered approach to enhance urban resilience, implications and application to improve outdoor comfort in dense urban spaces. Buildings 7(4) Available at: https://www.mdpi.com/2075-5309/7/4/113

Cohen-Shacham E, Walters G, Janzen C, Maginnis S (eds) (2016) Nature-based solutions to address global societal challenges. IUCN International Union for Conservation of Nature. https://doi.org/10.2305/IUCN.CH.2016.13.en

Corcelli F, Fiorentino G, Petit-Boix A, Rieradevall J, Gabarrell X (2019) Transforming rooftops into productive urban spaces in the Mediterranean. An LCA comparison of agri-urban production and photovoltaic energy generation. Resour Conserv Recycl 144:321–336. https://doi.org/10.1016/j.resconrec.2019.01.040

Cousins JJ (2021) Justice in nature-based solutions: research and pathways. Ecol Econ 180:106874. https://doi.org/10.1016/j.ecolecon.2020.106874

Croce S, Vettorato D (2021) Urban surface uses for climate resilient and sustainable cities: a catalogue of solutions. Sustain Cities Soc 75:103313. https://doi.org/10.1016/j.scs.2021.103313

Dumitru A, Frantzeskaki N, Collier M (2020) Identifying principles for the design of robust impact evaluation frameworks for nature-based solutions in cities. Environ Sci Policy 112:107–116. https://doi.org/10.1016/j.envsci.2020.05.024

Eggermont H, Balian E, Azevedo JMN, Beumer V, Brodin T, Claudet J, Fady B, Grube M, Keune H, Lamarque P, Reuter K, Smith M, van Ham C, Weisser WW, Le Roux X (2015) Nature-based solutions: new influence for environmental management and research in Europe. GAIA Ecol Perspect Sci Soc 24(4):243–248. https://doi.org/10.14512/gaia.24.4.9

Elderbrock E, Enright C, Lynch KA, Rempel AR (2020) A guide to public green space planning for urban ecosystem services. Land 9(10):391. https://doi.org/10.3390/land9100391

Eurac Research. (2018). Climate report. https://webassets.eurac.edu/31538/1630573573-rapporto-clima-2018-itnew.pdf

European Commission, D.-G. for R. and I (2015) Towards an EU research and innovation policy agenda for nature based solutions & renaturing cities. In: Final report of the Horizon 2020 expert group on 'Nature based solutions and renaturing cities'. Publications Office of the European Union. https://op.europa.eu/en/publication-detail/-/publication/fb117980-d5aa-46df-8edc-af367cddc202

European Commission, D.-G. for R. and I (2021) Evaluating the impact of nature-based solutions: a handbook for practitioners. Publications Office of the European Union. https://doi.org/10.2777/244577

Ferreira V, Barreira AP, Loures L, Antunes D, Panagopoulos T (2020) Stakeholders' engagement on nature-based solutions: a systematic literature review. Sustainability 12(2):640. https://doi.org/10.3390/su12020640

Fleck R, Gill R, Pettit TJ, Torpy FR, Irga PJ (2022) Bio-solar green roofs increase solar energy output: the sunny side of integrating sustainable technologies. Build Environ 226:109703. ISSN 0360-1323. https://doi.org/10.1016/j.buildenv.2022.109703

Foster S, Iaione C (2016) The city as commons. Yale Law Policy Rev 34(2):281. Connecticut: Yale Law School

Franck U, Krüger M, Schwarz N, Grossmann K, Röder S, Schlink U (2013) Heat stress in urban areas. Indoor and outdoor temperatures in different urban structure types and subjectively reported well-being during a heat wave in the city of Leipzig, Germany. Meteorol Zeitschrift. https://doi.org/10.1127/0941-2948/2013/0384

Gantioler S (2019) The right to ecological space in the city. TUM. University Press. https://www.ub.tum.de/en/node/7868

Gantioler S, Siclari I, McConaghy C, Croce S, Zambelli P, Zilio S, Giacovelli G, D'Alonzo V, Balest J, Mavrigiannaki A, van Lierop M, Gupta A, Chapman E, Vance C, Lim Y (2023) Conceptual and action framework on low carbon | high air quality nature-based solutions. JUSTNat Deliv 2(1). https://doi.org/10.5281/zenodo.7669321

Givoni B (1998) Climate considerations in building and urban design. John Wiley & Sons, Inc, USA

Henzler K, Horn R, Jäger M, Maier SD (2019) Sustainability of innovative urban surfaces—a new approach of assessment. IOP Conf Ser: Earth Environ Sci 323(1):012068. https://doi.org/10.1088/1755-1315/323/1/012068

Huttner S, Bruse M (2009) Numerical modeling of the urban climate–a preview on ENVI-met 4.0. In: 7th international conference on urban climate ICUC-7, Yokohama

Kellett J (2011) More than a roof over our head: can planning safeguard rooftop resources? Urban Policy Res 29(1):23–36. https://doi.org/10.1080/08111146.2010.514848

Kottek M, Grieser J, Beck C, Rudolf B, Rubel F (2006) World map of the Köppen-Geiger climate classification updated. Meteorol Zeitschrift 15(3):259–263. https://doi.org/10.1127/0941-2948/2006/0130

LAND. (2021, October 28). Piano del Verde di Bolzano: Analisi e quadro di sintesi interpretativo. Retrieved from https://opencity.comune.bolzano.it/Documenti-e-dati/Documenti-tecnici-di-supporto/Piano-del-verde-di-Bolzano

Leistner P, Kaufmann A, Koehler M, Würth M, Hofbauer WK, Dittrich S (2019) Building physics design of urban surfaces. IOP Conf Ser: Earth Environ Sci 323(1):012067. https://doi.org/10.1088/1755-1315/323/1/012067

Lobaccaro, G., Croce, S., Vettorato, D. & Carlucci, S. (2018) A holistic approach to assessing the exploitation of renewable energy sources for design interventions in the early design phases. Energy Build 175:235–256. Italy: Science Direct.

Municipality of Bolzano. (2020). Sustainable energy and climate action plan. https://www.comune.bolzano.it/UploadDocs/30270_PAESC_ITA.pdf

O'Sullivan F, Mell I, Clement S (2020) Novel solutions or rebranded approaches: evaluating the use of nature-based solutions (NBS) in Europe. Front Sustain Cities 2. https://doi.org/10.3389/frsc.2020.572527

Papathoma-Köhle M, Ulbrich T, Keiler M, Pedoth L, Totschnig R, Glade T, Schneiderbauer S, Eidswig U (2014) Vulnerability to heat waves, floods, and landslides in Mountainous Terrain. In: Assessment of vulnerability to natural hazards. Elsevier, pp 179–201. https://doi.org/10.1016/B978-0-12-410528-7.00008-4

Pappaccogli, G. (2019) Numerical modelling of urban atmospheric processes and energetics from city to building scale: case study in Bolzano [Free University of Bozen-Bolzano]. https://bia.unibz.it/esploro/outputs/doctoral/Numerical-modelling-of-urban-atmospheric-processes/991006593992601241?institution=39UBZ_INST

Pauleit S, Zölch T, Hansen R, Randrup TB, Konijnendijk van den Bosch C (2017) Nature-based solutions and climate change—four shades of green. In: Kabisch N, Korn H, Stadler J, Bonn A (eds) Nature-based solutions to climate change adaptation in urban areas: linkages between science, policy and practice. Springer International Publishing. https://doi.org/10.1007/978-3-319-56091-5_3

Peng, L.L.H. & Jim, C.Y. (2013) Green-roof effects on neighbourhood microclimate and human thermal sensation. Energies 6(2):598-618. https://doi.org/10.3390/en6020598

Pico T, Lim Y, van Lierop M, Chapman E, Meister J, Bukovski V (2023) State of the art report on good practice for co-governance of NbS. JUSTNat Deliv 7(1) Zenodo https://justnatureproject.eu/resources/https-justnatureproject-eu-resources-public-deliverables/d7-1-state-of-the-art-report-on-good-practice-for-co-governance-of-nbs

Programma Interreg IV Italia-Austria (2007–2013). Available at: https://www.provincia.bz.it/politica-diritto-relazioni-estere/europa/finanziamenti-ue/interreg-v-a-italia-austria.asp

Rakha, R., Zhand, P. & Reinhart, C. (2017) A framework for outdoor mean radiant temperature simulation: towards spatially resolved thermal comfort mapping in urban spaces. Available at: https://www.semanticscholar.org/paper/A-Framework-for-Outdoor-Mean-Radiant-Temperature-%3A-Rakha-Zhand/9a273b878b033233f3e8342b261d0f4f50345556

Redazione (2019) Agenda Bolzano, Assoimprenditori: bene gli interventi in Zona industriale. In: La Voce di Bolzano. Available at: https://www.lavocedibolzano.it/agenda-bolzano-assoimprenditori-bene-gli-interventi-in-zona-industriale/

Reid WV, Mooney HA, Cropper A, Capistrano D, Carpenter SR, Chopra K, Dasgupta P, Dietz T, Duraiappah AK, Hassan R, Kasperson R, Leemans R, May RM, McMichael AJ, Pingali P, Samper C, Scholes R, Watson RT, Zakri AH et al (2005) Ecosystems and human well-being—synthesis: a report of the millennium ecosystem assessment. Island Press. https://library.wur.nl/WebQuery/wurpubs/340442

Russo, A. (2008). Quantifying the local-scale ecosystem services provided by urban treed streetscapes in Bolzano. Bolzano.

Santamouris M, Ding L, Fiorito F, Oldfield P, Osmond P, Paolini R, Prasad D, Synnefa A (2017) Passive and active cooling for the outdoor built environment—analysis and assessment of the cooling potential of mitigation technologies using performance data from 220 large scale projects. Solar Energy 154:14–33. https://doi.org/10.1016/j.solener.2016.12.006

Somarakis, G., Stagakis, S., & Chrysoulakis, N. (2019). ThinkNature Nature-based Solutions handbook. ThinkNature project funded by the EU Horizon 2020 research and innovation programme under grant agreement No. 730338. https://platform.think-nature.eu/system/files/thinknature_handbook_final_print_0.pdf

Tapia F (2019) Green infrastructure delivered as nature-based solutions in an eco-industrial neighbourhood by service design approach. A case study in Bolzano. Archivio digitale delle tesi di laurea e di dottorato. Politecnico Di Milano. https://www.politesi.polimi.it/handle/10589/154344

Tsoka S, Theodosiou T, Tsikaloudaki K, Flourentzou F (2018) Modeling the performance of cool pavements and the effect of their aging on outdoor surface and air temperatures. Sustain Cities Soc 42:276–288. https://doi.org/10.1016/j.scs.2018.07.016

UNEP (2022) Resolution on nature-based Solutions for supporting sustainable development. United Nations Environment Programme UNEP/EA.5/Res.5, United Nations Environment Assembly UNEA 5.2. http://www.unep.org/environmentassembly/unea-5.2/proceedings-report-ministerial-declaration-resolutions-and-decisions-unea-5.2

United Nations Environment Programme UNEP/EA.5/Res.5, United Nations Environment Assembly UNEA 5.2 (n.d.) Resolution on Nature-based Solutions for supporting sustainable development. http://www.unep.org/environmentassembly/unea-5.2/proceedings-report-ministerial-declaration-resolutions-and-decisions-unea-5.2

Van der Meulen, G. (2019) Nature-based metropolitan solutions. TU Delft. OpenCourseWare. https://ocw.tudelft.nl/courses/nature-based-metropolitan-solutions/

Vuckovic M, Kiesel K, Mahdavi A (2017) Studies in the assessment of vegetation impact in the urban context. Energy Build 145:331–341. https://doi.org/10.1016/j.enbuild.2017.04.003

Wang Y, Zhou D, Wang Y, Fang Y, Yuan Y, Lv L (2019) Comparative study of urban residential design and microclimate characteristics based on ENVI-met simulation. Indoor Built Environ 28(9):1200–1216. https://doi.org/10.1177/1420326X19860884

Welch JL (1985) Research marketing problems and opportunities with focus groups. Ind Market Manage 14(4):245–253. Available at: https://www.sciencedirect.com/science/article/abs/pii/0019850185900173

Willems JJ, Molenveld A, Voorberg W, Brinkman G (2020) Diverging ambitions and instruments for citizen participation across different stages in green infrastructure projects. Urban Plan 5(1):22–32. https://doi.org/10.17645/up.v5i1.2613

Zheng G, Li K, Wang Y (2019) The effects of high-temperature weather on human sleep quality and appetite. Int J Environ Res Public Health 16(2):270. https://doi.org/10.3390/ijerph16020270

Zhou B, Rybski B, Kropp JP (2017) The role of city size and urban form in the surface urban heat island. Scientific Rep 7(1):4791. Available at: https://www.nature.com/articles/s41598-017-04242-2

Baltimore's Community Resiliency Hub Program: A Model for Advancing Climate Justice and Preparedness

Aubrey Germ

Abstract Since 2015, the City of Baltimore (the City) has been growing and facilitating a network of frontline, service-based, and mission-driven organizations that are dedicated to enhancing the resilience of the city's most climate-vulnerable neighborhoods. The City cultivates relationships with and aligns capacity-building resources for these community-trusted organizations providing daily services for vulnerable community members with the goal that in times of climate emergency, the organizations have the knowledge, training, confidence, partnerships, and capabilities to continue operating and providing essential resources for those who are most impacted. This concept has evolved into Baltimore's Community Resiliency Hub Program (CRH Program, or program), a nation-leading model for community-based resilience where local government and community-based organizations in underserved communities are working in close coordination to build place-based resilience before, during, and after disasters.

Community Resiliency Hub Partners (CRH Partners, or partners) receive support from the city in many forms including high-quality emergency preparedness supplies, energy efficiency upgrades to their buildings, back-up power capabilities (solar power and battery storage, if feasible), emergency preparedness and response training, connections to grant funding opportunities, and focused support, resources, and communications from the city. The City works with CRH Partners to enable them to serve as a safe space where vulnerable neighbors can gather in times of emergency; access reliable power for their essential devices; receive supplies, food, and drinking water; and store medications sensitive to temperature, among other things. CRH Partners can also serve as community-based staging areas for emergency and recovery personnel as well as conduits for critical resource distribution to the community. After successfully supporting community needs throughout heatwaves, floods, the COVID-19 Pandemic, and other local incidents, CRH Partners in Baltimore have proven to be well-positioned and highly successful at providing life-saving resources to vulnerable and hard-to-reach individuals, demonstrating

A. Germ (✉)
ICF, Reston, VA, USA

G. Iannaccone, E. V. Ellis (eds.), *Towards a Nature-Positive Built Environment*, Research for Development, https://doi.org/10.1007/978-3-031-95292-0_10

that they can be an effective solution to advancing climate justice and preparedness. Other cities and communities can learn from the successes and challenges of Baltimore's community-based resilience hub model and employ these lessons in their own communities.

Keywords Resilience hub · Hazard mitigation · Emergency preparedness · Emergency response · Climate resilience · Community-driven design

Baltimore's Community Resiliency Hub Program: An Overview

Since 2015, the City of Baltimore's Office of Sustainability, housed within the Department of Planning, has been leading an innovative program called the Community Resiliency Hub Program (CRH Program, or program) focused on increasing resilience within the City's most vulnerable communities by working hand in hand with local community partners. The overarching goals of the program are to build capacity within underserved communities and improve the City's responsiveness to community needs before, during, and after emergencies. Another core tenet of the program is to build trust between communities and city government through regular connection and resource sharing.

The CRH Partners are frontline, community-based non-profits and faith organizations that community members know and trust. Partners are willing to work with the City to expand their capacity to serve as well as advocate for community needs before, during, and after various climate hazards and other community disruptions. The City, in turn, commits to listening and facilitating access to additional resources to build their knowledge, confidence, and capacity to support continuity of operations and scale up their impact.

The program is not intended to place an additional burden on already burdened communities. The value of the program comes from the City's understanding that government is not always trusted by community members—especially communities of color, especially in Baltimore, and especially in emergencies. Decades of injustice, disinvestment, and inequity have led to very real and valid distrust of city government. Therefore, familiar faces from an existing service-based organization that is city-supported but not city-led can be very powerful in filling a gap by reaching hard-to-reach members of the population and providing life-saving services. The vision is to collectively work better together, government and community, by mutually listening to and supporting one another and slowly building trust by consistently showing up.

Another key objective of this model is to not only improve resource access in times of emergency, but day-to-day as well. The holistic resource investments that comprise Baltimore's CRH Program have been shown to drive additional investments over time. Through a lens of climate justice, Baltimore is reducing historic inequities present in underserved communities by providing more equitable access to emergency preparedness and response resources. The program provides a pathway for communities to voice their concerns directly to city government, enabling the delivery of resilience-building resources that reflect evolving community priorities and needs.

Background

Baltimore City is located along the Patapsco River, a tributary of the Chesapeake Bay. The city is vulnerable to a number of climate hazards including flooding, coastal storms, heavy precipitation, wind events, extreme temperatures, and sinkholes that can seriously impact infrastructure and communities. Combined with a projected rise in sea level, climate change will continue to exacerbate the impacts of these hazards and create disproportionate burdens on Baltimore's most vulnerable community members. Climate hazards often cause the most strain for low-income, disabled, elderly, unhoused, or socially vulnerable residents who face greater barriers to preparedness and recovery after a disruption. The legacy of historic redlining and other inequitable practices have led to stark disparities in resource access, public health, environmental health, and infrastructure investment, among others. (Baltimore City Department of Planning 2023)

Since 2013, Baltimore's Department of Planning has been overseeing the development and implementation of the city's Disaster Preparedness and Planning Project (DP3). The DP3 is Baltimore's Hazard Mitigation Plan, a FEMA-requisite for accessing federal mitigation funding. The DP3 details the city's vulnerabilities to hazards as well as opportunities for building resilience by strengthening buildings, utilities, nature-based solutions, and emergency preparedness, as well as intentionally improving social systems and resource access.

The foundation for Baltimore's Community Resiliency Hub Program (CRH Program) lies in the DP3. Specifically calling for establishing and growing the CRH Program in this foundational document has enabled the city to take a planning-based approach to program implementation. Embedding the CRH Program in a strategic plan that has significant community input as well local, state, and federal approval processes gives the program a strong base from which to build during plan implementation. It also positions the program, as well as other mitigation actions outlined in the plan, favorably for capturing FEMA mitigation funding. This strategy proved successful in 2023 when a $7.3 million FEMA award was granted to the city through FEMA's Hazard Mitigation Grant Program (HMGP).

The impacts of climate change are clear. Cities across the country are being impacted in unforgettable and devastating ways. In 2021 alone, there were extreme heatwaves in the Pacific Northwest and snowstorms accompanied by widespread power outages in Texas. Summer months continue setting global temperature records. It is commonly known that extreme climate events are happening and will likely worsen over time, continuing to disproportionately impact the same communities that have been overburdened in the past. Thus, it is imperative that cities and communities determine equitable and holistic approaches to building resilience both in terms of climate mitigation *and* climate adaptation. Baltimore's CRH Program sits at this precise intersection, serving to mitigate climate change by reducing greenhouse gas emissions through local renewable energy deployment as well as increasing the adaptive capacity of underserved communities across the city.

Baltimore's CRH Program Structure

The Baltimore Office of Sustainability is the lead agency that is responsible for growing and managing the Program as a key strategy of Baltimore's Disaster Preparedness Plan (DP3). The Office of Emergency Management (OEM) and Department of Health (BCHD) are also critical partners in the overall success of the program, largely through their role in emergency logistics coordination as well as the wide range of emergency preparedness, response, and recovery resources to which they can connect CRH Partners. Baltimore's CRH Program utilizes a network-based approach, serving to facilitate relationships amongst the CRH Partners themselves as well as build a partnership between the CRH Partners and the City. While the Office of Sustainability serves as the facilitator of the program, the program is designed to be community-driven. The program is centered around listening, adapting, and responding to the unique and evolving needs of the CRH Partners individually and as a network spanning the many diverse geographies, neighborhoods, and demographics across the city.

The CRH Partners range from community-serving faith organizations and food pantries to non-profits that provide after-school programming, job training, outdoor education, aging services, housing support, women's empowerment, case work, and re-entry programming, among many other services. These organizations are housed within a range of different building types such as historic churches, older school buildings, warehouses, and firehouses and are locally known and trusted. The daily work and mission of each organization creates critical touch points with highly vulnerable individuals on a regular basis to help meet the community's daily needs and improve their quality of life. As such, these buildings and organizations are considered 'natural fit' partners as they are essentially already serving as a resilience hub (in the broad definition of resilience) during 'blue skies,' or non-emergency situations. By officially partnering with the City and joining the CRH Program, organizations add the climate lens to their resilience efforts and have access to an additional level of support, connection, and resources to become better equipped to provide a temporary, expanded level of essential services during climate emergencies. They also have the 'ear of the city' and can more directly advocate for the evolving needs of their community members before, during, and after disasters.

The City commits to coordinating additional capacity-building resources for the CRH Partners including high-quality emergency preparedness supplies to keep on-site or loan out to community members, energy efficiency upgrades to their building, back-up power capabilities (solar power and battery storage, if feasible), emergency preparedness and response training, connections to and technical assistance for grant funding opportunities, and focused support, resources, and communications from the City. Aligning funding for each of these program components can be challenging and the city continues to take a multi-faceted approach by sourcing federal, state, and private local grant funding opportunities when city funds aren't necessarily available.

Since 2020, the City has held monthly virtual meetings to facilitate peer-to-peer learning amongst Hub leaders and provide regular touchpoints with City staff to share timely information and resources. Meetings are intended to be fun, inclusive, and informational, and topics vary based on seasonal or evolving needs of CRH Partners. Discussions begin with a community check-in where CRH Partners share community outreach events they are leading or supporting (such as food distributions or job training programs), any relevant happenings in their neighborhoods, and any immediate or long-term needs that either the City or another CRH Partner can support. The remainder of the meetings are discussion-based and include but are not limited to extreme heat season resources and water distribution coordination, extreme cold coordination, preparing for or sharing information about disaster preparedness and training activities, discussing ideal grant funding opportunities for program components, gathering program feedback, and discussing prospects for CRH Partner involvement in other city programs such as hosting city-sponsored resource distribution events. After each meeting, a comprehensive follow up email is sent with meeting notes and resources discussed for the CRH Partners that could not be in attendance. These 'blue skies' meetings are critical to overall CRH Program success and create structure and momentum as well as foster relationship-building so that, if and when a climate or other emergency happens, CRH Partners feel comfortable establishing contact with city staff and elevating community needs.

Partner Identification

As of 2024, over 20 organizations had joined the program as CRH Partners, reflecting its expanding impact. While several criteria are used to help identify strong CRH partners, there is no strict formula or blueprint. The City takes a flexible approach to CRH partnership, intentionally creating a low barrier to entry to maximize the program's accessibility and bridge the trust-gap. The City conducts outreach to ideal program partners based on resources available as well as considers requests by organizations interested in joining the program. The major determining factors are based around a series of questions asking whether the organization is:

- Already providing regular community-facing services that local residents know and trust;
- Located in a climate-vulnerable neighborhood;
- Working with highly vulnerable individuals who are most at risk during climate emergencies;

- Interested in learning more about climate change risks and expanding their role in climate resilience;
- Of the mindset that the CRH Program's is mission-aligned and that the Program's resources will be helpful to them and their community;
- Willing to actively participate in the program including attending meetings, trainings, and being open to back-up power and energy efficiency opportunities;
- Inclusionary in nature and committed to providing services for all members of their community.

Most 'natural fit' partners provide an affirmative answer to all questions, and many express that they are already dedicated to supporting residents no matter the circumstance, so connections to additional resources can only help. Ideally, the partnership should feel like a no-brainer and a win-win, as opposed to one that involves convincing a reluctant organization that the CRH Program is for them. Geographic distribution of CRH Partners across the city is also an important component to ensure that program resources are well distributed. The goal is to expand the program to support the establishment of CRH's in as many climate-vulnerable and underserved neighborhoods that could benefit from a local CRH.

Energy Resilience

A successful model for energy resilience in Baltimore's CRH Program has been the deployment of solar power and battery storage. Solar power and battery storage are integral components of building capacity within CRH Partners and can significantly enhance continuity of operations in the event of a disaster accompanied by a power outage. Beyond the back-up power capabilities that battery storage provides, solar power provides on-site electricity generation that reduces utility costs for CRH Partners. The reduced utility costs lead to savings for the organization and the funds are put back into their operations and community initiatives. Therefore, this renewable energy resource that provides backup power can result in multiple co-benefits. While current battery back-up technology cannot support backup power for the entire building, it does support certain essential loads like plug loads for charging cell phones or other essential devices, HVAC for heating and cooling, as well as refrigeration for food or temperature-sensitive medication.

Solar power and battery back-up systems that have been installed for CRH Partners have been fully grant funded and the city has successfully received funding from both state and federal grant programs, namely the Maryland Energy Administration (MEA) and FEMA. The city also maintains close partnership with a dedicated CRH 'solar project team' which consists of external solar companies and solar nonprofits that support the program's solar and battery offerings. These essential partners conduct solar and battery feasibility studies, provide critical technical assistance, manage grant implementation and solar and battery system installations, and even provide or align gap financing. The goal is for all solar and battery feasibility study and installation costs to be completed free of charge to the CRH Partner. Upon installation, CRH Partners can choose to own the systems themselves or enter into Power Purchase Agreements (PPAs) with a third-party owner, depending on what makes the most financial sense for them and results in maximum energy cost savings over the long term.

A strategic growth opportunity in the evolution of Baltimore's CRH Program took place from 2020-2021 with a grant award from MEA to conduct solar and battery feasibility studies for a significant number of CRH Partners and other non-profits interested in joining the program. This resulted in the alignment of organizations that had strong solar and battery feasibility and the development of preliminary system siting and designs. This created ideal pathways for the City to apply for implementation funds, including a $7.3 million FEMA HMGP grant awarded in 2023 to support an extensive network of solar and battery installations (or 'solar generators' for the purposes of the application) for sixteen of the City's CRH Partners. This funding also included capital for site enhancements necessary for the installation of solar and battery such as roof repairs or other structural enhancements. Part of the FEMA application included justification of CRH's as critical facilities and acknowledgement from the City's OEM that it considers CRH's as such. Baltimore's OEM was eager to support and provided the appropriate documentation for CRH's to be considered community-based critical facilities. This was significant because FEMA often requires critical facility designations for projects to be eligible for certain FEMA mitigation funding. This unique award came with support from, and close coordination with, the Maryland Department of Emergency Management (MDEM) and FEMA Region III. It is typically challenging for a city to receive federal funding for mitigation activities on behalf of non-city operated sites such as the nonprofits and churches of the CRH Program. This successful FEMA award sets a new bar for innovation and can inform what other cities and states are able to accomplish with FEMA funds.

It is important to note that a CRH Partner does not have to be able to receive solar and battery to participate in Baltimore's program. Many CRH Partners, especially early members of the program, are unable to receive solar and battery due to several reasons including, but not limited to, historic preservation rules, insufficient sun exposure, inadequate structural supports, and incompatible roof slope or materials. Even without solar and battery, CRH Partners can contribute uniquely to the program and support at-risk neighbors in many non-power outage situations and can still serve as critical resources and community gathering spaces. The CRH Program also seeks to support CRH Partners without solar and battery with other forms of resilient power such as solar-powered charging stations located in publicly accessible places near the organizations' buildings.

Energy Efficiency

An important complement to solar power and battery back-up systems is energy efficiency upgrades. Energy efficiency upgrades and weatherization, particularly for the types of older buildings that are home to many of the CRH Partners, are necessary to improve the operational resilience of CRH buildings and maximize the efficacy of the solar and battery systems. These upgrades also yield greater cost-savings for CRH Partners. The City has had success collaborating with a local neighborhood-improvement organization to conduct energy efficiency audits for CRH buildings and acquire MEA grants specifically for energy efficiency projects.

Preparedness Training

The objective of preparedness training as a component of the program includes providing knowledge-building resources to build confidence within organizations to prepare for and respond to emergencies. Knowing what to do, who to call, and how to help, can be empowering and reduce feelings of isolation and exhaustion during stressful times. The training offerings of Baltimore's CRH Program have evolved over time and are still evolving. However, three distinct, customized training modules have been formed based on input from CRH Partners on the type of training that would be most useful. Trainings take place via in-person or virtual workshops and are either facilitated by the city or by a consultant specializing in emergency preparedness. Embedded in each training are overviews of emergency management vocabulary to increase accessibility of the material and lower the barrier for communicating with emergency management professionals before, during, and after real emergencies.

Asset Mapping. One of the first training elements that the CRH Program developed was a customized neighborhood-based asset mapping activity to increase the organizations' awareness of the resilience of their own neighborhood. Asset mapping is a great way to help CRH Partners identify organizations, institutions, businesses, and other potential partners already located in their neighborhood that they could lean on or partner with. The goal is to prevent the CRH Partner from feeling like they are operating in isolation by facilitating additional awareness of the local assets that exist in their neighborhood as well as enhancing connections with city personnel and resources. The asset mapping activity is focused on the one-half mile radius around the Hub and asks CRH Partners to identify local organizations that they are or are not already connected to that could be useful partners in times of emergency. For example, there might be a corner store or a gas station down the street that they are not currently connected with that could be an asset in an emergency. Identifying these opportunities through asset mapping can help the CRH Partner build more hyperlocal partnerships during 'blue skies' so that they have even stronger relationships and resource access during disruptions.

Scenario Planning. Another training module developed by the CRH Program is based in scenario planning, which can be a powerful tool in understanding gaps and opportunities for improvement in preparedness and response measures. Scenario planning involves visualizing potential climate hazards of concern, often worst-case scenarios involving cascading events, and walking through the hypothetical operations and response activities. For example, what would a CRH Partner do to support their community during a severe snowstorm accompanied by a power outage? An extreme, prolonged heatwave and a power outage? At the end of the exercise, CRH Partners have a better sense of their own respective capacities and capabilities. They can identify what went well, gaps in resources or knowledge, and opportunities for improvement. The goal following the exercise is for each CRH Partner to fill out an After-Action Report that can inform their preparedness for if/when the real thing happens. Some example questions about CRH operational capacity during the scenarios included: How would your organization respond for your community? How would you get the message out to local residents? How would you communicate your situation to the City? How many people would you be able to take in? How would you handle animals? In short, walking CRH Partners and city personnel through various potential climate scenarios can identify gaps in emergency operations and coordination. It can also underscore opportunities for CRH Program improvements and highlight individual Hub needs. It is an effective way for the City to understand respective CRH Partner capacities and identify ways to continue growing and evolving the program to help fill gaps with additional resources.

Resiliency Hub Operations Planning. The CRH Program developed a customized Resiliency Hub Operations Plan Toolkit for CRH Partners to complete based on their own unique operational capacity and response goals. This resulted from CRH Partner feedback and the need for CRH leaders and volunteers to document how their Hub would operate in various situations. Traditional emergency operations plans are large, complex documents that are inaccessible to smaller organizations like churches and community-based non-profits. They are also often inward facing and focused on how to protect an organizations' assets, as opposed to how to expand operations to provide additional outward-facing services. Therefore, a Resiliency Hub-specific operations plan template was determined to be helpful guidance for CRH leaders and volunteers to build out operations plans in a structured, well-documented way.

Hosting Existing Training Programs for Staff, Volunteers, and Community Members. Other successful training opportunities have involved individual CRH Partners working with organizations like the local Office of Emergency Management (OEM), local Voluntary Organizations Active in Disaster (VOAD) groups, the American Red Cross, or FBI to host free or low-fee trainings onsite at a CRH. Community Emergency Response Training (CERT), CPR, Basic First Aid, and Stop the Bleed can be extremely valuable programs for bringing knowledge, awareness, and capacity to communities. An option for faith-based organizations can be to host other forms of training such as Active Shooter and Hate Crimes training, which are often offered by Department of Homeland Security-affiliated agencies. These can serve to both train staff and volunteers in holistic emergency preparedness and response skills as well as offer additional opportunities for community members to frequent the Hub, build connection with staff and volunteers, and better-understand the CRH's role in their neighborhood.

Emergency Supplies

Baltimore works with CRH Partners to have emergency equipment and supplies on hand that can support either the Hub or community members during various emergencies. For example, these can include snowblowers, jumper cables, sump pumps, fans, warming blankets, ice packs, canopy tents, AEDs, first aid kits, sandbags, and walkers and wheelchairs for seniors. It can also include items the Hub specifically thinks are useful for their own unique operations and community needs. Additionally, the City proactively provides supplies requested by CRH Partners before potential climate emergencies. For example, the CRH Program provides all partners with cases of bottled water in advance of extreme heat season for distribution to community members on Code Red Extreme Heat days, or sandbags to lend out to community members in advance of flood season for CRH Partners that are in flood-prone areas. The CRH Partners are encouraged to let the City know if they are running low on any supplies. Supplies and equipment are often purchased with grant funds but can also be supported by city funds, depending on the situation.

Technical Assistance

The CRH Program regularly shares relevant and timely grant opportunities with CRH Partners. The program also offers technical assistance and grant writing support for CRH Partners that have limited administrative capacity or feel that additional resources would be helpful for the operation of the Hub and the resilience of their community before, during, and after disruptions. The City is also willing to provide letters of support and grant technical assistance for sustainability and resilience-related funding opportunities that are pursued by CRH Partners. For example, the CRH Program helped connect one CRH Partner with grant funding for the development of a youth and adult learning center when the Hub identified learning and education as a critical component of their neighborhood's resilience during the COVID-19 Pandemic. The CRH Program also directed short-term emergency relief funding towards CRH Partners and provided on-call grant writing support during the COVID-19 Pandemic to enable the CRH Partners to continue operating and providing essential resources such as food, Personal Protective Equipment (PPE), and COVID-19 testing and vaccination resources to community members when it was most needed. The technical assistance component of the program has proven highly effective and helpful for CRH Partners and greatly improves the accessibility of resilience funding for underinvested communities across the city.

Challenges and Opportunities

Though Baltimore's CRH Program serves as an impactful and successful model in many ways, managing and funding Baltimore's CRH Program comes with challenges that threaten the long-term sustainability of the program. Funding for the various components of the program is fragmented and inconsistent. The majority of funding for the program's larger capital projects comes from state and federal grant programs, which can be extremely technical, competitive, and limited in availability. Many of the solar and battery grant opportunities were successfully pursued by the generosity of the city's 'solar project team' members in offering their technical expertise, as the Office of Sustainability does not carry this capacity internally. Outside of staff time dedicated by the Office of Sustainability, OEM, and BCHD, the City of Baltimore itself has not yet allocated any program-based funding beyond emergency resource provision. As the program continues to grow, accessing consistent funding for all CRH Program components will become more challenging unless a different funding model is developed. Another key issue is staff capacity. While CRH's are largely locally managed and self-directed, the program hinges upon the dedication and commitment of key city staff in listening to CRH Partners and working to change the existing paradigm. Changes or gaps in CRH Program leadership from the city-side could lead to program disruptions and uncertainty in services. A more substantial commitment to the program by city leadership through the allocation of additional staff or dedicated funding could help the program gain further traction and ensure its long-term success.

Conclusion

Through their regular services, frontline community-based organizations in urban areas already save lives daily. When partnered with local government and equipped as resilience hubs, their operational capacities are expanded to include the unique scenarios that come with climate emergencies, leading to even more lives being protected. Government-led preparedness, response, and recovery efforts are not always nimble, but a commitment to listening to and working hand-in-hand with community partners can build trust and improve emergency preparedness and responsiveness in underserved communities. Faith and service-based organizations are critical resources and pillars of strength for their urban communities. Building capacity within their existing community-driven efforts can cultivate climate resilience where it is needed most. By providing holistic resource investments and regular touchpoints for elevating community concerns before, during, and after emergencies, Baltimore's CRH Program proactively builds capacity and trust between the city and its vulnerable communities in advance of disasters. If managed well, the city-facilitated, community-based resilience hub model can be a powerful tool for the equitable advancement of climate justice and community preparedness within underserved urban areas most vulnerable to the impacts of climate change.

Additional Resources

For additional information regarding Baltimore's CRH Program, including case-studies, please reference the following:

Stoler, L., Velani, S., & Yu, A. (2023, January 12). Street Notes 4: The case for (Balti)more resilience hubs. Parachute. Retrieved May 20, 2024, from https://parachuteearth.substack.com/p/street-notes-4-the-case-for-baltimore

Muir, M., Oesterle, M., Pew, J., & George, A. (2021). Detroit Resilience Hub Framework: Best practices for the implementation of resilience hubs in Detroit, Michigan. MA Thesis, University of Michigan. 33–42. https://deepblue.lib.umich.edu/bitstream/handle/2027.42/172183/EcoWorks%20Detroit%20Resilience%20Hub_399.pdf?sequence=1

For additional information regarding resilience hub concepts, please reference the Urban Sustainability Director's Network (USDN)'s Resilience Hubs resource page.

Resilience Hubs. (2024). Urban Sustainability Director's Network (USDN). Retrieved May 20

Reference

Baltimore City Department of Planning (2023). 2023 Disaster Preparedness and Planning Project (DP3). Baltimore City. Retrieved May 20, 2024, from https://www.baltimoresustainability.org/wp-content/uploads/2024/01/2024-01-30_Baltimore2023DP3.pdf

Decarbonization Strategies for Nature-Positive Cities

From Neighborhoods Regeneration to Cities Decarbonization

Andrea Giuseppe di Stefano, Tecla Caroli, Alessio Costantino Mirabella, and Serena Girani

Abstract In order to achieve decarbonization in cities, it is crucial to spatially limit the first phase of intervention: starting with neighborhoods is fundamental for planning, social, governance, and practical reasons.

Pilot projects involving neighborhoods, whether newly built or existing, allow for initiatives with public and private actors on a smaller scale, endow policy makers and all stakeholders with an experimental attitude in measuring the benefits of different actions, and permit actors to tackle challenges and overcome obstacles with more ease and with tailored solutions. The scale of the neighborhood is appropriate for testing the readiness of communities to change and for ensuring that the change is real, rooted in the community, and supported by public and private actors.

Indeed, for the operation to be truly effective and the decarbonization targets to be reached as quickly as possible, a systemic effort is needed: a commitment by all citizens, associations, workers and professionals, the city, and even higher levels of governance along with tangible actions to make decarbonization simpler and more feasible.

Keywords Urban sustainability · Energy-efficient design · Neighborhood-scale decarbonization · Climate-neutral urban development · Sustainable urban planning · Renewable energy communities

A. G. di Stefano (✉)
Department of Architecture, Built environment and Construction Engineering, Politecnico di Milano, Milan, Italy

Sustainable Development Team, Arup, Milan, Italy
e-mail: andreagiuseppe.distefano@polimi.it

T. Caroli · A. C. Mirabella
Sustainable Development Team, Arup, Milan, Italy
e-mail: tecla.caroli@arup.com; alessio.mirabella@arup.com

S. Girani
Masterplanning and Urban Design Team, Arup, Milan, Italy
e-mail: serena.girani@arup.com

G. Iannaccone, E. V. Ellis (eds.), *Towards a Nature-Positive Built Environment*, Research for Development, https://doi.org/10.1007/978-3-031-95292-0_11

Introduction

From the onset of the industrial revolution, cities have played a pivotal role in shaping energy consumption trends. Today, with urbanization progressing unabated, more than half of the world's population lives in urban areas. These areas are not only crucial for generating over three-quarters of the world's total economic output but also contribute significantly to environmental concerns, accounting for approximately 75% of worldwide carbon emissions. This highlights the dual role of cities as both engines of economic vitality and major contributors to environmental challenges (Intergovernmental Panel on Climate Change (IPCC) 2015). The significant carbon footprint associated with urban centers highlights their complex role as both substantial contributors to climate change and key agents in addressing it. This is evidenced by the pledge of more than 100 cities globally to attain net-zero carbon emissions by 2050 (International Institute for Sustainable Development 2019). In this context, the role of urban structures is particularly noteworthy, serving as critical elements in the equation of carbon emissions. Despite the acknowledged significance of buildings in this context, the path towards achieving meaningful reductions in carbon emissions is fraught with challenges and complexities (Di Stefano et al. 2023).

The current century poses a crucial challenge to the built environment: the creation of energy-efficient urban spaces and the implementation of existing ones to align with contemporary standards of living and environmental sustainability. Tackling this issue requires a detailed comprehension of building energy usage and the identification of effective strategies to reduce this consumption. A collaborative effort involving building designers, policymakers, utility providers, managers, and occupants is essential. By combining their knowledge and resources, they can develop and apply measures that are both impactful and economically viable (Ang 2022).

City developments often have a strong focus on how to create positive outcomes on each plot, while leaving the surrounding neighborhood and communities too rarely integrated into a holistic thinking and urban development strategies. According to Birgisdóttir et al. (2023) "this results in positive impact left unrealized and a risk of adverse outcomes because of the isolated, reductionist approach" (Birgisdóttir et al. 2023) reducing also the positive feedback from the surrounding communities, suffering the impact of works for longer periods without immediate and tangible outcomes for them except, in particular cases, for increases in property values.

Regulatory interest since 2015 has improved some metrics but not overall energy consumption and carbon intensity (both embodied and operational). As depicted in Fig. 1 from the UN's 2022 Global Status Report (United Nations Environment Programme 2022) for Buildings and Construction, emission and energy intensities have slightly declined, while gross floor area, references to buildings in national policies, building energy codes, and investments have all increased, signaling growing awareness and action toward sustainable buildings. Looking at the Emission and

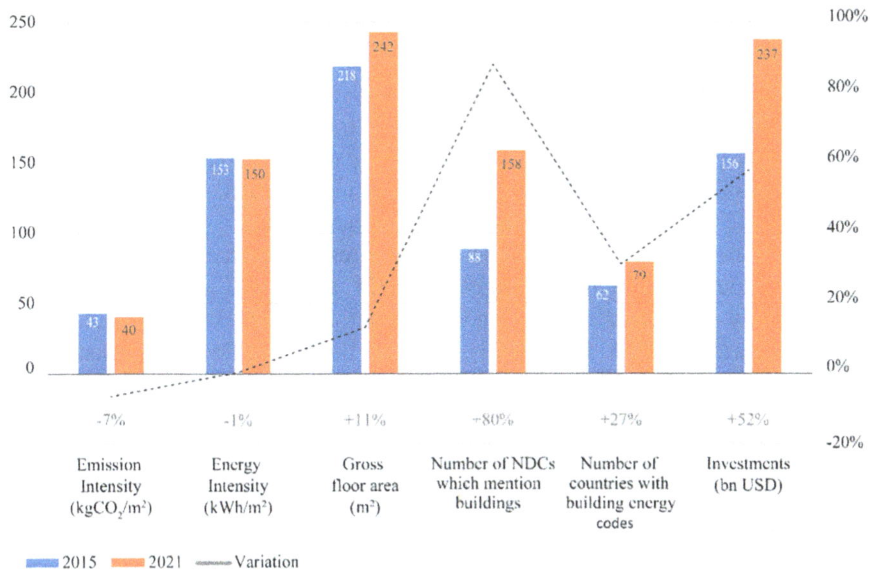

Fig. 1 Comparative analysis of sustainability metrics in real estate between 2015 and 2021. The dotted line indicates the percentage change for each metric over the six-year period, illustrating trends in the industry's efforts towards sustainability

Energy Intensity, it is possible to note a small decrease from 2015 to 2021, marked by −7% and −1% changes respectively, while Gross Floor Area shows a relatively modest increase of +1.1%. The Number of Nationally Determined Contributions (NDCs) which mention buildings has a significant increase of +80%, and the Number of countries with building energy codes has increased by +27%. Finally, the Investments have a substantial increase of +52%.

The increases in the gross floor area, the number of NDCs mentioning buildings, the number of countries with building energy codes, and the investments, suggest a growing awareness and proactive measures in sustainable building practices. Conversely, the slight reductions in emission and energy intensities indicate efficiency improvements, though modest. According to our approach, this scenario can be shifted by an increased focus on neighborhoods and urban areas in a holistic approach.

For existing cities, which are expanding due to population increase worldwide, attention must also turn to the existing building stock. To limit building stock emissions below 580 $GtCO_{2e}$ by 2050, the global renovation rate must jump from 1% to 5% annually, and all new constructions must be carbon neutral by 2030 (Weber et al. 2021).

Challenges and Limitations: The Italian and European Scenario

In the UK, reaching a 5% renovation rate, means retrofitting 1.5 homes every minute until 2050 (Timperley 2018). This urgency is translated into political actions all over Europe. For example, in Italy, in the three-year period 2020–2023 a huge program of public incentives has been introduced. It updated existing incentives (Facade Bonus, *Bonus Facciate*) and introduced new ones like the *SuperEcobonus*, complete with new refund schemes. These measures aim to accelerate the refurbishment of residential units and buildings. By 2024, the expenditure for these incentives, namely the *SuperEcobonus* and the *Bonus Facciate,* is approximately 130 billion Euros (Nicoletti Altimari 2023). Although the precise financial impact on the state treasury is still under assessment, the *SuperEcobonus*' estimated influence on the renovation of residential buildings is about 4% of the total building stock, accounting for 110 billion Euros. This figure corresponds to 470,000 formal declarations out of a total of 12 million residential buildings over a period of approximately 30 months (ISTAT—Istituto Statistico Italiano 2015; Agenzia Nazionale per le nuove tecnologie, l'energia e lo sviluppo economico sostenibile et al. 2023). This data underscores the challenges inherent in establishing a stable and financially viable refurbishment program.

At the European level, the updated Energy Performance of Buildings Directive (EPBD), ratified at the outset of 2024, establishes ambitious benchmarks to promote the decarbonization of the building sector within the European Union. Under this directive, all new public buildings are mandated to achieve zero emissions by 2028. This requirement extends to new private buildings beginning in 2030, accompanied by an escalated rate of refurbishments for existing structures, particularly targeting those with higher energy consumption (European Parliament and the Council of European Union 2024). The overarching objective is to realize a fully decarbonized building sector by 2050.

In this scenario, it appears obvious that designing new and maintaining old urban spaces is our century's defining challenge, necessitating resource-conscious design from the earliest stages. Urban buildings, as significant emission sources, are prime targets for mitigation efforts. Given the limited availability of energy and other critical resources, it is imperative for architects and building engineers to prioritize efficient resource management right from the initial design phase. This necessitates a thorough understanding of the energy consumption of buildings, including both direct and indirect usage, alongside the expenses and enhancements necessary for reducing energy use in both renovations and new projects. Additionally, it is crucial to work closely with policymakers, administrators, and building owners to advocate for the implementation of the most effective and economical strategies. This can be achieved by developing policies and programs designed to incentivize the adoption of such measures (Ang 2022). Future building stock models must ensure long-term stability and sustainability while ensuring that context changes can be absorbed quickly and at a low cost (model flexibility) to optimize design processes in an ever-changing environmental, climatic, social, and economic context. In short, we need a new, practical decision support framework for buildings, applicable from the outset of design.

Cities worldwide, in their ambition to achieve net-zero emissions, must endeavor to reduce greenhouse gas (GHG) emissions while also offsetting residual emissions. Notably, reducing emissions offers several advantages such as improved community health and well-being, enhanced air and water quality, reduced urban heat, and better resource management. In Europe, there are two main strategies adopted to mitigate climate interference: improving cities' efficiency to reduce GHG emissions and enhancing GHG sinks. While projects focusing on carbon sinks are necessary given the constraints of the carbon allowance worldwide, the priority is still to reduce global emissions through a just energy transition (Bourban 2022).

Tools and Instruments for Cities Decarbonization

Building Performance Simulations (BPS) have historically served as a foundational element in the decision-making process of building design, playing a pivotal role from the early planning stages through to occupancy, with their impact being most significant during the design phase. As regulatory frameworks have shifted towards promoting buildings that are both more environmentally friendly and efficient, BPS has evolved, incorporating increasingly sophisticated and reliable data streams into the design process. The advent of the digital era has marked a period of unparalleled access to data, facilitating the integration of machine learning and other data-driven technologies into BPS. These developments not only promise to refine simulation and modeling techniques but also to facilitate their integration from the very beginning of the design process. The recent update to the EPBD reflects this trend, advocating for the incorporation of data-driven and performance-oriented tools into BPS. However, navigating the extensive variety of modeling methods within BPS can present significant challenges, often leaving a range of stakeholders, from researchers to policymakers, navigating a complex labyrinth (Attia and De Herde 2011). On the other hand, the data, the lifeblood of these models, are often fragmented, unstructured, or locked behind gates of inaccessibility.

The energy consumption of buildings is significantly influenced by urban typology, which includes factors such as geometry, form, and typology (Pisello et al. 2013), along with shading, daylighting, and the urban heat island effect (Monien et al. 2017). The development of cities and neighborhoods involves iterative and multidisciplinary processes that encompass design, engineering, economics, and politics. In this multidimensional context, urban planners and policymakers can employ Urban Building Energy Modelling (UBEM) to swiftly assess the energy and carbon implications of various urban forms and their characteristics. UBEM facilitates the examination of a broad spectrum of urban planning and technical design considerations, including land use, zoning, urban economics, transportation, and compliance with standards, codes, as well as ensuring safety and security (Ang 2022). This approach can and must be implemented when working on a built environment, especially in the case of larger developments, as they are able to impact relevant parts of cities and a higher number of inhabitants. The analyses must be undertaken with advanced tools, considering different types of input and providing numerous types of outcomes, allowing for a deep and complete understanding of the relationship between the different topics related to urban and built environment, highlighting and comparing the outcomes from different scenarios, providing comparable results with a quantitative approach since early design stage, even with higher degree of uncertainties.

Significance of Starting with Neighborhoods

In the context of current climate urgency and the intricate urban environments we inhabit, it is increasingly vital to consider architectural design in terms of its neighborhood and urban-scale impacts. This shift reflects an evolving understanding of architecture's role within the broader perspective on sustainable urban development carried on by various organizations worldwide.

The strategic focus on neighborhoods in the decarbonization process is pivotal for several reasons. At the grassroots level, neighborhoods offer a manageable scale for implementing and testing innovative decarbonization strategies. This scale is significant enough to demonstrate tangible impacts, yet small enough to adapt and respond to specific local needs and conditions. Starting at the neighborhood level provides a unique opportunity to engage local communities directly, fostering a sense of ownership and participation in sustainability efforts. Neighborhoods serve as microcosms of larger urban systems, making them ideal for trialing diverse approaches to reduce carbon emissions. These may include implementing local renewable energy sources (Li et al. 2022), enhancing building efficiency while dealing with climate change (Isinkaralar 2023), promoting sustainable transportation (Zhang et al. 2023), developing green spaces, and integrating district level energy systems (Haase and Baer 2022). These options push the single building scale optimization to a further level, achieving ever-higher sustainability levels. Moreover, the learnings and successes from these neighborhood-scale interventions can then be scaled up and adapted for wider urban areas.

The social dynamics within neighborhoods also play a critical role in the decarbonization process. Neighborhood-based interventions can strengthen community bonds and foster a shared sense of purpose, encouraging residents to take collective action towards common sustainability goals and fostering emulation of best practices between residents in the same area. Moreover, these initiatives provide a platform for community education and awareness, crucial for adapting behaviors related to energy use and environmental stewardship (Ye and Niyogi 2022).

From a governance perspective, neighborhood-level interventions allow for more targeted and efficient use of resources. Local authorities can work closely with residents and stakeholders to identify specific needs and tailor interventions accordingly. This localized approach enables more effective monitoring and evaluation of decarbonization strategies, leading to continuous improvement and adaptation.

They allow for a more integrated and multi-scale approach to urban planning, where sustainability is embedded in every aspect of neighborhood development. This approach can lead to more resilient and adaptive urban environments, better equipped to meet the challenges of climate change and future growth.

In summary, focusing on neighborhoods in the decarbonization process offers a pragmatic and effective approach. It leverages the power of local communities and governance structures to pilot, refine, and eventually scale sustainable urban solutions. This bottom-up approach is not only strategic in achieving broader decarbonization goals but also essential in fostering resilient and inclusive urban communities.

Regulatory Drivers in Neighborhood Decarbonization

Recent legislative developments are crucial in shaping the urban decarbonization landscape, establishing a trajectory towards energy efficiency and sustainable development. Building codes are essential in reducing emissions in the building sector and providing clear guidelines for energy performance. According to the United Nations Environment Programme (UNEP), as of September 2022, approximately 40% of countries have implemented mandatory or voluntary building energy performance regulations or codes. However, this figure drops to 26% when considering countries with mandatory regulations for both residential and non-residential buildings (United Nations Environment Programme 2022).

According to the European Green Deal, the policy response to climate change and its industry consequences must be bold and comprehensive, seeking to maximize benefits for health, quality of life, resilience and competitiveness (European Commission 2019), allowing private citizens, companies and public entities to point out to the greater center on consumption, allowing for more efficient interventions. Moreover, according to the regulation 2020/852 of the European Parliament and of the Council on the establishment of a framework to facilitate sustainable investment (European Parliament and the Council of European Union 2020), a large part of the EU's investment has the objective of incentivize the energy efficiency of the built environment[1], also in line with the recent EU Taxonomy (EU Technical Expert Group of Sustainable Finance 2020) and the revision of the EPBD (European Commission 2021), part of the 2021 Commission Work Programme "Fit for 55"

[1] "For the purposes of this Regulation, the term 'energy efficiency' is used in a broad sense and should be construed by taking into account relevant Union law, including Regulation (EU) 2017/1369 of the European Parliament and of the Council and Directives 2012/27/EU and (EU) 2018/844 of the European Parliament and of the Council, as well as the implementing measures adopted pursuant to Directive 2009/125/EC of the European Parliament and of the Council".

(European Council 2022). These directives are already seeing their first fruits thanks to the implementation directives of the individual states, showing that energy efficiency in buildings is a top priority. In Italy, for example, to obtain funding from the PNRR (Ministero dello Sviluppo Economico 2021), according to Circular No. 32 of the "Ragioneria Generale dello Stato" of 30 December 2021 (Ministero dell'Economia e delle Finanze 2021), for new buildings, it is necessary to have a consumption of 20% below the NZEB standard.

The main objectives of the EPBD revision (IV) and its regional applications are reducing buildings' GHG emissions and final energy consumption by 2030, setting a long-term vision for buildings towards EU-wide climate neutrality in 2050. Article 4 (ex-Article 3) on the methodology for calculating the energy performance of buildings, in conjunction with Annex I, is updated to introduce the possible use of metered energy use to calculate energy performance, thus envisioning the integration of data-informed and performance-driven tools for BPSs.

In 2024, a new revision of the EPBD Directive (V) has been formalized. This Revision sets rigorous energy savings targets for both residential and non-residential sectors. It also indicates the necessity of a masterplan-level approach for significant interventions, emphasizing standardized strategies to enhance energy efficiency. Accordingly, Directive 2023/1791—Energy Efficiency Directive (EED recast) mandates annual energy savings targets and among the many requirements, setting the obligation for cities with over 45,000 inhabitants to develop district heating and cooling plans using geolocation and energy demand data. The directive's implementation is challenged by municipalities' dependency on external companies for planning and design, a concern noted by Mazzucato and Collington (2023). Further issues include limited public access to energy data collected by public portals. The directive also introduces efficiency requirements for data centers, enabling them to potentially earn Guarantees of Origin (GoO) through excess heat export. Energy audits, as part of this directive and mandatory for larger industrial and commercial companies could facilitate heat recovery, contributing to district heating systems. Directive 2023/2413—Promotion of Energy from Renewable (RED III) advocates for the integration of renewable energy, aligning with urban decarbonization objectives. It emphasizes the transition to renewable energy sources to meet sustainability targets. Also, different States (e.g., Italy) have been integrating Renewable Energy Communities (REC) legislative framework (referring to Directive 2018/2001 on promotion of the use of energy from renewable sources - RED II) in national legislation, with less stringent regulations, increasing maximum size for each renewable plant, increasing territorial boundary of subject being able to join the Community, enabling the proper start of shared renewable production on neighborhood level, allowing for a higher coverage of consumption and fostering shifting towards electrification and matching of production and demand curves, also with monetary incentives for energy virtually self-consumed within the Communities.

Collectively, these directives form a comprehensive regulatory framework for neighborhood decarbonization. While ambitious in their goals and innovative in approach, challenges in data accessibility, implementation, and retrofitting scale-up necessitate ongoing adaptation and multi-stakeholder collaboration. Addressing these challenges is crucial for cities to progress towards their decarbonization ambitions and contribute to a sustainable urban future.

The research of IEA ANNEX-30 shows that the critical design parameters are determined in the early design stage, and more than 40% of the building energy-saving potential comes from the early design stage (Attia et al. 2013). Therefore, it is necessary to optimize the key design parameters from the beginning to improve building performance (Lin et al. 2021) and adhere to the regulatory objectives. It is necessary to study the building physics from a holistic point of view and include all the physical phenomena and their relationship to users' behavior (Hensen and Lamberts 2011), studying in detail the different energy demand nature and use schedule of the neighborhood users being private, public, commercial, or industrial ones.

Case Studies: Translating Theory to Practice

In our design approach, we envision the neighborhood as an opportunity, defining it as a distinct area within a city, referred to variably as precincts or communities across different geographies. Regardless of size variations, neighborhoods typically encompass a blend of residential, commercial, and retail structures, green and public spaces, as well as supporting infrastructure. Implementing Net Zero solutions at the neighborhood level offers numerous opportunities for community engagement to identify and reinforce residents' needs, facilitating a transition to behavioral changes.

A neighborhood project serves as an avenue to involve and enhance the skills of a diverse array of individuals, such as urban planners, developers, construction workforce, and community groups. This exposure allows them to explore emerging practices, recognize their capabilities, and identify opportunities to contribute. The planning, design, and operation of a sustainable neighborhood should prioritize minimizing emissions throughout its entire lifecycle, meeting the needs of residents, workers, and visitors while creating a high-quality and resilient environment that fosters community thriving.

A sustainable neighborhood is achieved by addressing both the imperative of emissions reduction and the promotion of quality of life on a neighborhood scale. Through an integrated approach, the neighborhood becomes an appealing destination for residents, workers, and visitors, supported by a green business ecosystem. It showcases effective emissions reduction practices, positioning itself as a model for other neighborhoods to emulate. Additionally, a carbon-zero neighborhood should strive to minimize emissions throughout the project lifecycle and attain net zero by effectively offsetting any remaining emissions in a robust and transparent manner.

Frameworks for categorizing emissions include operational (energy use), embodied (material use for constructions), and consumption-based (goods purchased during use by users) emissions. Neighborhoods should target achieving net zero operational and embodied emissions, undertaking ambitious actions on consumption-based emissions by focusing on specific measurable initiatives that impact lifestyles and behavior, such as locally grown or plant-based food, transport or logistics hubs, and 'sharing economy' initiatives.

While achieving net zero emissions is a gradual process, neighborhoods should adopt short-term targets to drive rapid emissions reduction, aligning with global commitments on emissions reduction. According to Arup vision, specific operational, embodied, and consumption-based emissions targets should be set to guide the neighborhood's development.

- *Operational emissions target:* All new buildings in the neighborhood should achieve net zero operational emissions. The neighborhood should strategize to ensure all existing buildings attain net zero operational emissions, with an ambitious target to complete renovations within the period of neighborhood redevelopment, no later than 2050, aligning with global commitments.
- *Embodied emissions target*: The neighborhood should implement a process requiring lifecycle emissions assessments for all construction works, setting a minimum reduction target for embodied emissions against a 'Business-as-usual' approach for a comparable neighborhood. A reasonable target would be a minimum 50% reduction over the 'Business-as-usual' approach.
- *Consumption-based emissions target:* Due to the challenges in accurately calculating a wide range of consumption-based emissions, neighborhoods are advised to focus on a select number of ambitious actions with measurable impacts on consumption-based emissions, such as promoting locally grown or plant-based food, establishing transport or logistics hubs, and encouraging 'sharing economy' initiatives.

Focusing on the energy component, the goal of urban energy design is that all buildings in the neighborhood should have net zero operational emissions. In collaboration with C40 Cities, Arup defined a strategy to achieve the net-zero scope, following three consequentially principles: minimize energy demand, invest in energy infrastructure, and decarbonize energy supply with onsite and offsite contribution.

Focusing on the first principle, both new and existing buildings should prioritize energy demand reduction, adopting passive design principles that prioritize well-insulated fabric, natural ventilation and the size and orientation of glazing. The spatial design and internal layout of new buildings is important to encourage natural ventilation. Green roofs can be installed to reduce internal solar gains, as well as improving urban biodiversity and reducing heat-island effect. Building services systems should be sized correctly, avoiding outdated rules of thumb, ensuring best performance at partial loads. All systems should be equipped with sensor-based smart controls that can monitor energy consumption and report against energy use intensity (EUI) targets, maintenance should be targeted using performance data, rather than merely planned or reactive, and all building services systems should be designed for end-users, with clear guidance on how they can influence energy consumption, highlighting the parameters that have the greatest impact on energy consumption.

Many neighborhood-scale decisions can reduce building energy demands, such as building massing combined with the topography and orientation of green spaces and streets, bringing benefit from prevailing winds and minimizing solar exposure on glazed facades. Tree planting, green spaces and water bodies should be incorporated to help reduce surface temperatures and urban heat island effects, also reducing the risk of surface water flooding. Well-controlled energy demand will improve the comfort and livability of workplaces, reduce energy bills and reduce peak loads on energy infrastructure. A mixture of regulation, policies and incentives can help to ensure that existing buildings are in-line with these ambitions (Rode and Flynn 2021). It is important to provide suitable financial support for vulnerable groups, so that all existing buildings are extensively upgraded, at least for city users that require it most.

Energy supply infrastructure should be strategically planned. If the buildings will be all-electric in anticipation of grid decarbonization, it will be important to ensure sufficient network capacity and reliability. If there is a major infrastructure transition (e.g., to hydrogen), the neighborhood will need sufficient power to ensure that all buildings adopt these strategic aims. The energy supply infrastructure should be sized correctly, with provision to expand and adapt if demands change, be integrated and flexible, accommodating local renewable energy generation and capturing waste heat from local sources, such as chiller heat rejection, tunnel extract ventilation or even data centers. The creation of energy infrastructure covering different uses featuring different demand profiles (e.g., different working and operating hours) and different type of demand (e.g., cooling versus heating) would also allow an overall increase of efficiency and reduction of cost, allowing heat recovery on a district level and the use of the same equipment for different buildings, even with a distributed approach with different energy centers following the development phasing of the neighborhood.

Investing in digital-enabled systems will support distributed energy generation and enable flexible demand management. This "two-way" infrastructure requires an integrated approach, so that different stakeholders can participate in the energy system, selling excess energy, providing thermal or electrical storage or making demand reductions on request. Neighborhood-wide energy balancing can be enabled through real-time monitoring and extended to include vehicle-to-grid (V2G) solutions, or peer-to-peer trading between local micro-generators or batteries, increasing diffusion of electrical vehicles being also able to reduce pollutant emissions. To reduce operational emissions, the efficiency of new and existing supply infrastructure should be increased. It is relevant to reduce losses in heat networks and thermal storage, especially above-ground and in the final runs to building heat interface units. Also, in case water from the ground is used for thermal purposes (e.g. ground source heat pump) only a neighborhood or urban approach would allow the integration of buffer basin large enough to add flexibility and resilience respect to usual approaches with supply and return wells with no relevant water storage.

Net Zero energy supply requires efficient and affordable renewable energy. From the outset, the renewable energy sources and storage potential should be mapped and utilized. Priority should always be on local renewable generation, in preference to off-site agreements. Local renewable energy generation can bring about wider benefits including energy security, job creation and lower transition risk.

When on-site generation is maximized, off-site renewable energy should be procured. A stable, long-term arrangement is likely to be preferable, with a scheme such as Power Purchase Agreement (PPA) while buying on an urban scale should reduce price impacts. The governance needs to be considered; particularly where local market conditions allow end-users to change suppliers. The neighborhood will need robust monitoring and reporting to calculate annual operational emissions. Although the return on investment for local renewable energy systems is often positive, they usually require high upfront capital investment. Financial incentives and innovative business models can help overcome this barrier and should be adopted to ensure that all residents can have access to clean energy, regardless of their ability to invest. The introduction of Renewable Energy Communities will help create more efficient energy infrastructures, as one of the main aspects is related to the alignment between energy supply and demand, possible only with an extensive network of sensors to track in real-time renewable production and consumption together with user's direct actions to change their own behavior in term of energy consumption (already common at building scale for users with renewable sources integrated into their home). The change of behavior will also be fostered by monetary incentives for virtual self-consumption, already present in some countries, provided by public subjects to avoid issues on the national grid.

While the concept of a net-zero neighborhood introduces a renewed emphasis on the urgency of transparent emissions reduction alongside comprehensive livability, there are instances worldwide where neighborhoods have already initiated measures to reduce emissions and foster vibrant communities. One such illustration is the L'Innesto project (Fig. 2).

L'Innesto, the first Climate Neutral social housing project in Italy strives to serve as the exemplar for Milan's freshly integrated and multi-scale sustainability approach. L'Innesto encompasses the establishment of a cutting-edge 4th generation district heating system, fueled by renewable sources alongside an urban wastewater heat-recovery system. The architectural conception of Nearly Zero Energy Buildings is harmonized with prefabricated construction technology and an optimal combination of bio-sourced materials.

The objective is to reduce both CO_2 emissions and waste, with the intention of enabling the structure to undergo disassembly and achieve 100% recyclability upon reaching the end of its life cycle. Minimization of soil excavation is prioritized, and on-site treatment utilizing bio-remediation techniques is implemented to facilitate reuse for landscaping purposes. The incorporation of vegetable gardens, nursery gardens, nature-based roofs, and edible landscapes establishes a centralized hub for food and agriculture. A comprehensive sustainable mobility strategy ensures citizens' complete accessibility. The initiative advocates for active mobility, collaborative systems, and a substantial reduction in the reliance on private vehicles by actively discouraging resident car ownership and providing alternative mobility solutions (Fig. 3).

Fig. 2 L'Innesto masterplan aerial view (Credit: ARUP)

As reported in the LCA (Life Cycle Assessment) study developed by Politecnico di Milano (Famiglietti et al. 2023), the aim to reach climate neutrality for the project is feasible only if the energy strategy includes the urban scale: using district heating networks (of the latest generation, with reduced supply temperature confronting to usual district systems) at the urban level instead of at the district level could optimize space, increasing the photovoltaic panels surface, in order to reach the target set. In other words, the goal of climate neutrality must be pursued from the point of neighborhood definition to leave adequate space for photovoltaic modules and analyze appropriate solutions. In addition, the connection of existing buildings with traditional building services to the decarbonized district heating network would allow an even greater reduction in carbon emissions, balancing the embodied carbon due to the construction phase of the new development, while also tackling the pollutants emitted by building services using fossil fuel. The project testifies the need of strategic planning and development programs including a clear approach

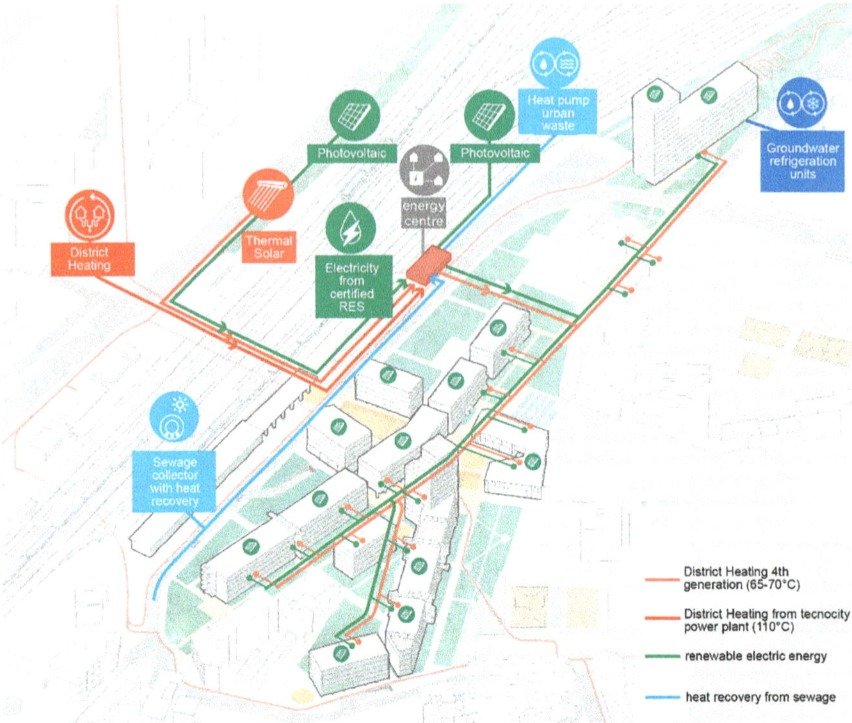

Fig. 3 L'Innesto active energy strategies. The district heating and cooling system is highlighted, showing the heat-cool exchange between buildings (Credit: ARUP)

and methodology to calculating emissions, monitoring progress and demonstrating net zero outcomes. To this end, it is fundamental that the strategic approach includes the involvement of private and public sector developer organizations, national government bodies, residents and communities, and local businesses.

An example of this bottom-up approach is the masterplan for the University Campus of Heidelberg in Germany (Fig. 4).

Starting from the sustainability goal of the Federal Government to drastically reduce energy consumption, the masterplan focused on densification and organic growth of the city center, with new facilities, existing buildings will be provided with new connecting spaces to accommodate the increase in interdisciplinary activities and the resulting synergies between hospital, teaching, research and private facilities. The energy concept is a key pillar for achieving climate protection goals for the campus. The goal of the Heidenheimer Feld masterplan is a climate-neutral energy supply by 2050, in line with the city's ambitions and the German government's climate protection program.

The first pillar of the project's energy strategy is the premise of passive design for the entire campus and individual campus buildings. The second pillar is efficient design, i.e. increasing the efficiency of the building standard. The third pillar is the

Fig. 4 Masterplan im Neuenheimer Feld (INF) Heidelberg. Arup's focus lies on sustainability, infrastructure and mobility strategies (Credit: ARUP)

coverage of the remaining energy needs using renewable and climate-neutral energy. The focus of energy development is on significantly increasing the energy efficiency of existing buildings on the one hand and new buildings on the other. To cover the remaining energy requirements, the third pillar of the energy concept, Arup conceived the energy infrastructure (urban scale) according to the following fundamental components energy.

A strategy for the maximization of energy efficiency is also integrated, including the use of solar radiation for energy production, the use of prevailing easterly winds to cool the campus in the summer months, while reducing energy consumption and using geothermal energy.

Efficiently instituting a climate-neutral neighborhood necessitates a methodical and all-encompassing approach. Solely investing only in technical solutions proves inadequate. To address the imperative of emission reduction and the augmentation of quality of life, it is imperative that the municipality and pertinent local governmental entities establish an enabling milieu. This entails the integration of the neighborhood into pertinent land use strategies and developmental policies, the execution of capacity-building programs, and the judicious allocation of funds.

Consequently, the formulation of a decarbonization scheme for the urban scale mandates the collaboration of diverse stakeholders cognizant of the significance of climate action and united by a shared vision (Rode and Flynn 2021). It is essential to delineate the inclusion and collaboration of pivotal stakeholders throughout the duration of the project. Instituting a lucid engagement process from the project's inception ensures a comprehensive understanding among all stakeholders regarding how they can contribute most effectively.

Various methodologies exist for discerning and engaging pertinent stakeholders, and the Inclusive Climate Action program by C40 Cities encompasses engagement guidebooks developed in partnership with Arup (Arup and C40 Cities 2021). The stakeholders are likely to be categorized as follows:

- *Municipality/Regional Authority*: Assumes a pivotal role in guiding, facilitating, or executing the development of a neighborhood, encompassing land use policies and the provisioning of infrastructure.
- *Private Developers:* Determine the way land is utilized for development purposes.
- *Building Firms*: Shape decisions related to material selection and construction methodologies.
- *Utilities Providers*: Supply energy, water, waste, and ITC services to the neighborhood and its broader vicinity.
- *Local Enterprises*: Play a crucial part in mitigating consumption-related emissions by offering goods and services, as well as making informed choices regarding their supply chain.
- *Community Organizations & Inhabitants*: Can serve as advocates for an eco-friendly and flourishing neighborhood when adequately involved in the planning stages.
- *Non-Governmental Organizations (NGOs)/Development Agencies*: Can contribute financial resources, expertise, and capacity.
- *National Government*: Might possess, operate, regulate, or fund essential infrastructure, assets, and services critical to the neighborhood.

Climate-neutral communities offer a distinctive chance to alter the production process, foster creativity, and establish fresh demand for low-carbon products in the market. Effective leadership plays a crucial role in transmitting explicit messages that propel this transformation, and such influence can be heightened through the gathering influence of the mayor or local government. Engaging community stakeholders at an early stage is advantageous, guaranteeing their input in the neighborhood planning process. This connection will strengthen over time as mutual comprehension and trust evolve. Supportive community organizations have the potential to strengthen and assume responsibility for prolonged initiatives aimed at promoting low-carbon lifestyles.

Reporting our experience in Milan, the energy and environmental strategy of the masterplan considers elements that will have a positive influence on the performance of buildings to achieve the requirements related to the General Urban Plan (PGT) in force in Milan and in particular the objectives for the carbon neutrality of new construction and climate impact reduction (RIC).

At the neighborhood level, the application of these principles is an opportunity to scale up and amplify economic, social and environmental benefits. Indeed, this approach can create a built environment that is resilient to volatile commodity prices, maintains essential natural ecosystem services, creates more livable, productive and affordable urban areas, and contributes to achieving decarbonization targets. It needs to be said that, especially within existing districts and neighborhood, large refurbishments are not easy to be designed and started due to the numerous requirements and obstacles related to authorizing procedures on both public (e.g., landscape authorization) and private (e.g., condo assembly) level.

Going further, the urban requalification projects, in which we are involved, fit into and dialogues with an urban context that is extremely proactive on the decarbonization front, establishing challenges and opportunities for new developments and future city districts to be at the forefront of innovation in this field. The collaboration with private and public stakeholders is fundamental for the generation of transformation on a larger scale. Specifically, the urban context is framing its commitments to energy efficiency using various policies and instruments, from policy awareness, tax and fiscal incentives, subsidies, space concessions, and European funds. A neighborhood-scale approach offers the opportunity to instill the principles of the climate neutral approach to reduce climate emissions and to create a long-term value for the building sector.

Conclusions

In conclusion, this paper has underscored the critical role that urbanization plays in global energy consumption and carbon emissions, presenting both a significant challenge and a unique opportunity for sustainable development. As urban areas continue to expand, the integration of energy-efficient design and retrofitting existing infrastructure emerges as a pivotal strategy for mitigating environmental impact. Collaborative efforts across disciplines, from architects and engineers to policymakers and community stakeholders, are essential in developing and implementing effective solutions.

The evolution of BPS and UBEM reflects the increasing sophistication of tools available for designing sustainable urban spaces. These advancements, coupled with a growing regulatory focus on energy efficiency and carbon neutrality, signal a promising shift towards more sustainable building practices. However, challenges such as data accessibility and the complexity of modeling methods persist, necessitating ongoing innovation and collaboration.

The emphasis on neighborhood-scale initiatives is particularly noteworthy, offering a scalable and community-oriented approach to decarbonization. These efforts, supported by legislative developments and technological advancements, highlight the potential for comprehensive and effective urban planning strategies that prioritize sustainability.

Furthermore, the significant investments in retrofitting and the establishment of ambitious emissions targets underscore the urgency and scope of the task at hand. The European Union's revised EPBD, along with various national incentives, exemplifies the commitment to transforming the built environment.

Ultimately, achieving a sustainable urban future requires a multi-faceted approach that integrates technological innovation, regulatory support, and community engagement. By focusing on both the macro and micro scales of urban development, from city-wide policies to neighborhood-level projects, we can navigate the complexities of urban sustainability and work towards a more resilient and carbon-neutral future. Now is the time to accelerate action and increase our ambition. Sustainable neighborhood design provides a new focus to respond to the urgency of the global climate agenda. These reflect the priorities of the "Green and Just Recovery" Agenda released by the C40 Global Mayors Task Force and supporting the UN "Global Green New Deal", while also contributing directly towards the UN Sustainable Development Goals on climate action (SDG 13) and sustainable cities and communities (SDG 11).

References

Agenzia Nazionale per le nuove tecnologie, l'energia e lo sviluppo economico sostenibile, Agenzia Nazionale Efficienza Energetica, Ministero dell'Ambiente e della Sicurezza Energetic (2023) Dati SuperEco Bonus 110% - art. 119 Decreto Legge 34/2020

Ang YQ (2022) Using urban building energy modeling to develop carbon reduction pathways for cities. Ph.D. thesis,. Massachusetts Institute of Technology

Attia SG, De Herde A (2011) Early design simulation tools for net zero energy buildings: a comparison of ten tools. In: Proceedings of Building Simulation 2011. Sydney, pp 94–101

Attia S, Hamdy M, O'Brien W, Carlucci S (2013) Assessing gaps and needs for integrating building performance optimization tools in net zero energy buildings design. Energy Build 60:110–124. https://doi.org/10.1016/j.enbuild.2013.01.016

Birgisdóttir H, Bjørn A, Branny A et al (2023) Doughnut for urban development: a manual, 2nd edn. The Danish Architectural Press, Copenhagen

Bourban M (2022) Ethics, energy transition, and ecological citizenship. In: Comprehensive renewable energy. Elsevier, pp 204–220

Di Stefano AG, Ruta M, Masera G (2023) Advanced digital tools for data-informed and performance-driven design: a review of building energy consumption forecasting models based on machine learning. Appl Sci 13:12981. https://doi.org/10.3390/app132412981

EU Technical Expert Group of Sustainable Finance (2020) Taxonomy: final report of the technical expert group on sustainable finance

European Commission (2019) The European Green Deal

European Commission (2021) Directive of the European Parliament and of the Council on the energy performance of building (recast)

European Council (2022) Fit for 55. https://www.consilium.europa.eu/en/policies/green-deal/fit-for-55-the-eu-plan-for-a-green-transition/. Accessed 19 Sep 2022

European Parliament and the Council of European Union (2020) Regulation (EU) 2020/852 of the European Parliament and of the Council of 18 June 2020 on the establishment of a framework to facilitate sustainable investment, and amending Regulation (EU) 2019/2088

European Parliament and the Council of European Union (2024) Energy Performance of Buildings Directive. https://energy.ec.europa.eu/topics/energy-efficiency/energy-efficient-buildings/energy-performance-buildings-directive_en. Accessed 11 Mar 2024

Famiglietti J, Aprile M, Spirito G, Motta M (2023) Net-zero climate emissions districts: potentials and constraints for social housing in Milan. Energies 16:1504. https://doi.org/10.3390/en16031504

Haase M, Baer D (2022) Implementation of positive energy district concepts and energy master plans for decarbonization of districts. Acta Polytech CTU Proc 38. https://doi.org/10.14311/APP.2022.38.0546
Hensen J, Lamberts R (eds) (2011) Building performance simulation for design and operation. Spon Press, London, New York
Intergovernmental Panel on Climate Change (IPCC) (2015) Climate change 2014: synthesis report. In: Contribution of Working Groups I, II and III to the fifth assessment report of the intergovernmental panel on climate change. IPCC, Geneva, Switzerland
International Institute for Sustainable Development (2019) 77 Countries, 100+ cities commit to net zero carbon emissions by 2050 at climate summit. In: IISD—SDG Knowl. Hub. https://sdg.iisd.org/news/77-countries-100-cities-commit-to-net-zero-carbon-emissions-by-2050-at-climate-summit/. Accessed 25 Sep 2022
Isinkaralar O (2023) Bioclimatic comfort in urban planning and modeling spatial change during 2020–2100 according to climate change scenarios in Kocaeli, Türkiye. Int J Environ Sci Technol 20:7775–7786. https://doi.org/10.1007/s13762-023-04992-9
ISTAT—Istituto Statistico Italiano (2015) Annuario Statistico Italiano 2015—Costruzioni
Li F, Schell KR, Yang H, Tsamis A (2022) Infomorphism: an urban planning framework for local renewable energy integration
Lin B, Chen H, Liu Y et al (2021) A preference-based multi-objective building performance optimization method for early design stage. Build Simul 14:477–494. https://doi.org/10.1007/s12273-020-0673-7
Mazzucato M, Collington R (2023) The big con: how the consulting industry weakens our businesses, infantilizes our governments and warps our economies. Allen Lane, London
Ministero dell'Economia e delle Finanze (2021) Circolare del 30 dicembre 2021 n.32
Ministero dello Sviluppo Economico (2021) Piano Nazionale di Ripresa e Resilienza
Monien D, Strzalka A, Koukofikis A et al (2017) Comparison of building modelling assumptions and methods for urban scale heat demand forecasting. Future Cities Environ 3:2. https://doi.org/10.1186/s40984-017-0025-7
Nicoletti Altimari S (2023) Nota di aggiornamento del Documento di economia e finanza 2023
Pisello AL, Taylor JE, Cotana F (2013) Simulating the effect of urban morphology on indoor thermal behavior: an Italian case study. In: 2013 Winter Simulations Conference (WSC). IEEE, Washington, DC, pp 2008–2019
Rode P, Flynn R (2021) Multilevel emergency governance: enablinc adaptive and agile responses. Emergency Governance for Cities and Regions
Timperley J (2018) UK homes need 'deep efficiency retrofit' to meet climate goals. In: Carbon brief. https://www.carbonbrief.org/uk-homes-need-deep-efficiency-retrofit-meet-climate-goals/. Accessed 25 Sep 2022
United Nations Environment Programme (2022) 2022 Global status report for buildings and construction: towards a zero-emission. Efficient and Resilient Buildings and Construction Sector, Nairobi
Weber R, Mueller C, Reinhart C (2021) Building for zero, the grand challenge of architecture without carbon. SSRN Electron J. https://doi.org/10.2139/ssrn.3939009
Ye X, Niyogi D (2022) Resilience of human settlements to climate change needs the convergence of urban planning and urban climate science. Comput Urban Sci 2:6. https://doi.org/10.1007/s43762-022-00035-0
Zhang J, Li S, Wang Y (2023) Shaping a smart transportation system for sustainable value co-creation. Inf Syst Front 25:365–380. https://doi.org/10.1007/s10796-021-10139-3

Reducing Greenhouse Gas Emissions: Lessons from Case Studies

Mahsa Safari, Hugh Johnson, Joel Krakower, Jin Wen, and Patrick L. Gurian

Abstract Efforts to ensure a nature-positive built environment must identify plausible scenarios for greenhouse gas (GHG) emissions reductions. There is a rich literature of planning studies seeking to identify and evaluate various scenarios for GHG emissions reductions. These planning studies are conducted on a range of scales from international, to national, to state, to local. This chapter reviews a set of GHG emissions reduction planning studies from the period of 2014 through 2019. Most planning efforts have concluded that dramatic emissions reductions are feasible. Efforts to reduce greenhouse gas emissions (referred to as "carbon abatement") rely on some mixture of the following strategies: (1) demand reduction via energy efficiency measures such as more efficient equipment and more insulated buildings, (2) electrification of cooking, space heating, and water heating, coupled with low GHG electricity generation, (3) adoption of low GHG fuels such as hydrogen produced from renewables. The mixture of these three strategies varies from study to study. A detailed cost study in Philadelphia concluded that GHG abatement costs for energy efficiency measures vary dramatically by building with some cases showing negative abatement costs (i.e., utility saving were sufficient to pay for upfront investment without consideration of greenhouse gas emissions reductions) and others requiring large investments that would likely not be feasible for their owners. In particular, low and mid-rise residences were not particularly attractive for retrofits with abatement costs exceeding $200 per tonne of carbon dioxide. Switching cur-

M. Safari
The Cadmo Group, Arlington, VA, USA

H. Johnson
The Environmental Collaboratory, Drexel University, Philadelphia, USA

J. Krakower
Department of Mechanical Engineering and Mechanics, Drexel University, Philadelphia, PA, USA

J. Wen · P. L. Gurian (✉)
Department of Civil Architectural Environmental Engineering, Drexel University, Philadelphia, PA, USA
e-mail: pgurian@drexel.edu

G. Iannaccone, E. V. Ellis (eds.), *Towards a Nature-Positive Built Environment*, Research for Development, https://doi.org/10.1007/978-3-031-95292-0_12

rent electricity uses to low GHG electricity had an estimated cost of $23/tonne abated. However, reducing GHG emissions from building heating presents a greater challenge as most buildings currently rely on gas heating and hence electrification requires a substantial capital investment in heat pumps, as well as the capital investment in low GHG electricity generation and storage capabilities. A study found that in New York City, abatement costs for switching to heat pumps ranged from roughly $125 to $160/tonne. The role for low GHG fuels in emissions reductions remains uncertain and contested with many studies seeing electrification as providing the bulk of emissions reductions with an important niche role for low GHG fuels. While the optimal pathway remains uncertain, a wide range of studies concur in their conclusions that substantial reductions in emissions are technically and economically feasible.

Keywords Climate change mitigation planning · Greenhouse gas emissions reduction · Carbon abatement costs

Introduction

The development of nature-positive built environments must include consideration of greenhouse gas emissions. Buildings are embedded in electricity, water, gas, and transportation infrastructure systems, and building emissions are strongly influenced by these infrastructure systems. This chapter considers overall strategies for greenhouse gas emissions reductions, including options for directly reducing building greenhouse gas emissions as well as strategies to reduce greenhouse gas emissions from the larger-scale infrastructure in which buildings are embedded. There has been considerable attention to this issue in the literature with many planning studies that have considered a wide range of scenarios. These planning efforts range in scale from international, to national, to regional. They seek to envision a set of possible future options that provide similar or enhanced standards of living while dramatically reducing greenhouse gas emissions. They develop demand reduction scenarios as well as scenarios in which low emissions energy sources are substituted for higher emission energy sources. They must consider emissions in sectors and then integrate how demand reduction and source substitution strategies in different sectors interact (Fig. 1). Most focus primarily on carbon dioxide, often using “carbon” more or less synonymously with “carbon dioxide equivalents”, a measure of overall GHG emissions, but they at least implicitly acknowledge the role of other greenhouse gases. The studies employ a variety of mathematical models, that are often complex and hard to replicate, to estimate the performance of these hypothetical future infrastructures. They set a variety of goals for reductions in GHG emissions over a variety of timelines. Given the diversity of methods and goals, it is not surprising that these studies reach differing conclusions. Cross comparing the results of such studies provides insight into the discourse around greenhouse gas emissions reduction efforts by identifying the range of views expressed to policy

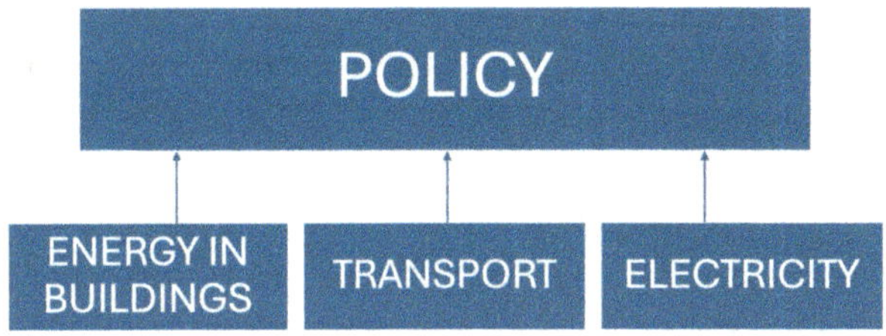

Fig. 1 Approach in which energy demand in buildings, transport emissions, and electrical generation are first considered separately and then integrated

makers by these studies. This chapter examines the results of 18 of these emissions reduction planning studies conducted in the time period 2014–2019 with special attention to how they envision the future of built infrastructure.

Feasibility

One of the first questions to be addressed regarding such far-ranging infrastructure changes is whether they are feasible at all. Such reports typically conclude that dramatic reductions are feasible. For example, out of eighteen studies conducted between 2014 and 2019 (NYC 2016; EPRI 2018; Foti et al. 2015; PGE 2018; Stirchele et al. 2018; National Grid 2016, 2017; PÖYRY 2018; NREL 2017; EFI 2019; E3 et al. 2014; PA-DEP 2019; Toronto 2017; E3 2018; NRDC 2017; Sepulveda et al. 2018; CEC 2018), fifteen identified plans for reducing emissions 80% from 1990 levels by 2050 that were presented as feasible courses of action (NYC 2016; Foti et al. 2015; PGE 2018; Stirchele et al. 2018; National Grid 2016; National Grid 2017; PÖYRY 2018; EFI 2019; E3 et al. 2014; Toronto 2017; E3 2018; NRDC 2017; Sepulveda et al. 2018; CEC 2018). Feasibility is often not precisely defined but appears to indicate that the technologies exist and can be implemented without additional innovations. For example, one might note the use of "currently" in Foti et al.'s statement that "options for achieving greenhouse gas emissions reductions are currently feasible and do not involve foregoing the benefits of modern technology". The breadth of the usage of the term "feasible" can be seen by noting that one of the fifteen reports indicated that achieving deep emissions reductions would be "extremely challenging" but "possible" (EFI 2019). Two of the fifteen reports identified plans for achieving a 95% reduction of 1990 emissions by 2050 (Stirchele et al. 2018; PÖYRY 2018) and one modeled 54%, 77%, 89%, 99%, and 100% reduction of 2005 emissions in U.S. electricity sector (Sepulveda et al. 2018). There were three other reports that identified plans for achieving less than 80% reductions (EPRI 2018; NREL 2017; PA-DEP 2019). One report focused on the full electrification of the U.S. economy and found that 70% of 2015 emissions could be reduced by 2050 (EPRI 2018). Another report focused on U.S. electrification alongside power sector decarbonization (i.e., GHG reduction) and identified options for a 74%

reduction in emissions from 2005 levels (NREL 2017). The third report was based on the Pennsylvania Climate Change Act and found that Pennsylvania could reduce 36% of 2005 emission levels by 2050. None of these reports declared that deeper reductions were infeasible.

Technological Options

These emissions reduction studies employ a mixture of several elements including: (1) demand reduction through increases in energy efficiency, (2) replacement of fossil fuel electricity generation with low GHG sources and subsequent electrification of end uses that have historically involved fossil fuel combustion, such as building heating and transportation, and (3) use of low GHG fuels, such as hydrogen (generated from renewable sources or from fossil fuels with sequestration of the carbon) or renewably generated methane (in which case the carbon uptake in producing the methane is considered to be balanced by the carbon emissions from the combustion of the methane). The mixture of strategies used varies from study to study. Often studies compare different approaches in order to assess which ones are most promising. Several of these studies are briefly described below.

A European study (PÖYRY 2018) considered a zero-carbon gas pathway and an all-electric pathway. Both pathways require sufficient insulation of buildings so that heat pumps are feasible without creating extensive winter electricity peaks and costs, the large-scale integration of the electricity grid (e.g., improved long range transmission capabilities), and the ability to use electric vehicle batteries for electricity storage (i.e., electric vehicle batteries capable of being discharged to meet grid needs). The zero-carbon gas path allows technologies to be adopted based on economics. This pathway envisions multiple gaseous fuels including biogas/biomethane, hydrogen from electrolysis and methane reforming with carbon capture, utilization, and storage (CCUS), and natural gas with CCUS. Methane reforming was the major source of hydrogen, with electrolysis used in regions with high renewables potential and limited grid flexibility. In this pathway, nuclear plants were phased out over time due to being more costly than alternatives. The all-electric pathway constrains the choice of technologies such that gas technologies, including CCUS, are excluded, and electrification is the only approach used for decarbonization. The scenario envisions support for the increased use of nuclear power. The integration of electricity grids across the continent enables the additional nuclear power to be located in countries that permit nuclear power (e.g., France) but used by other countries as well. While both pathways were considered feasible, the all-electric pathway was more expensive than the zero-carbon gas pathway. The levelized costs (i.e., full cost of production accounting for both fixed and variable costs) for various potential heating technologies (including fuel and appliances) are also presented and range from 82.7€/MWh (for a gas hybrid heat pump using biomethane) to 166.6 €/MWh (for electric resistance heating).

A Portland decarbonization study (PGE 2018) identified three feasible pathways that did not involve nuclear or CCUS including (1) High Electrification pathway which relies on electrifying space and water heating in buildings and deploying bulk energy storage to balance high levels of renewable generation, (2) Low Electrification pathway which decarbonizes the energy supply with a variety of renewable fuels, and electrolysis and power-to-gas facilities, and (3) High Distributed Energy Resources (DER) pathway which is highly electrified and distributed, with increased rooftop solar PV and distributed energy storage in buildings and industry. There were three common elements to each pathway: (1) a decline in per capita final energy consumption by approximately 40%; (2) a decrease in the carbon intensity of electricity generation to near zero; and (3) an increase in the share of energy coming from electricity (including fuels produced from electricity) from approximately one-quarter today to at least half by 2050. The study did not give overall system costs but did provide some information on the range of annual incremental household costs experienced by consumers (including transportation). The low electrification pathway had cost increases of up to $100 per month per customer (although some would see declines of up to $150 per month), while the two high electrification pathways had price increases of up to $50 per month (and declines of up to $200 per month for some customers). The range of uncertainties reflects both net increases and decreases (savings) in a range of alternative fossil fuel price and end-use electric technology cost projections by 2050. In contrast the European study (PÖYRY 2018) estimated higher overall system cost (including transport, heat, residual power, network and other costs) for the all-electric pathway than for the low carbon fuels pathway, the opposite of the trend in consumer prices reported by the PGE study. Comparisons across studies are complicated by the different settings and assumptions of the studies, such as the exclusion of CCUS from the PGE study and the complete exclusion of gas technologies from the all-electric pathway of PÖYRY (2018).

A U.S. national study (E3 et al. 2014) identified four different pathways to achieve 80% reductions in greenhouse gas emissions by 2050, referred to as the 80 × 50 goal. These pathways were: high nuclear, high renewables, high CCUS, and a mixed scenario. These pathways share the features that (1) energy efficiency is increased, (2) electricity generation is decarbonized, (3) end uses are electrified when possible, and (4) where electrification is not possible, low carbon fuels are used, specifically renewable gas and hydrogen (derived either from electrolysis of water with renewable electricity or reformulation of methane with sequestration of carbon). These four pathways all involve extensive electrification of buildings. All four pathways are judged to be feasible with costs for the nuclear and CCUS estimated at roughly 0.8% of GDP. However, the high renewables pathway was roughly 3 times more costly (with the mixed scenario falling between these extremes). Technology cost assumptions are described as conservative in that modest declines are forecast in the unit costs of new technologies as they are adopted over time. The 75th percentile of the uncertainty distribution of costs (i.e., a relatively conservative estimate) corresponds to “little to no” decline in costs over four decades and is over twice the baseline estimate (i.e., without innovation costs would be about 1.8% of GDP).

The E3 et al. (2014) study also identifies investment risks but indicates that these risks can be managed if there is appropriate long-range planning and policy making for 80 × 50 goals. Specifically, if initial efforts are focused exclusively on efficiency gains (where returns are often highest) then ongoing capital investments in greenhouse gas emitting generation sources would be stranded (obsolete before the capital investment is recovered) as further gains could only be achieved by decarbonization of electricity generation. For this reason, a gradual approach of replacing carbon emitting generation sources as they reach the end of their lifetimes with low carbon generation capacity is recommended. The report identified two societal decisions that must be made to guide the selection of a pathway: (1) the extent of nuclear generation, and (2) the extent of CCUS. In addition, the extent of biomass available for energy needs to be studied and better quantified.

Foti et al. (2015) estimated carbon abatement costs in Philadelphia for "ambitious building retrofits aimed at achieving 30–50% reductions in energy use". Abatement costs varied greatly by sector. For hospitals, schools, and retail stores, the utility cost savings were sufficient to justify the required capital investment given a 3% interest rate and 35-year project period without consideration of the greenhouse gas emissions reductions (that is, abatement costs were negative). Other sectors had markedly higher abatement costs. Low and mid-rise residences had abatement costs exceeding $200 per tonne of carbon dioxide and abatement costs for warehouses were estimated at $140/tonne. The use of low carbon electricity had an abatement cost of $23/tonne, which indicates that switching building electricity use to low carbon sources would be more cost-effective than deep reductions in demand in some sectors.

Comparing Emissions Reduction Strategies

Common features of the studies include continued, though reduced, use of fossil fuels, generally enabled by CCUS (PÖYRY 2018, NREL 2017; EFI 2019; E3 et al. 2014; H21 2018) and some role for nuclear power (PÖYRY 2018; NREL 2017; E3 et al. 2014), although not all studies include nuclear generation (PGE 2018; Stirchele et al. 2018; EFI 2019). PÖYRY (2018) developed a pathway that paired electrification with an expanded role for nuclear power, as the use of nuclear power provides a source of baseload for electrification. Many studies include hydrogen as an energy carrier (Stirchele et al. 2018; National Grid 2017; 2018; NREL 2017; E3 et al. 2014; H21 2018) produced either by electrolysis with low carbon electricity or through natural gas reforming. The pairing of high renewable generation with hydrogen generation enables the hydrogen to be used as a form of storage for surplus renewable power (National Grid 2018). E3 et al. (2014) found that the importance of such "electric fuels" (hydrogen by electrolysis and synthetic natural gas) varied greatly by scenario. In the high renewables scenario these fuels accounted for 8.8% of total energy, in the high nuclear scenario 12.3%, in the mixed scenario 8.5%, but only 0.9% in the high CCUS scenario.

CCUS enables the production of low carbon fuels by natural gas reforming which maintains a cost advantage over "electric fuels". PÖYRY (2018) developed a pathway that paired low carbon fuels with CCUS to enable low-carbon hydrogen production predominantly by methane reformulation. In contrast, E3 et al. (2014) found little role for hydrogen in the high CCUS scenario. In the high CCUS scenario of E3 et al. (2014), buildings were largely electrified, and transportation relied on a mixture of electrification, biofuels, and fossil fuels, such that overall little hydrogen was used.

While biogas is often part of the energy mix, some sources raise questions as to its ability to substantially replace existing gas supplies. AGF (2011) estimates 4–10% of current U.S. gas consumption could be replaced with renewable gas. If other low carbon options are used for electricity generation, then available biogas resources could supply a larger fraction of building demand. National Grid (2010) indicates that the technical potential of biogas in Massachusetts, New Hampshire, New York, and Rhode Island is 25% of current demand, excluding demand from electricity generation. E3 et al. (2014) sees substantial roles for biomass for gas and liquid fuels in all four scenarios, but it is difficult to compare usage rates from this report with availability from other reports. If biogas supplies are more constrained, then it is likely that the analysis of report (E3 et al. 2014) would be modified to include more reformulation of methane to hydrogen with CCUS, which would bring it more in line with PÖYRY (2018), which pairs hydrogen reformulation with CCUS (see above). However, further electrification could also be considered as an alternative to biogas.

Sepulveda (2018) also stresses the importance of the role of "firm", or "dispatchable" low-CO_2 electricity resources. These are technologies that can be counted on to meet demand when needed in all seasons and over long durations (e.g. weeks or longer). Firm low-CO_2 resources include nuclear power plants capable of flexible operations, hydro plants with high capacity reservoirs, coal and natural gas plants with CCUS and capable of flexible operations, geothermal power, and biomass and biogas fueled power plants). The results indicate that the cost of a decarbonization pathway in the northern part of the United States, where renewables potential is more limited, particularly in winter, could be between 42% to 163% higher without the use of firm low-CO_2 resources. Such resources provide baseload generation to compliment the variable output of renewable technologies and offer protection from the risk that a significant decrease in the cost of variable renewable energy resources like onshore wind and solar PV may not occur. Modeling was performed for the state of California in the EFI report (2019) that reinforces the danger of steering away from firm low-CO_2 electricity resources when no change in the cost of renewables occurs. California will lose about 10,000 MW of capacity by 2025 due to natural gas and nuclear plant closures. The need to compensate for this loss compounds the cost of decarbonization by requiring either building more renewable capacity or importing energy. In this report the modeling assumed no change in technology cost, which resulted in a marginal cost of more than $1000 per ton of CO_2 abated as their greenhouse gas reduction percentage approached 80%.

Firm low-CO_2 resources, specifically nuclear and natural gas with CCUS, provide a reliable base of power as both the demand and the power generated by renewables fluctuate. As the emissions reduction percentage increases, i.e. as the amount of CO_2 emissions allowed approaches zero, variable renewable technologies like wind and solar lose a significant amount of reliability as electricity generation resources. Due to the variable nature of weather, the renewable energy sources often do not generate at their maximum capacity at a given time. The result of which is systems with no firm low-CO_2 resources requiring enough variable renewable energy resources to be built so that peak demand can be met even when the theoretical minimum amount of renewable generation is produced by the installed capacity. In other words, in order to replace a specific amount of capacity provided by firm low-CO_2 resources, significantly more capacity of variable renewable energy and battery storage must be built in order to maintain the same·level of reliability. In systems with no firm low-CO_2 electricity resources, Sepulveda et al. (2018) state that generation and storage capacity is five to eight times the peak system demand compared to a system with firm low-CO2 resources, where the ratio is 1.3–1.6 times peak demand.

Cost Estimates

Detailed cost information are not provided for most of the decarbonization studies and the few that have reported cost results (NREL 2017; E3 et al. 2014; Toronto 2017; E3 2018) used a variety of different cost metrics: 2014 dollars per ton CO_{2e} in 2050 (NREL 2017), percentage of the U.S. GDP level (E3 et al. 2014), marginal abatement costs per action (Toronto 2017), and 2017 dollar per ton CO_{2e} in 2050 (E3 2018). Pathway costs might reflect the electricity sector only and exclude additional costs of electrifying end-use sectors, as well as the cost savings associated with higher energy efficiency or fuel switching (NREL 2017). In the study of E3 et al. (2014) costs of different pathways are reported as a percent of GDP and broken down by end use sector. In another study the total reported costs include amortized capital costs for energy infrastructure in each sector, associated operations and maintenance costs, and fuel costs without reflecting macroeconomic effects (e.g. jobs or structural changes to the economy) or avoided externality costs like the social cost of carbon or changes in health outcomes or costs due to changes in regional air quality (E3 2018). Co-benefits and co-harms of different actions are difficult to quantify and are not included in most of the studies. However, in (Toronto 2017) a multi-criteria analysis was used to account for social equity, public health and economic considerations or criteria.

Abatement costs are often expressed in dollars per ton to enable comparisons across studies and technologies. The 0.8% of GDP estimated by E3 et al. (2014) corresponds very roughly to \$30–40/ton of avoided emissions, and the lower bound estimate of (E3 2018) is \$40/ton. While NREL's analysis (NREL 2017) identified reduction of only 74%, rather than 80%, these reductions cost less than \$50/ton.

However, more conservative assumptions result in substantially higher costs. The 75th percentile cost estimate of (E3 et al. 2014) is roughly $80/ton, and the upper bound of (E3 2018) is $190/ton. The lower to mid-range of reported abatement costs are on the order of typical estimates of the social cost of carbon. EPA mean estimates for the SCC over time are in the $42–69/ton range.

Cost estimates are sensitive to the role of nuclear power and CCUS, particularly given that E3 et al. (2014) indicate that a high renewables pathway costs roughly three times as much as the high nuclear and high CCUS pathways. Assumptions regarding the extent to which costs of new technologies will decline are important as E3 et al. indicate that using current technology costs (rather than assuming that costs will decline as technologies are more widely adopted) roughly doubles the overall abatement costs. These results further emphasize the conclusions of E3 et al. (2014) that decisions need to be made on the role of nuclear power and CCUS. They also suggest that policies based on an expectation that costs for a variety of new technologies will substantially decline in the coming decades need to have alternatives identified for the eventuality in which these forecast declines are not achieved.

Building Electrification

These studies make clear that substantial changes are needed in the built environment in order to achieve greenhouse gas emissions reduction goals. Foti et al. (2015) note that even if all current building electricity use were converted to low carbon sources, natural gas use would still need to be reduced by 41% to achieve an 80% overall reduction in building emissions. Heat pumps powered by low carbon electricity offer an opportunity to replace natural gas for space and water heating in buildings. The economics of heat pumps compared to gas heating vary by region. In the southeastern U.S. heat pumps are considered competitive or even preferable to gas heating and dedicated air conditioning. In more northernly climates heat pumps have generally been considered less economically attractive as they are less efficient at low temperatures and the option to provide air conditioning which is provided by most heat pumps is less valuable. However, higher efficiency heat pumps and potential regulatory incentives for building decarbonization are driving consideration of heat pumps in more northernly climates. Stief et al. (2021) developed an integrated building energy use and economic model to compare lifecycle costs of gas heating and heat pumps for a multifamily building in the northeastern United States. They found that gas heating was economically preferred under current market conditions. If low carbon electricity is used, then Stief et al. estimated that a carbon price of $125/tonne is required to equalize the lifecycle costs of gas heating and heat pumps. This study did not consider the use of electrical resistance heating as a back up. The use of back up resistance heating can reduce the required capacity of the heat pump and lower capital costs. As resistance heating is not as efficient as heat pumps, this approach will increase the electrical consumption on cold days when the resistance heating is required to meet a portion of the heating demand. However, the system can be designed such that the heat pump meets the bulk of the heating demand with only limited use of resistance heating. Hence the impacts on operational costs may be modest. Another factor that would favor heat pump adoption would be valuing the air conditioning capability which is provided by heat pumps. Stief et al. noted that many buildings in the northeast do not have air conditioning and hence did not include air conditioning in the capital costs for the gas heating system. Had air conditioning been included, this would have increased the capital costs and lessened the economic advantage of the gas system.

Building electrification may present substantial challenges to the current electrical supply infrastructure with respect to both peak demand and resilience. Existing electricity supply infrastructure has generally been designed around meeting peak demand in summer. The replacement of fossil fuel heating by heat pumps could substantially increase winter demand to the point where peak demand occurs in winter rather than summer. This presents a challenge as solar radiation is less intense during the winter. Fawcett et al. (2014) estimate that 100% use of heat pumps in UK would increase national electricity demand by 25%, and peak electricity demand by 65%. Globally, this study estimates that 100% heat pump adoption would require 11% of current world electricity use and increase peak demand by 65%. Clearly a systems approach is needed in which electrification is combined with load shifting and shedding strategies. In addition, technologies for energy storage are rapidly progressing and offer potential tools to mitigate the impact of shifting demand patterns on the electrical supply infrastructure.

Conclusions

Developing a nature-positive built environment will require reducing GHG emissions. This involves not only buildings but also consideration of the complex infrastructures systems in which buildings are embedded. To identify options to reduce GHG emissions, a wide range of future scenarios have been considered by literature studies. While these studies vary greatly in assumptions, scale, and conclusions, on the balance they suggest that dramatic reductions in greenhouse gas emissions are possible. Cost estimates are uncertain and depend greatly on (1) the extent to which costs are assumed to decline as technologies are more widely adopted and (2) the degree to which firm (or dispatchable) sources of electrical power generation are used vs. relying on intermittent sources with battery storage. Most of the studies considered here include continued reliance on firm sources of power, such as nuclear power and fossil fuels with CCUS. The mixture of demand reduction, use of low carbon electricity and electrification of end uses, and use of low carbon fuels varies from study to study. Overall, these studies suggest that a wide variety of pathways to a low carbon energy infrastructure exist, but that there is no consensus yet as to the preferred pathway. Ongoing technological developments, particularly in the area of energy storage, may open up further options that enable the sustainable development of a nature-positive built environment.

Acknowledgements Atiq Rahman assisted with the literature review for this chapter. Gena Ellis and Bob Ramirez provided comments on drafts. Numerous colleagues assisted with the identification of relevant planning studies.

References

AGF (2011) The potential for renewable gas: biogas derived from biomass feedstocks and upgraded to pipeline quality. American Gas Foundation, Washington, DC

CEC (2018) Deep decarbonization in a high renewables future. California Energy Commission Sacramento, California

E3 (2018) Pacific Northwest pathways to 2050, energy and environmental economics, San Francisco, California.

E3, LBNL, PNNL (2014) Pathways to deep decarbonization in the United States, Energy and Environmental Economics (E3). Lawrence Berkeley National Laboratory, Pacific Northwest National Laboratory, San Francisco, CA

EFI (2019) Optionality, flexibility, and innovation: pathways for deep decarbonization in California. Energy Futures Initiative, Washington, DC

EPRI (2018) U.S. National Electrification Assessment. Electric Power Research Institute, Washington, DC

Fawcett T, Layberry R, Eyre N (2014) Electrification of heating: the role of heat pumps. In: BIEE conference Oxford, UK

Foti R, Gurian P, Wen J, Hunold C et al (2015) Options for achieving deep reductions in carbon emissions in Philadelphia by 2050. Drexel University, Philadelphia, PA. Available online at: https://www.researchgate.net/publication/283503711_Options_for_Achieving_Deep_Reductions_in_Carbon_Emissions_in_Philadelphia_by_2050?channel=doi&linkId=563b9b0808ae45b5d28693e7&showFulltext=true

H21 (2018) H21 North of England, Northern gas networks in partnership with Cadent and Equinor, United Kingdom, North of England, Leeds, UK.

National Grid (2010) Renewable gas—vision for a sustainable gas system. National Grid, Waltham, MA

National Grid (2016) National Grid UK's future energy scenarios 2016. National Grid System Operator, Warwick, UK

National Grid (2017) National Grid UK's future energy scenarios 2017. National Grid System Operator, Warwick, UK

National Grid (2018) National Grid UK's future energy scenarios 2018. National Grid System Operator, Warwick, UK

NRDC (2017) America's clean energy frontier: the pathway to a safer climate future. Natural Resources Defense Council, Washington, DC

NREL (2017) Electrification and decarbonization: exploring US energy use and greenhouse gas emissions in scenarios with widespread electrification and power sector decarbonization. National Renewable Energy Laboratory, Golden, Colorado

NYC (2016) New York City's roadmap to 80 × 50. NYC Mayo's Office of Sustainability, New York City

PA-DEP (2019) Pennsylvania Climate Action Plan 2018. PA Department of Environmental Protection Harrisburg, Pennsylvania

PGE (2018) Exploring pathways to deep decarbonization for the Portland General Electric Service Territory. Portland General Electric, Portland, OR

PÖYRY (2018) Fully decarbonizing Europe's energy system by 2050. PÖYRY Engineering Firm, Helsinki, Finland

Sepulveda NA, Jenkins JD, de Sisternes FJ, Lester RK (2018) The role of firm low-carbon electricity resources in deep decarbonization of power generation. Joule 2(11):2403–2420. https://doi.org/10.1016/j.joule.2018.08.006

Stief T, Safari M, Johnson H, Wen J, Gurian P (2021) Economic analysis and comparison of electric and conventional heating systems. ASHRAE Trans 127(Part 2):287–295

Stirchele P, Palzer A, Hening H (2018) Electrify everything? IEEE Power Energy Mag 16(4):24–33. https://doi.org/10.1109/MPE.2018.2824100

Toronto (2017) Transform TO: the pathway to a low carbon future, City of Toronto, Toronto, Ontario, Canada.

Nature-Based Solutions to Overheating for Equitable Heat Health Protection

Simi Hoque, Leah Schinasi, Molly Hesson, and Ramyar Tajik

Abstract The majority of individuals spend more than 90% of their time inside buildings, underscoring the critical importance of healthy indoor environments. Excessive heat can have a significant impact on human well-being, affecting both physical and mental health, and prolonged exposure to high temperatures can lead to heat-related illnesses, ranging from heat exhaustion to heatstroke. Vulnerable populations, such as the elderly, young children, and individuals with pre-existing health conditions, are particularly susceptible to the adverse effects of excessive heat. These individuals often spend a significant amount of time indoors, making their homes the primary location where heat-related health issues can arise. Indeed, most heat-related fatalities tend to occur in residential settings, particularly within people's homes. Although mechanical air conditioning is a common choice for regulating indoor climate, it represents an unsustainable approach that will exacerbate health disparities, increase energy consumption, and add to greenhouse gas emissions. Instead, nature-based solutions, including energy-efficient building designs and landcover adaptation methods, can offer more cost-effective and climate-resilient solutions to help protect public health from the effects of extreme heat.

In this chapter, we present nature-based solutions that can mitigate the mortality and morbidity associated with heat stress from indoor overheating. We review the literature from passive design, building science, landscape architecture, green infrastructure, and public health to identify behavioral and built environment strategies to address indoor overheating in residential buildings. Behavioral adaptations for cooling include actions and practices that a home occupant can implement to remain cool, such as reducing metabolic load, strategic ventilation, and night flush cooling.

S. Hoque (✉) · M. Hesson · R. Tajik
Department of Civil, Architectural and Environmental Engineering, Drexel University, Philadelphia, PA, USA
e-mail: sth55@drexel.edu

L. Schinasi
Department of Environmental and Occupational Health, Drexel University, Philadelphia, PA, USA
e-mail: lhs36@drexel.edu

G. Iannaccone, E. V. Ellis (eds.), *Towards a Nature-Positive Built Environment*, Research for Development, https://doi.org/10.1007/978-3-031-95292-0_13

Building innovations include the use of reflective roofing and thermal mass to optimize cooling, as well as careful consideration of window orientation and shading as well as material selection. And on a broader scale, community or regional interventions aimed at reducing urban heat islands involve cultivating green spaces and tree planting and enhancing the reflectivity (albedo) of pavements and rooftops.

Ultimately by embracing behavioral, building and site design strategies tailored to specific regions and climates, future heat stress-related challenges can be avoided. Moreover, by applying these strategies to reduce reliance on energy-intensive mechanical cooling, we can not only drive down greenhouse gas emissions and forestall the ongoing climate crisis, but also derive multiple co-benefits such as reduced stress, enhanced wellbeing, as well as increased community engagement and social equity.

Keywords Indoor environment · Heat stress · Indoor overheating · Nature-based solutions · Natural cooling

Introduction

Accelerating urbanization and unprecedented global warming have created critical environmental challenges. Every summer has been a "hottest summer on record," exacerbating health risks and threatening human health and well-being. The escalating frequency and intensity of heatwaves, often disproportionately impacting vulnerable communities, demand urgent and innovative interventions to mitigate the adverse impacts of rising temperatures. The experience of staying at home while living through heat waves will amplify the well-documented racial/ethnic and socioeconomic disparities in heat related morbidity and mortality (Gronlund 2014; Benmarhnia et al. 2015). This situation highlights the importance of identifying optimal and energy efficient, long-term building adaptation strategies to maintain health and prevent premature heat-related deaths. The exact total number of deaths attributable to heat is unclear, but one thing is certain: the number of cases of this preventable cause of death is unnecessarily high. A recent analysis estimated that, in 384 locations across the world, 42% of deaths were attributable to high temperatures (Benmarhnia et al. 2015). In another analysis of data from 297 locations across the U.S. between years 1997–2006, attributed an estimated 5608 annual deaths to heat (Weinberger et al. 2020). The risks from heat are not equitably distributed; heat disproportionately impacts low socioeconomic position and ethnic/racial minority populations. There are a multitude of explanations for these disparities (Gronlund 2014), but they may be explained, at least in part, to disproportionately higher overall heat exposure inside the home.

Indeed, in developed countries, the majority of heat-related deaths occur indoors (Fouillet et al. 2006). The quickest solution to overheating inside the home, and to reducing disparities, is to expand access to mechanical air conditioning (Centers for Disease Control and Prevention (CDC) 1994). Yet differential access to mechanical air conditioning has been shown to explain racial disparities in heat related mortality (O'Neill 2005), and tremendous disparities in access have been documented. In the U.S., for example, more than three-quarters of households with incomes higher than $100,000 have central air conditioning, while less than 40% of households living below the poverty line have mechanical cooling systems (U.S. Energy Information Administration 2011). Mechanical air conditioning is a non-sustainable band-aid solution that, in the long run, will amplify disparities, exacerbate energy burden, and contribute to greenhouse gas emissions. Air conditioner and electric fan use accounts for 10% of global energy consumption (IEA 2018). In the United Kingdom, energy use from air conditioning was estimated to account for approximately 26% of the country's total carbon dioxide emissions (Wilkinson et al. 2009). An analysis in Mexico found that each additional day above 90 °F (32.2 °C) increased monthly electricity consumption by 3.2% (Davis and Gertler 2015). Often, lower income households cannot afford higher energy costs associated with mechanical cooling. Based on the most recent Residential Energy Consumption Survey by the U.S. Energy Information Administration, 31% of households in the U.S. reported that they had trouble paying energy bills or adequately heating or cooling their homes (U.S. Energy Information Administration 2017). In qualitative interviews of residents of several major U.S. cities, participants reported utility costs as a barrier to air conditioning use (Sampson et al. 2013). Even if made affordable, peak energy demand during hot days overloads the electrical grid; this problem will intensify with increased global warming (Auffhammer et al. 2017) This, in turn, will contribute to power outages and potentially more dangerous conditions for vulnerable populations (Farbotko and Waitt 2011). There is ample evidence of the negative consequences of relying on mechanical cooling (Isaac and Van Vuuren 2009; Buranyi 2019). Areas across the world have experienced power outages due to excessive energy demands from air conditioner use during heat wave events (Klinger et al. 2014).

We will present various mechanisms through which nature-based solutions can effectively mitigate overheating, enhance thermal comfort, and minimize heat-related health risks in diverse residential building types in different climate regions. For example, passive cooling techniques such as natural ventilation, shading, thermal mass, and the use of cool roofing materials as well as building orientation and glazing design are effective strategies to minimize indoor heat gain. These techniques are further enhanced by landscape architecture interventions such as green roofs and tree canopies. Reducing grey (built-up) space through green infrastructure elements such as permeable paving around buildings also contributes to temperature moderation and mitigates urban heat island effects. Building envelope technologies such as advanced insulation materials, smart glazing, and phase change materials reduce the need for energy intensive cooling systems. And finally, the promotion of adaptive behavior, education, and awareness campaigns aimed at encouraging heat-resilience have played a key role in reducing heat-related health risks and enhancing community resilience. Our aim is to provide valuable insights into the opportunities and best practices for the equitable integration of nature-based approaches into overheating protection strategies for residential dwellers. Ultimately, we are advocating for an inclusive approach to leveraging nature-based solutions as a cornerstone for achieving equitable heat health protection in the face of a rapidly warming world.

Passive Cooling

Designing for passive cooling is not trivial. In most residential building design projects in Europe and North America, developer driven housing designs predominate and thermal comfort is largely achieved by the use of mechanical HVAC systems. Passive design is a concept that is applied by a select few architects rather than a mainstream design approach. Because passive heating and cooling design is not an integral part of a designer's traditional workflow, the ways that these strategies are learned and applied are often ad hoc (Toroxel and Silva 2024). Passive strategies, drawn from vernacular housing, historical precedents, and sustainable building design standards, such as the Leadership in Energy and Environmental Design (LEED), Passivhaus, and the WELL building standard, are well known to be an energy efficient means to reduce energy loads. Indeed, the vernacular and historical homes from the last century offer valuable lessons about how cooling and heating were managed before the advent of mechanical systems (Vural et al. 2007). However, as the climate is changing, the benefits of these strategies are less apparent and many of the historical and vernacular homes in different cities are more vulnerable to climate-related environmental stresses like overheating. Examples of what strategies to apply are now no longer straightforward. For example, a historical home that was passively cooled in the past in the City of Philadelphia may no longer be cool today, because Philadelphia summers are now hotter and more humid (ICF International 2014; Santamouris 2020).

Designers often look to vernacular or historical passive design—derived from precedents of housing typologies before the advent of mechanical systems—to implement effective passive cooling design strategies. For example, Philadelphia's typical vernacular housing typology—the rowhome—is now now overheating in a hotter climate. The envelope systems of Philadelphia rowhomes are simple, especially the older ones, built with a masonry wall (typically brick but sometimes with brownstone) exterior and minimal insulation. The shared walls between homes minimize external exposure making them easier to insulate, and there are ways to improve the insulation by adding it to the roof and exterior walls to meet modern energy-efficiency standards. Rowhomes also typically have flat black, tar roofs, which absorb sunlight to heat the building easily and efficiently for the heating season. However, for a warming climate, these types of buildings overheat because they are designed and built to store thermal energy. This is appropriate for what used to be the dominant season—winter—in Philadelphia. Nowadays, this vernacular building type can become uncomfortably hot in the summer to the point of overheating with severe consequences. For Philadelphia, addressing overheating may require looking at vernacular traditions from regions beyond its climate zone, to find passive cooling strategies that have been implemented in hotter and more humid regions around the globe.

For instance, passive strategies for cooling from in Akure, Nigeria, where the mean relative humidity is 77.1% and temperatures range from 21.4 °C to 34.19 °C, might be more appropriate to apply to new housing design in Philadelphia. In Akure, roofing colors are commonly pink, red, or brown; these are colors that reflect rather than absorb infrared radiation (Adegun and Ayoola 2019). The shape of the roof is not flat (like the typical rowhome) but rather pitched to provide a buffer of air between the roof and interior. And the interior surfaces are generally light in color, allowing for thermal storage and keeping the interiors cool on hot days. On the ground floor, ceramic tiles, a popular flooring material, provide thermal mass where radiant heat can be absorbed by the ground underneath the building. These architectural features help to protect homes from overheating. Similarly historical housing can offer insights about passive design because it was built before the advent of mechanical heating and cooling systems. In another example, the Milam House, designed by Paul Rudolph in 1962, is located in the sub-tropical climate of Jacksonville, Florida. The original design takes into account solar angle, shading, and thermal mass to prevent overheating. While the Milam house was built with air conditioning and does not have operable windows, it has been shown that if the windows were operable, the need for mechanical cooling would be negated (Fiocchi et al. 2011). Similarly, a simulation study in the Midlands, United Kingdom, estimated that installing shutters for use during the daytime would reduce population heat-related mortality by 32–38% (Taylor et al. 2018).

Landscape Architecture

Expanding beyond the home, characteristics of the urban landscape can also be modified to reduce overall temperatures, improve thermal comfort, and reduce energy use and demand. These neighborhood or regional level interventions aim to reduce urban heat islands, or the phenomenon of urban centers being hotter than suburban or rural settings (Davies et al. 2008). The urban heat island effect is a pressing concern, particularly within the context of our rapidly urbanizing world (Tuholske et al. 2021), and urban expansion processes, alone, have been shown to induce warming, independent of carbon emissions (Georgescu et al. 2014). The phenomenon is caused by lack of vegetation and shade trees (Katunský and Lopušniak 2012), low albedo caused by the presence of dark concrete and asphalt (Virk et al. 2014), and sub-standard building envelopes (Kandya and Mohan 2018) that are found in urban settings.

The urban heat island effect has been shown to be even more substantial in low income and racial/ethnic minority areas and may contribute to disparities in heat related mortality (Jenerette et al. 2011; Jesdale et al. 2013). Site specific considerations such as landscaping elements, such as blue and green infrastructure are also utilized to help reduce indoor temperatures (Sun et al. 2017). For example, common approaches to reducing the urban heat island effect include planting shade trees near buildings, installing vegetation in parks or along roadways, and/or increasing the albedo of surfaces (i.e., solar reflectivity). Replacing dark with light roofs or pavement increases sunlight reflectivity, reduces net radiation, and decreases downward heat flux during the day and upward heat flux at night. Vegetation and water bodies reduce air temperature through evaporation and transpiration. Trees provide shading to decrease net radiation from the underlying surface. Positive impacts of these interventions on energy use, air temperature, and thermal comfort, have been observed at the household, neighborhood, and regional levels (Dhar and Khirfan 2017).

A study with the Department of Conservation Resources in Massachusetts found that the urban canopy of deciduous street trees reduced cooling energy use at the household level by up to a factor of two or an average of 600 kWh (Morzuch 2013). The study focused on site specific micro-climates and the impact of shade trees on both neighborhood and regional energy consumption. Other simulation-based studies confirm these findings. At the neighborhood level, in Los Angeles, CA, adopting cool pavements led to surface air temperature cooling of up to 2.0 °C, and planting new trees reduced surface air temperatures by up to 1.0 °C (Taleghani et al. 2013). At the regional level, in a simulation study, increasing albedo in some urban areas led to summertime temperature reductions of 0.11°–0.53 °C. However, increasing albedo in rural locations, most of which had lower soil moisture, fewer or thinner clouds, and less precipitation, led to temperature increases, highlighting the need for area-appropriate interventions (Millstein and Menon 2011).

Empirical epidemiological studies have demonstrated positive impacts of urban greenspace on heat-mortality/morbidity relationships. For example, a recent meta-analysis of five epidemiology studies showed higher risk of heat-related mortality or morbidity among those living in less green areas (Schinasi et al. 2018). In Seoul, Korea, the risk of total mortality in association with increases in higher temperatures were higher in areas with low vs. high densities of greenness (Son et al. 2016). In NYC, compared to those living in less green areas, residents of greener areas were less likely to die during heat waves (Madrigano et al. 2015). However, in thinking about these landcover and built environment modifications, it is important to consider the tradeoff between seasonal energy demands; while interventions such as cool roofs decrease energy demand during the summer, they increase heat-energy demands in the winter (i.e., cool vs. warm season) (Georgescu et al. 2014).

Building Envelope

The components that comprise a building envelope are the walls, ground floor, roof, shading, windows, and internal heat loads, respectively. Combating overheating requires careful material selection and placement, such as the choice and order of materials for a wall system from the exterior to interior. It is also important to choose the appropriate type of glazing and placement for windows, which can be a significant source of heat gain during the summer months when solar radiation is at its peak. And finally, the roofing type, its insulation, and morphology are important to design with specific regard to heat gain. Several research studies have explored the concept of an "optimal building envelope" which would help to mitigate the effects of heat gain in the summer and heat loss in the winter (Manni and Nicolini 2022). However, an optimal envelope system in the hot season may not be optimal for the cold season. Envelopes that have high heat capacity for the walls and higher sun absorption on the roof will work well during colder temperatures to hold in the heat and take advantage of passive heating (Gupta and Tiwari 2016). Lightweight building walls that are well insulated, low sunlight absorbance on the roof, and a lack of insulation on the ground floor slab may be best suited for high temperatures and humidity (Sadineni et al. 2011). And for hot and dry climates, exterior envelopes that absorb and maintain heat might work more efficiently to keep the space cool during the hottest time of the day instead of letting the heat pass through into the building interior (Yılmaz 2007). For different climates, it is increasingly clear that business-as-usual approach to designing building envelopes per building code or even precedent may no longer apply. Climate change is creating anomalous outdoor conditions such that regions that were typically warm and wet may becoming hotter and dryer, or vice versa. And these types of extreme heat events are especially dangerous given that building envelopes have not been designed for those types of loads (Pajek and Košir 2021).

Roofs are also a part of how a building can overheat as they are warmed by direct sunlight. In warmer climates it is better to have a thermo-reflective roof so as to not absorb sunlight and therefore heat. The optimal solar absorbance must be low, at a value of 0.3, to reduce heat loads through the roof (Baglivo et al. 2017). Philadelphia rowhomes are built with flat black tar roofs, maximizing solar absorbance which is extremely detrimental to overheating in residential buildings.

When the focus is on thermal aspects of a building envelope, windows represent the weakest element of the system. The optimal design of windows relies on their type, position, size, and shading. Low emissivity glazed windows can reduce significantly building energy consumption depending on the placement and size of the windows (Sorooshnia et al. 2023). Low emissivity surfaces minimize the amount of UV solar rays that pass through and therefore the amount of thermal energy that passes through. Low emissivity windows as well as double paned windows are a better insulator than simple paned windows as they block the amount of heat passing through and are low solar gain which is preferable in a warm climate to decrease overheating in the summer. Specifically, the optimal window design found in the study is a window with double glass, a *G*-value of 0.333 meaning it transmits very little thermal energy, an argon cavity, and an aluminum frame.

Looking beyond materiality, window placement, size, and shading systems play a crucial role in mitigating overheating. Zhao and Du (2020) and Zhai et al. (2019) emphasize the significance of optimizing window placement and size, along with implementing effective shading strategies, to reduce solar heat gain. Kim et al. (2018) and Tajik et al. (2024) highlight how dynamic shading systems can be an effective way to adapt to changing environmental conditions, allowing for better control of daylighting and thermal performance. By integrating these considerations into building design and retrofitting practices, overheating challenges can be addressed for a variety of building types in different climate zones. There is no single optimal solution to mitigating overheating.

The third most prominent component in building envelope design to mitigate overheating is insulation of the ground floor slab. For a warm climate, no ground slab insulation is a beneficial design choice although it is less beneficial for winter conditions (Baglivo et al. 2017). Having a ground floor with no layer of insulation will avoid the risk of overheating and provide a lower, more comfortable, and healthy indoor temperature when it is hot outside. The challenge is to determine which is the dominant season to protect indoor occupants from. In temperate climates like Philadelphia, which now experiences much longer and hotter humid summers and cool winters, the dominant season to protect against is now the summer (Staszczuk and Kuczyński 2019).

Adaptive Behavior

One of the main obstacles to addressing overheating in residential buildings is the awareness and engagement of building occupants. Despite the availability of passive design strategies and technological solutions, occupant behavior and attitudes towards energy use and comfort significantly influence the effectiveness of these interventions. Without active participation and cooperation, such as adjusting window openings, using shading devices, or managing internal heat gains, the benefits of overheating mitigation measures may be limited. Therefore, comprehensive approach is required, encompassing targeted education and awareness campaigns, collaborations with stakeholders, and leveraging innovative technological solutions.

Addressing occupant behavior involves developing targeted education and awareness campaigns. By implementing these campaigns, households can gain knowledge and understanding of overheating drivers, potential health risks, and strategies to improve thermal comfort. Additionally, ensuring these educational efforts are introduced to empower participants, fostering their active engagement, will ensure long term behavior adaptation and adoption. These require a collaborative approach in which community organizations, local authorities, and relevant agencies may offer valuable assistance and resources. Collaborators can contribute expertise, funding, and community outreach, enhancing participants' involvement and broader engagement. Strong partnerships and collaborative relationships ensure collective efforts that will ultimately increase community engagement and translate into effective interventions.

Additionally, there are several economic and logistical challenges to mitigating overheating that can present significant obstacles to households. Implementing shading devices, improving roof insulation, or upgrading cooling systems often require upfront investments that are prohibitive to homeowners or tenants. Additionally, the availability of materials, technologies, and skilled professionals for retrofitting existing buildings or integrating passive design strategies into new construction may vary regionally and pose logistical challenges. Efficient coordination, clear communication channels, and flexibility in scheduling are key to overcoming some of these challenges. Streamlining the regulatory framework, offering financial incentives, and establishing effective policies to ensure the health, comfort, and resilience of residential occupants will help to overcome logistic and economic barriers.

Discussion and Conclusion

What are the solutions as we face projections of higher frequency and duration of heat waves and more record high temperatures between now and the end of the 21st century (Intergovernmental Panel on Climate Change (IPCC) 2023)? We must advocate a shift in focus to affordable, energy efficient adaptation strategies for protecting against adverse health impacts of extreme heat. The concept of nature-based solutions (NBS) is a promising avenue for fostering climate resilience, promoting sustainable urban development, and protecting public health. Despite the growing recognition of the multifaceted benefits of NBS, there remains a pressing need to deepen our understanding of their efficacy in curbing overheating and their potential for ensuring equitable heat health protection, particularly in marginalized and underserved communities.

Design standards that focus specifically on energy conservation are no longer enough to prevent overheating due to climate change. For example, in 1991, German designers developed the "Passivhaus" system, which defines optimal housing based on building orientation and shape, thermal insulation, airtightness, and mechanical ventilation. This system was originally invented with the idea of preventing excessive energy use from heating during the winter, which is the dominant season in Germany. Passivhaus certified homes have extremely low energy costs, using less than 120 kWh/m^2 per year of energy to operate. Unfortunately, Passivhaus design standards have limited success in the summer and can result in overheating (Mitchell and Natarajan 2019). This is because the primary strategies to reduce heating loads in the winter season are not similarly effective in reducing cooling loads in the summer. Climate regions with a dominant winter season would benefit from Passivhaus but, as global temperatures rise, these strategies must be adapted for warmer and longer summers. This highlights the importance of developing and applying low energy housing design strategies that are both climatically and culturally specific.

Few epidemiological studies have empirically quantified the impact of housing characteristics on heat related mortality or morbidity. Housing characteristics might correlate with less energy efficient housing; however, more epidemiological research on housing factors as determinants of heat-vulnerability is needed. There is an urgent need to critically assess the role of nature-based interventions in mitigating overheating in residential buildings, while emphasizing the importance of equitable distribution and access to these solutions for ensuring social justice and health equity.

References

Adegun OB, Ayoola HA (2019) Adaptation to heat stress within housing estates in Akure, Nigeria. Covenant J Res Built Environ 7. https://doi.org/10.20370/45y0-hs421

Auffhammer M, Baylis P, Hausman CH (2017) Climate change is projected to have severe impacts on the frequency and intensity of peak electricity demand across the United States. Proc Natl Acad Sci U S A 114:1886–1891. https://doi.org/10.1073/pnas.1613193114

Baglivo C, Congedo P, Di Cataldo M et al (2017) Envelope design optimization by thermal modelling of a building in a warm climate. Energies 10:1808. https://doi.org/10.3390/en10111808

Benmarhnia T, Deguen S, Kaufman JS, Smargiassi A (2015) Review article: vulnerability to heat-related mortality. Epidemiology 26:781–793. https://doi.org/10.1097/EDE.0000000000000375

Buranyi S (2019) The air conditioning trap: how cold air is heating the world. The Guardian

Centers for Disease Control and Prevention (CDC) (1994) Heat-related deaths—Philadelphia and United States, 1993–1994. MMWR Morb Mortal Wkly Rep 43:453–455

Davies M, Steadman P, Oreszczyn T (2008) Strategies for the modification of the urban climate and the consequent impact on building energy use. Energy Policy 36:4548–4551. https://doi.org/10.1016/j.enpol.2008.09.013

Davis LW, Gertler PJ (2015) Contribution of air conditioning adoption to future energy use under global warming. Proc Natl Acad Sci U S A 112:5962–5967. https://doi.org/10.1073/pnas.1423558112

Dhar TK, Khirfan L (2017) A multi-scale and multi-dimensional framework for enhancing the resilience of urban form to climate change. Urban Climate 19:72–91. https://doi.org/10.1016/j.uclim.2016.12.004

Farbotko C, Waitt G (2011) Residential air-conditioning and climate change: voices of the vulnerable. Health Promot J Austr 22:13–15

Fiocchi C, Hoque S, Shahadat M (2011) Climate responsive design and the milam residence. Sustainability 3:2289–2306. https://doi.org/10.3390/su3112289

Fouillet A, Rey G, Laurent F, Pavillon G, Bellec S, Guihenneuc-Jouyaux C, Hémon D (2006). Excess mortality related to the August 2003 heat wave in France. *International archives of occupational and environmental health, 80*(1), 16–24.

Georgescu M, Morefield PE, Bierwagen BG, Weaver CP (2014) Urban adaptation can roll back warming of emerging megapolitan regions. Proc Natl Acad Sci U S A 111:2909–2914. https://doi.org/10.1073/pnas.1322280111

Gronlund CJ (2014) Racial and socioeconomic disparities in heat-related health effects and their mechanisms: a review. Curr Epidemiol Rep 1:165–173. https://doi.org/10.1007/s40471-014-0014-4

Gupta N, Tiwari GN (2016) Review of passive heating/cooling systems of buildings. Energy Sci Eng 4:305–333. https://doi.org/10.1002/ese3.129

ICF International (2014) Useful climate information for Philadelphia: past and future. ICF International, Philadelphia

IEA (2018) World energy outlook 2018. IEA, Paris

Intergovernmental Panel on Climate Change (IPCC) (2023) Climate change information for regional impact and for risk assessment. In: Climate change 2021—the physical science basis: working group I contribution to the sixth assessment report of the intergovernmental panel on climate change. Cambridge University Press, Cambridge, pp 1767–1926

Isaac M, Van Vuuren DP (2009) Modeling global residential sector energy demand for heating and air conditioning in the context of climate change. Energy Policy 37:507–521. https://doi.org/10.1016/j.enpol.2008.09.051

Jenerette GD, Harlan SL, Stefanov WL, Martin CA (2011) Ecosystem services and urban heat riskscape moderation: water, green spaces, and social inequality in Phoenix, USA. Ecol Appl 21:2637–2651. https://doi.org/10.1890/10-1493.1

Jesdale BM, Morello-Frosch R, Cushing L (2013) The racial/ethnic distribution of heat risk-related land cover in relation to residential segregation. Environ Health Perspect 121:811–817. https://doi.org/10.1289/ehp.1205919

Kandya A, Mohan M (2018) Mitigating the Urban Heat Island effect through building envelope modifications. Energy Build 164:266–277. https://doi.org/10.1016/j.enbuild.2018.01.014

Katunský D, Lopušniak M (2012) Impact of shading structure on energy demand and on risk of summer overheating in a low energy building. Energy Procedia 14:1311–1316. https://doi.org/10.1016/j.egypro.2011.12.1094

Kim J, Schiavon S, Brager G (2018) Personal comfort models—a new paradigm in thermal comfort for occupant-centric environmental control. Build Environ 132:114–124. https://doi.org/10.1016/j.buildenv.2018.01.023

Klinger C, Landeg O, Murray V (2014) Power outages, extreme events and health: a systematic review of the literature from 2011–2012. PLoS Curr. https://doi.org/10.1371/currents.dis.04eb1dc5e73dd1377e05a10e9edde673

Madrigano J, Ito K, Johnson S et al (2015) A case-only study of vulnerability to heat wave–related mortality in New York City (2000–2011). Environ Health Perspect 123:672–678. https://doi.org/10.1289/ehp.1408178

Manni M, Nicolini A (2022) Multi-objective optimization models to design a responsive built environment: a synthetic review. Energies 15:486. https://doi.org/10.3390/en15020486

Millstein D, Menon S (2011) Regional climate consequences of large-scale cool roof and photovoltaic array deployment. Environ Res Lett 6:034001. https://doi.org/10.1088/1748-9326/6/3/034001

Mitchell R, Natarajan S (2019) Overheating risk in Passivhaus dwellings. Build Serv Eng Res and Technol 40:446–469. https://doi.org/10.1177/0143624419842006

Morzuch EL (2013) The energy benefits of trees: investigating shading, microclimate and wind shielding effects in worcester and Springfield, Massachusetts. https://doi.org/10.7275/3827749

O'Neill MS (2005) Disparities by race in heat-related mortality in four US cities: the role of air conditioning prevalence. J Urban Health: Bullet NY Acad Med 82:191–197. https://doi.org/10.1093/jurban/jti043

Pajek L, Košir M (2021) Exploring climate-change impacts on energy efficiency and overheating vulnerability of bioclimatic residential buildings under Central European Climate. Sustainability 13:6791. https://doi.org/10.3390/su13126791

Sadineni SB, Madala S, Boehm RF (2011) Passive building energy savings: a review of building envelope components. Renew Sustain Energy Rev 15:3617–3631. https://doi.org/10.1016/j.rser.2011.07.014

Sampson NR, Gronlund CJ, Buxton MA et al (2013) Staying cool in a changing climate: reaching vulnerable populations during heat events. Glob Environ Change 23:475–484. https://doi.org/10.1016/j.gloenvcha.2012.12.011

Santamouris M (2020) Recent progress on urban overheating and heat island research. Integrated assessment of the energy, environmental, vulnerability and health impact. Synergies with the global climate change. Energy Build 207:109482. https://doi.org/10.1016/j.enbuild.2019.109482

Schinasi LH, Benmarhnia T, De Roos AJ (2018) Modification of the association between high ambient temperature and health by urban microclimate indicators: a systematic review and meta-analysis. Environ Res 161:168–180. https://doi.org/10.1016/j.envres.2017.11.004

Son J-Y, Lane KJ, Lee J-T, Bell ML (2016) Urban vegetation and heat-related mortality in Seoul, Korea. Environ Res 151:728–733. https://doi.org/10.1016/j.envres.2016.09.001

Sorooshnia E, Rashidi M, Rahnamayiezekavat P et al (2023) A novel approach for optimized design of low-E windows and visual comfort for residential spaces. Energy Built Environ. S2666123323000776. https://doi.org/10.1016/j.enbenv.2023.08.002

Staszczuk A, Kuczyński T (2019) The impact of floor thermal capacity on air temperature and energy consumption in buildings in temperate climate. Energy 181:908–915. https://doi.org/10.1016/j.energy.2019.05.202

Sun S, Xu X, Lao Z et al (2017) Evaluating the impact of urban green space and landscape design parameters on thermal comfort in hot summer by numerical simulation. Build Environ 123:277–288. https://doi.org/10.1016/j.buildenv.2017.07.010

Tajik R, Soltanmohammadlou S, Kianfar A et al (2024) A case study: intelligent shading retrofit to existing home-office using multi-objective optimization. J Green Build 19:123–156. https://doi.org/10.3992/jgb.19.1.123

Taleghani M, Tenpierik M, Van Den Dobbelsteen A, De Dear R (2013) Energy use impact of and thermal comfort in different urban block types in the Netherlands. Energy Build 67:166–175. https://doi.org/10.1016/j.enbuild.2013.08.024

Taylor J, Wilkinson P, Picetti R et al (2018) Comparison of built environment adaptations to heat exposure and mortality during hot weather, West Midlands region, UK. Environ Int 111:287–294. https://doi.org/10.1016/j.envint.2017.11.005

Toroxel JL, Silva SM (2024) A review of passive solar heating and cooling technologies based on bioclimatic and vernacular architecture. Energies 17:1006. https://doi.org/10.3390/en17051006

Tuholske C, Caylor K, Funk C et al (2021) Global urban population exposure to extreme heat. Proc Natl Acad Sci U S A 118:e2024792118. https://doi.org/10.1073/pnas.2024792118

U.S. Energy Information Administration (2011) Residential energy consumption survey (RECS). 2009: Air Conditioning

U.S. Energy Information Administration (2017) Residential energy consumption survey (RECS). 2015

Virk G, Jansz A, Mavrogianni A et al (2014) The effectiveness of retrofitted green and cool roofs at reducing overheating in a naturally ventilated office in London: direct and indirect effects in current and future climates. Indoor Built Environ 23:504–520. https://doi.org/10.1177/1420326X14527976

Vural N, Vural S, Engin N, Reşat Sümerkan M (2007) Eastern Black Sea region—a sample of modular design in the vernacular architecture. Build Environ 42:2746–2761. https://doi.org/10.1016/j.buildenv.2006.07.017

Weinberger DM, Chen J, Cohen T et al (2020) Estimation of excess deaths associated with the COVID-19 pandemic in the United States, March to May 2020. JAMA Intern Med 180:1336. https://doi.org/10.1001/jamainternmed.2020.3391

Wilkinson P, Smith KR, Davies M et al (2009) Public health benefits of strategies to reduce greenhouse-gas emissions: household energy. Lancet 374:1917–1929. https://doi.org/10.1016/S0140-6736(09)61713-X

Yılmaz Z (2007) Evaluation of energy efficient design strategies for different climatic zones: comparison of thermal performance of buildings in temperate-humid and hot-dry climate. Energy Build 39:306–316. https://doi.org/10.1016/j.enbuild.2006.08.004

Zhai Y, Wang Y, Huang Y, Meng X (2019) A multi-objective optimization methodology for window design considering energy consumption, thermal environment and visual performance. Renew Energy 134:1190–1199. https://doi.org/10.1016/j.renene.2018.09.024

Zhao J, Du Y (2020) Multi-objective optimization design for windows and shading configuration considering energy consumption and thermal comfort: a case study for office building in different climatic regions of China. Solar Energy 206:997–1017. https://doi.org/10.1016/j.solener.2020.05.090

Reducing Greenhouse Gas Emissions from the Existing Building Stock: A Scenario Analysis for the European Union

Giuliana Iannaccone, Francesco Pittau, and Guillaume Habert

Abstract The construction sector plays a decisive role worldwide in achieving the global objectives of decarbonizing the economy by 2050, being responsible for 39% of total emissions. In the Global North, the efforts to cut emissions by buildings are currently geared towards drastically reducing energy demand for heating. The European Union, in view of the widespread ageing of the building stock in the Member States, has identified its energy refurbishment as one of the priorities in view of the objectives linked to climate change agreements. Thermal façade insulation solutions offer a great potential for energy savings by significantly reducing heat loss over large areas. However, once the primary energy demand of a building is reduced through retrofitting, the contribution to overall carbon emissions caused by the production processes of insulation materials increases. On a European scale, the demand for building materials to be used in the renovation of the building stock is expected to increase significantly in the coming years; therefore, the carbon emissions generated by material production and construction processes would significantly slow down the desired ecological transition of our society. At the same time, in developing countries of the Global South a huge growth in population has been estimated by 2050, with an increasing demand for new houses and infrastructure. Therefore, the energy needs of countries that need to develop their economies and provide materials for new buildings and infrastructure are likely to dominate global building sector emissions and impact the transition to a zero-carbon society. It is therefore necessary to define a carbon budget for construction and construction-related activities in such a way as to limit emissions in each country and stay within the available global budget limits. In this scenario, the adoption of materials with a

G. Iannaccone · F. Pittau (✉)
Department of Architecture, Built Environment and Construction Engineering, Politecnico di Milano, Italy
e-mail: giuliana.iannaccone@polimi.it; francesco.pittau@polimi.it

G. Habert
Department of Civil, Environmental and Geomatic Engineering, ETH Zurich, Zurich, Switzerland
e-mail: habert@ibi.baug.ethz.ch

G. Iannaccone, E. V. Ellis (eds.), *Towards a Nature-Positive Built Environment*, Research for Development, https://doi.org/10.1007/978-3-031-95292-0_14

low environmental impact and capable of storing carbon in the long term is an opportunity that should be addressed urgently in order to meet climate change mitigation goals.

Keywords Biogenic materials · Bio-based materials · Carbon budget · Carbon neutrality · Construction · Buildings · Embodied carbon · Environmental impacts · Greenhouse gas (GHG) emissions

Introduction

The emission of greenhouse gases into the atmosphere due to increasing anthropogenic activity began to grow exponentially following the industrial revolution, reaching alarming levels in recent decades. In the last century, the world population has increased from 1.5 to 6.1 billion people. According to UN estimates, there are now more than 8 billion people on earth (Ritchie et al. 2023), and this overall growth is accompanied by a steady migration of population from rural to urban areas. Today, more than 55% of the world's population lives in urban areas, a figure that is expected to increase to 68% by 2050 (IRP 2018). Consequently, the increasing population growth on the one hand, and rising levels of affluence in many developing regions of the world on the other, are likely to lead to an ever-increasing consumption of resources, also leading to an increase in the demand for energy and fossil fuels, as well as the production of an unsustainable amount of waste (Dutil et al. 2011). Currently, fossil emissions from the building and construction sector account for approximately 37% of total emissions (United Nations Environment Programme 2024). As shown in Fig. 1, they are divided into indirect emissions, which come from electricity consumption, and direct emissions, which come from the combustion of fossil fuels for heating/cooling buildings. In addition to these, there is the contribution of the construction industry, with its related emissions: direct emissions, generated by the production of building materials and installation, and indirect emissions, from the heating of workspaces and the use of electricity.

The risk of a 1.5 °C increase in the average annual global temperature is real and, according to several studies, may already occur within a decade if we do not quickly reverse course and drastically reduce fossil emissions (Lee and Romero 2023). At the Conference of the Parties (COP21) in Paris in December 2015, held in conjunction with the Sustainable Innovation Forum, the parties to the United Nations Framework Convention on Climate Change (UNFCCC) reached an historic agreement to combat climate change and to accelerate and intensify the actions and investments needed for sustainable low-carbon living. The Intergovernmental Panel on Climate Change (IPCC) has developed climate models that show that if we continue to use resources in the same way as today, and thus emit a high amount of greenhouse gases, there will be a temperature increase of 1.4 °C to 5.4 °C by the end

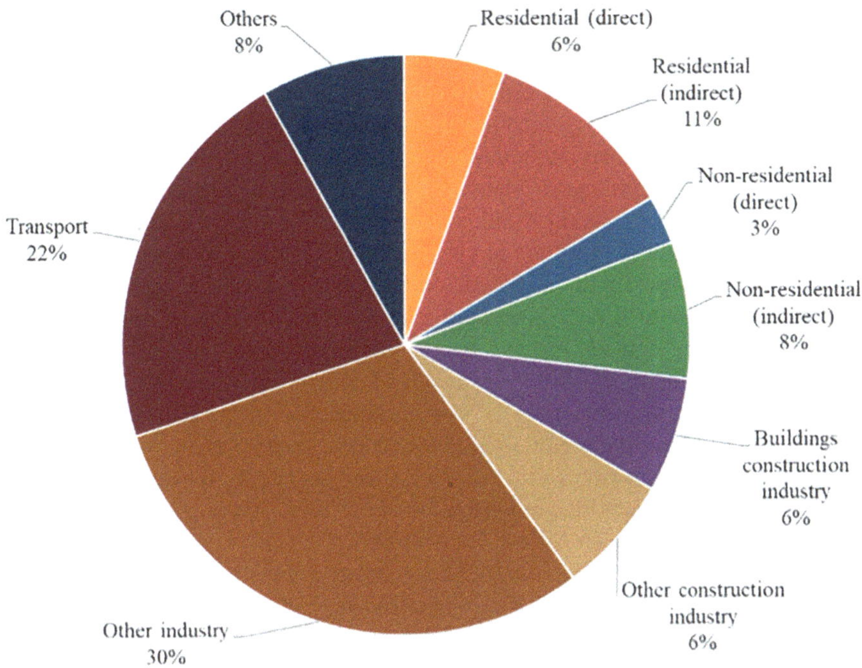

Fig. 1 Global energy and process emissions from buildings, including embodied emissions from new construction. Adapted from (IEA 2022)

of 2100. The increase in temperature on the earth's surface has a climatic impact that directly influences human life in various fields, such as an economic impact, an agricultural impact, effects on aquatic systems and effects on the hydrological cycle. To combat this prospect of severe global warming, new scenarios have been created (Prather et al. 2012). They represent the maximum CO_2-eq value that can be emitted in the coming years with a 66% probability of reaching it. The Representative Concentration Pathways (RCP) 2.6 is the name of one of the analyzed scenarios that consider a maximum global warming between 1.5 and 2 °C by the end of 2100 (Van Vuuren et al. 2011). According to the report and the Paris Agreement, the carbon budget of RCP Scenario 2.6, which is equivalent to about 800 Gt CO_2-eq from 2016, should last until 2100 to complete the energy transition. Unfortunately, the various scenarios recently developed by the IPCC (Calvin et al. 2023) show that it is extremely unlikely that we will be able to reduce emissions to zero by 2100. Therefore, the only realistic option is to rapidly develop CO_2 absorption and storage (Carbon Capture and Storage - CCS) technologies that ensure a sustainable pathway through the transition. The uncertainties, however, about the long-term effectiveness of such systems and the enormous economic cost of their implementation are currently the focus of heated debate within the scientific community (Smith et al. 2016).

The Carbon Budget for Buildings

In Western countries, energy use in dwellings is dominated by heating and many policy efforts are currently geared towards drastically reducing energy demand for heating buildings. In contrast, in developing countries in the global south, energy for heating/cooling buildings is relatively low, while most of the share is required to produce building materials (also called 'grey energy'). By 2050, a colossal increase in population to almost 11 billion has been estimated, and as shown in Fig. 2, most of the growth is expected in the global South, especially in Africa, India and South-East Asia. In most of these countries, the level of wealth is also expected to rise, with an increasing demand for new houses and infrastructure needed to cover the housing deficit. Therefore, the energy needs of countries that need to develop their economies and provide materials for new buildings and infrastructure are likely to dominate global building sector emissions and slow the transition to a zero-carbon society. It is therefore necessary to define a carbon budget for construction and construction-related activities in such a way as to limit emissions in each country and stay within the available global budget limits.

This estimation is possible through a top-down approach in order to obtain the maximum amount of CO_2-eq that can be released from a building in the period of transition. The eleven macro-regions, subdivided as in Fig. 2, and the corresponding carbon budget available for building and construction can be analyzed according to two different scenarios. The first, represented with the letter A, is based on calculating the budget as a function only of the number of buildings and their volume that must be constructed or renovated in the next 30 years to support population growth in the various geographic areas. This scenario therefore does not consider the past

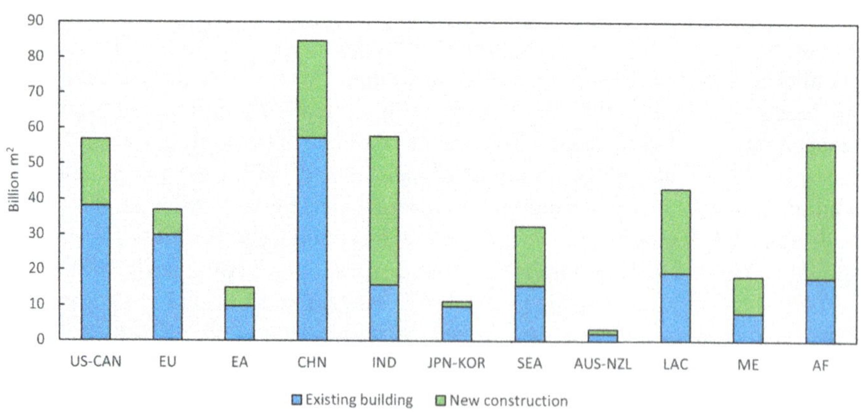

Fig. 2 Growth in building area over the period 2020–2050 by macro-regions: North America (US-CAN), Western Europe (EU), Eurasia (EA), China (CHN), India (IND), Japan and Korea (JPN-KOR), South East Asia (SEA), Australia and New Zealand (AUS-NZL), Latin America and the Caribbean (LAC), the Middle East (ME) and Africa (AF)

contribution of western countries on air emissions (otherwise called 'carbon debt'). The second scenario, represented by letter B, instead takes into account not only the need to renovate existing buildings—mostly in Europe—and to construct new ones, but also the debt on greenhouse gas emissions that Western countries have accumulated in past years, from 1975 onwards, while rewarding, with an extra budget, those countries that have not contributed to emissions with 'carbon credits' proportional to how small their share is compared to the world average.

Considering the share of new construction and the past contribution of each macro-region to climate change, the carbon budget of construction can be defined for each macro-region and divided per m^2 of gross built floor area, as shown in Fig. 3.

As shown in the figure above, when taking into account the contribution of past emissions to climate change (scenario B), Western countries do not have a budget to spend, as most of the carbon is expected to go to emerging economies, particularly India, China and Africa. Moreover, the countries responsible for the current climate change emergency are expected to invest additional resources in developing and promoting carbon capture and storage (CCS) technologies and achieve negative emissions from construction. Within this framework, Western countries are responsible for reabsorbing fossil carbon emitted in the past, while Southern countries are asked to limit their future emissions to develop their growing economies and ensure a transition to a carbon-neutral society.

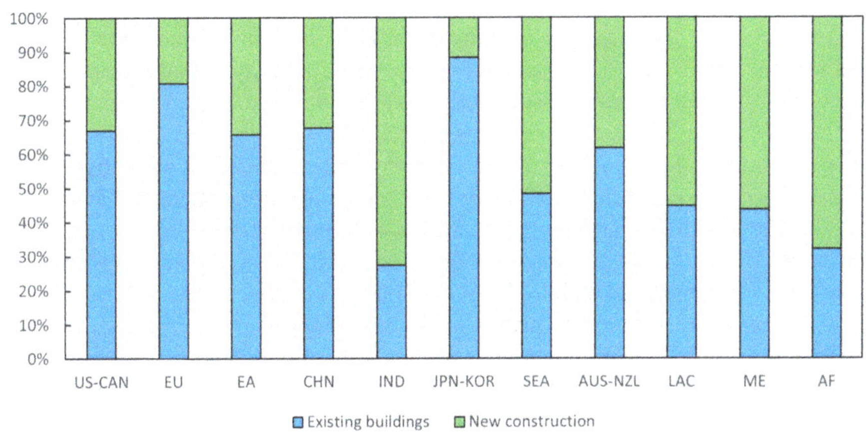

Fig. 3 Expected share in 2050 of buildings built before 2020 (existing buildings) and after 2020 (new construction) by macro-regions: North America (US-CAN), Western Europe (EU), Eurasia (EA), China (CHN), India (IND), Japan and Korea (JPN-KOR), South East Asia (SEA), Australia and New Zealand (AUS-NZL), Latin America and the Caribbean (LAC), the Middle East (ME) and Africa (AF)

The Contribution of Biogenic Building Materials: A Valuable Opportunity for Climate Regeneration and Sustainable Growth

Quantitative analysis of global resource requirements for future urbanization shows that, without a new approach, material consumption by the world's cities will rise from 40 billion tonnes in 2010 to around 90 billion tonnes by 2050 (IRP 2018).

Numerous studies on the life cycle analysis (LCA) of buildings in economically already developed countries show how the climate change contribution, normally measured through the Global Warming Potential (GWP) indicator according to the IPCC method (IPCC 2014), of building materials has become increasingly important in recent years (Hoxha et al. 2017). Whereas in the past, emissions from the use and heating of buildings (so-called operational emissions) were dominant in the life cycle, their contribution has become minimal today, thanks to increasingly stringent energy-saving requirements. In contrast, material-related specific emissions, measured in $kgCO_2$-eq/kg or $kgCO_2$-eq/m^3, have not seen significant reductions over the years, other than those indirectly related to the decarbonization of the energy grid. Reducing emissions for the production of materials is in fact much more complex than containing use-related emissions, where it is often simply necessary to increase the level of thermal insulation of the building envelope and generate thermal energy with energy-efficient systems, possibly powered by renewable energy (Röck et al. 2020).

In addition to the difficulty of reducing specific emissions, one therefore also finds oneself in the position of having to increase the quantity of materials per m^2 of building, especially thermal insulation, increasing the pressure on the production sector even more. If one considers, finally, that in developing countries the energy problem is much less sensitive due to often tropical or milder climates that do not require energy inputs comparable to those of the global north, one immediately realizes how critical the contribution of materials on anthropogenic emissions is. Huge quantities of new building materials will be needed in the coming decades to compensate for the housing deficit in the global south, as well as for the construction of new infrastructure, much of which consists of carbon-intensive materials such as cement, steel and aluminum (Müller et al. 2013)

For these reasons, it is necessary to turn these challenges into opportunities. The buildings of the future can be designed to reduce emissions or even store CO_2, if based on a careful selection of building materials (Churkina et al. 2020). Buildings can also be designed as repositories of materials to be reused in the future, or even act as catalysts to reactivate a neighborhoods' social and economic networks (Brunner 2011).

The widespread use of bio-based materials in construction is a valuable opportunity that can bring several advantages. Indeed, not only are they renewable in nature, but they naturally capture carbon dioxide from the atmosphere as they grow. The sequestered CO_2 is then stored as biogenic carbon during their use as a building product. The removal of CO_2 from the atmosphere and storage in biogenic materials over a long period of time has a positive influence on the development of the earth's climate with a regenerative effect on climate change. This effect is particularly pronounced if the biogenic material regenerates rapidly in the soil, e.g. through the regrowth of forests or plantations.

The delayed emission of carbon previously stored in fast-growing biogenic materials - typically grass, such as straw, hemp, cane, bamboo, etc. - is completely captured by the regrowth of crops within a few years, which makes these materials more effective in mitigating climate change than wood products, which instead require long regrowth periods. In addition, fast-growing biogenic materials have many other advantages, such as their affordability, local availability, reduced grey energy and exceptional hygrothermal properties, which can contribute to improving comfort within a building. The availability and types of biogenic materials are related to the specific geographical context and can be used as innovative components in different building elements, as shown in Fig. 4 below (Fig. 5).

Some of these materials, e.g. bio-concrete, mycelium blocks, etc., are innovative proposals and their possible use and application in building practice is currently being studied, while many other products have already been used in vernacular construction for centuries. However, their widespread use in a contemporary society also brings with it several challenges. While their inherent variability due to natural growth makes standardization difficult, this can be offset by an increase in labor and manual labor, which in emerging economies is often a positive factor providing employment and specialized skills. Unlike conventional alternatives, moreover,

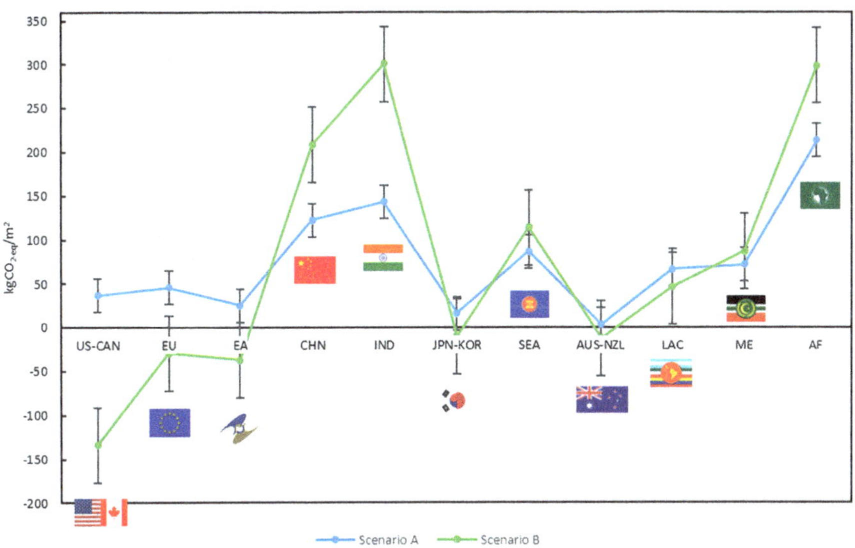

Fig. 4 Carbon budget for the building and construction sector in each macro-region, according to the two reference scenarios. Scenario A represents the allowable emission share per m^2 of built area in the case where past carbon debts/credits are not counted; Scenario B represents the allowable emission share in the case where the contribution of historical emissions is taken into account

Fig. 5 Bioconcrete with wood particles (left), algae insulation board (centre left), silk plaster (centre right), mycelium block structure (right)

Fig. 6 Rural house in Nepal made by limestone and straw (श्रेष्ठ भूपेन्द्र, CC BY-SA 4.0 via Wikimedia Commons)

unconventional materials are typically valued for their cultural, environmental and technological advantages (Suffian et al. 2016). A growing interest in alternative and vernacular construction, deeply rooted in the local culture and traditions of a place, can be observed in parallel with the increase in research on unconventional materials, typically of natural origin, obtained or produced locally (Fig. 6).

The construction industry, originally organized into trade guilds based on empirical and traditional knowledge, has undergone a shift in the last century from a purely qualitative system to a quantitative one based on sophisticated scientific research. The invention of new materials and construction systems, such as frame systems, allowed the creation of many extraordinary structures, but also required the specialization of building professionals. The builder-engineers stopped learning and passing on the practice of 'building know-how' on construction sites and started acquiring more and more specific scientific knowledge in university classrooms. Traditional knowledge, passed down from generation to generation, was gradually replaced by materials science, structural engineering and analytical mathematics, modern subjects that have often confirmed, sometimes without being able to explain them, the effectiveness of traditional knowledge, based on empirical evidence and incremental improvements following mistakes.

The worldwide dominance of reinforced concrete constructions and the general atrophy of today's architectural language show that, in addition to traditional empirical knowledge, site-specific culture has also been lost. Vernacular architecture contrasts with an increasingly identical and globalized architecture, stimulating, in modern times, a growing interest in traditional ancient construction and unconventional materials (Langenbach 2016) which can become a valuable opportunity: for emerging countries, to develop alternative low-carbon constructions and, for Europe, to rediscover traditional materials for the recovery and renovation of existing buildings.

New Vegetarian Diets for Carbon Neutral Buildings

In order to efficiently reduce carbon emissions from the construction sector, there is an urgent need to change the current 'materials diet' (Rovers 2019). The most reasonable option implies that new buildings should be carbon neutral at every single stage of the life cycle, i.e. zero net emissions and zero impact on emissions from building materials. To achieve this ambitious goal, a complete shift towards bio-based materials is needed to reduce the carbon intensity of materials processing, store biogenic CO_2 in the elements, and quickly offset the initial emission through carbon restoration in soils (Pittau et al. 2018).

As shown in Fig. 7, these must be balanced and predominant in the composition of a building, in order to compensate for the positive emissions from high

Fig. 7 The building materials pyramid (right), which inherits the concept of the food pyramid (left) for a balanced plant-based diet. At the top, high-carbon materials; at the bottom, low-carbon regenerative materials, e.g. wood and bio-based materials

CO_2-emitting materials, e.g. cement and concrete, which are required in any case for the foundations and basements of buildings (Carcassi et al. 2022).

On the one hand, biogenic materials are commonly used in vernacular architecture in many areas of the world, and provide a practical way to address extreme poverty, as most people live in remote and rural areas rich in natural resources. Unfortunately, the use of concrete and steel has become a kind of economic 'status symbol' in these areas (Ghavami 2016). However, such materials must be imported, as they require massive infrastructures for their production and processing that are not currently available there. Many natural materials could perform the same structural function as conventional alternatives, at a lower economic and environmental cost, while respecting cultural heritage. Furthermore, many newly developed materials take advantage of the excellent mechanical properties of bio-based materials, such as bricks reinforced with natural fibers, which provide greater resilience against earthquakes due to their higher specific tensile strength (Suffian et al. 2016). On the other hand, bio-based materials are generally characterized by reduced durability when used in their natural state. For this reason, there is a need to develop material processing and treatment technologies to generate advanced natural products, such as in the case of bamboo, where the raw material is commonly treated with borax and boric acid to repel insects and increase the durability of the final product (Xiao 2020).

Finally, particular attention should be paid to the effectiveness of carbon storage, as the increasing industrialization of bio-based materials may lead to overestimating the capacity of materials that are not so good at removing carbon from the atmosphere. In fact, the more a bio-based material is processed, the less effective it is at offsetting carbon emissions. This can be seen in the comparison between bamboo culms and laminated bamboo panels (Lugt and Vogtlander 2015) or in the comparison of more industrialized wood products, such as OSB and MDF, with sawn wood (Höglmeier et al. 2015).

This is due to the increased energy requirements of more elaborate processing and the inclusion of other materials such as glue and resin, which contribute to reducing the relative storage potential of the product, generally not considered when calculating the carbon footprint according to the standard ISO 16745:2017. Consequently, materials such as lime-canapulo used as insulation may not always be more environmentally friendly than other conventional alternatives, due to the high amount of mineral binders used to hold the plant particles together (Arrigoni et al. 2017). Therefore, one should always opt for a material that requires a non-intensive industrial process and, ideally, needs basic processing to meet essential performance requirements, in addition to replicability and economic feasibility.

In Europe, the renovation of the building stock has been identified as the main key to achieving climate neutrality by 2050. To achieve this goal, the energy efficiency of existing buildings should be increased by 75 per cent by 2030 and the renovation rate should be increased to at least 3% (Sesana et al. 2020). Therefore, in order to improve the energy efficiency of the building stock, it will be necessary to introduce a large number of building materials into the market, in particular for thermal insulation (Heeren and Hellweg 2018). If conventional fossil-based materials with high embedded carbon values are used, there is a risk of drastically consuming the remaining budget. Conversely, if the entire demand for additional insulation is covered by biogenic solutions, especially fast-growing species (e.g. straw, hemp, etc.), buildings can act as carbon sinks: the rapid restoration of biomass in crops removes CO_2 from the air and makes extra budget available to spend on energy transition and infrastructure development (Pittau et al. 2019).

References

Arrigoni A, Pelosato R, Melià P et al (2017) Life cycle assessment of natural building materials: the role of carbonation, mixture components and transport in the environmental impacts of hempcrete blocks. J Clean Prod 149:1051–1061. https://doi.org/10.1016/j.jclepro.2017.02.161

Brunner PH (2011) Urban mining a contribution to reindustrializing the city. J Ind Ecol 15:339–341. https://doi.org/10.1111/j.1530-9290.2011.00345.x

Calvin K, Dasgupta D, Krinner G et al (2023) IPCC, 2023: climate change 2023: synthesis report. In: Core Writing Team, Lee H, Romero J (eds) Contribution of working groups I, II and III to the sixth assessment report of the intergovernmental panel on climate change. IPCC, Geneva, Switzerland. Intergovernmental Panel on Climate Change (IPCC)

Carcassi OB, Habert G, Malighetti LE, Pittau F (2022) Material diets for climate-neutral construction. Environ Sci Technol 56:5213–5223. https://doi.org/10.1021/acs.est.1c05895

Churkina G, Organschi A, Reyer CPO et al (2020) Buildings as a global carbon sink. Nat Sustain 3:269–276. https://doi.org/10.1038/s41893-019-0462-4

Dutil Y, Rousse D, Quesada G (2011) Sustainable buildings: an ever evolving target. Sustainability 3:443–464. https://doi.org/10.3390/su3020443

Ghavami K (2016) Introduction to nonconventional materials and an historic retrospective of the field. In: Nonconventional and vernacular construction materials. Elsevier, pp 37–61

Heeren N, Hellweg S (2018) Tracking construction material over space and time: prospective and geo-referenced modeling of building stocks and construction material flows. J Ind Ecol. https://doi.org/10.1111/jiec.12739

Höglmeier K, Steubing B, Weber-Blaschke G, Richter K (2015) LCA-based optimization of wood utilization under special consideration of a cascading use of wood. J Environ Manage 152. https://doi.org/10.1016/j.jenvman.2015.01.018

Hoxha E, Habert G, Lasvaux S et al (2017) Influence of construction material uncertainties on residential building LCA reliability. J Clean Prod 144:33–47. https://doi.org/10.1016/j.jclepro.2016.12.068

IEA (2022) Global energy and process emissions from buildings, including embodied emissions from new construction 2021

IPCC (2014) Climate change 2013: the physical science basis. Cambridge University Press, New York

IRP (2018) The weight of cities: resource requirements of future urbanization. Swilling, M., Hajer, M., Baynes, T., Bergesen, J., Labbé, F., Musango, J.K., Ramaswami, A., Robinson, B., Salat, S., Suh, S., Currie, P., Fang, A., Hanson, A. Kruit, K., Reiner, M., Smit, S., Tabory, S. Nairobi, Kenya

Langenbach R (2016) What we learn from vernacular construction. In: Nonconventional and vernacular construction materials. Elsevier, pp 3–36

Lee H, Romero J (eds) (2023) Climate Change 2023: synthesis report. Intergovernmental Panel on Climate Change, Geneva, Switzerland

Lugt PVD, Vogtlander JG (2015) The environmental impact of industrial bamboo products—life-cycle assessment and carbon sequestration. https://doi.org/10.13140/RG.2.2.20797.46560

Müller DB, Liu G, Løvik AN et al (2013) Carbon emissions of infrastructure development. Environ Sci Technol 47:11739–11746. https://doi.org/10.1021/es402618m

Pittau F, Krause F, Lumia G, Habert G (2018) Fast-growing bio-based materials as an opportunity for storing carbon in exterior walls. Build Environ 129:117–129. https://doi.org/10.1016/j.buildenv.2017.12.006

Pittau F, Lumia G, Heeren N et al (2019) Retrofit as a carbon sink: the carbon storage potentials of the EU housing stock. J Clean Prod. https://doi.org/10.1016/j.jclepro.2018.12.304

Prather MJ, Holmes CD, Hsu J (2012) Reactive greenhouse gas scenarios: systematic exploration of uncertainties and the role of atmospheric chemistry. Geophys Res Lett 39:2012GL051440. https://doi.org/10.1029/2012GL051440

Ritchie H, Rodés-Guirao L, Mathieu E, et al (2023) Population growth. https://ourworldindata.org/population-growth. Accessed 12 Mar 2024

Röck M, Saade MRM, Balouktsi M et al (2020) Embodied GHG emissions of buildings—the hidden challenge for effective climate change mitigation. Appl Energy 258:114107. https://doi.org/10.1016/j.apenergy.2019.114107

Rovers R (2019) People vs resources: restoring a world out of balance. Eburon, Utrecht

Sesana MM, Rivallain M, Salvalai G (2020) Overview of the available knowledge for the data model definition of a building renovation passport for non-residential buildings: the ALDREN project experience. https://doi.org/10.3390/su12020642

Smith P, Davis SJ, Creutzig F et al (2016) Biophysical and economic limits to negative CO2 emissions. Nat Clim Change 6:42–50. https://doi.org/10.1038/nclimate2870

Suffian S, Dzombak R, Mehta K (2016) Future directions for nonconventional and vernacular material research and applications. In: Nonconventional and vernacular construction materials. Elsevier, pp 63–80

United Nations Environment Programme (2024) 2023 Global status report for buildings and construction: beyond foundations—mainstreaming sustainable solutions to cut emissions from the buildings sector. United Nations Environment Programme

Van Vuuren DP, Stehfest E, Den Elzen MGJ et al (2011) RCP2.6: exploring the possibility to keep global mean temperature increase below 2 °C. Clim Change 109:95–116. https://doi.org/10.1007/s10584-011-0152-3

Xiao Y (2020) Engineered bamboo in China. In: Nonconventional and vernacular construction materials. Elsevier, pp 625–643

www.ingramcontent.com/pod-product-compliance
Lightning Source LLC
Chambersburg PA
CBHW071700300126
38800CB00056B/122
* 9 7 8 3 0 3 1 9 5 2 9 1 3 *